ANNALS OF
THE NEW YORK ACADEMY
OF SCIENCES
Volume 529

EDITORIAL STAFF

Executive Editor
BILL BOLAND

Managing Editor
JUSTINE CULLINAN

Associate Editor
MARION L. GARRY

The New York Academy of Sciences
2 East 63rd Street
New York, New York 10021

FOURTH COLLOQUIUM IN BIOLOGICAL SCIENCES:
BLOOD-BRAIN TRANSFER

ANNALS OF THE NEW YORK ACADEMY OF SCIENCES
Volume 529

FOURTH COLLOQUIUM IN BIOLOGICAL SCIENCES: BLOOD-BRAIN TRANSFER

Edited by Fleur L. Strand

The New York Academy of Sciences
New York, New York
1988

Library of Congress Cataloging-in-Publication Data

Colloquium in Biological Sciences (4th : 1986 : New
    York, N.Y.)
Fourth Colloquium in Biological Sciences.

    (Annals of the New York Academy of Sciences,
ISSN 0077-8923 ; v. 529)
    Colloquium held in New York City on Nov. 3, 1986 by
the New York Academy of Sciences.
    Includes bibliographies and index.
    1. Blood-brain barrier—Congresses.   I. Strand,
Fleur L.   II. New York Academy of Sciences.   III. Title.
IV. Title: Blood-brain transfer.   V. Series. [DNLM:
1. Biological Transport—congresses.   2. Biology—
congresses.   3. Blood-Brain Barrier—congresses.
W1 AN626YL v.529 / QH 301 C714 1986f]
Q11.N5   vol. 529   500 s   88-17911
[QP375.5]   [612.8]
ISBN 0-89766-454-X
ISBN 0-89766-455-8 (pbk.)

PCP
Printed in the United States of America
ISBN 0-89766-454-X (cloth)
ISBN 0-89766-455-8 (paper)
ISSN 0077-8923

ANNALS OF THE NEW YORK ACADEMY OF SCIENCES

Volume 529
June 14, 1988

FOURTH COLLOQUIUM IN BIOLOGICAL SCIENCES: BLOOD-BRAIN TRANSFER[a]

Editor
FLEUR L. STRAND

Advisory Council
CRAIG D. BURRELL, HERBERT J. KAYDEN, and WALTER N. SCOTT

Area Chairs
ERIC J. SIMON, ELEANOR B. MCGOWAN, KEVIN L. KEIM, LAWRENCE G. PALMER, MARCELINO F. SIERA, and LINDA HALL

CONTENTS

[a]This volume is the result of a conference entitled the Fourth Colloquium in Biological Sciences, held in New York City on November 3, 1986 by the New York Academy of Sciences.

SECTION ON BIOMEDICINE

Preface

FLEUR L. STRAND

Center for Neural Science
New York University
Washington Square
New York, New York 10003

The six plenary lectures delivered at the Fourth Colloquium in the Biological Sciences organized by the New York Academy of Sciences comprise a unique collection of papers on blood-brain transfer. The Colloquium attracted distinguished investigators who first discuss the overall significance of this complex and important topic. Once this coherent background is laid, each paper carefully presents the significant advances in techniques and approaches that have been made over the past few years. The coherence of these papers is remarkable, and the final paper by William M. Pardridge specifies exciting new directions for blood-brain research.

The timeliness and relevance of the topic also drew a large number of excellent posters on this theme, as well as many fine posters on topics of high interest to the many members of the Academy who participate in the areas of the biological sciences. The value of this *Annal* is consequently increased by the inclusion of the papers based on the poster presentations.

This Colloquium, as did the previous three Colloquia, permitted the Academy to achieve one of its major goals: that of bringing established investigators together with young scientists in an exciting and intellectually stimulating environment. I am happy that this also has resulted in the publication of a particularly fine *Annal*. This success again depended upon the care and hard work of our Advisory Council and the chairs and vice-chairs of the six sections of the Division of Biological Sciences of the Academy: Biochemistry, Biological Sciences, Biomedical Sciences, Biophysics, and Microbiology and Neuroscience.

Brain Endothelium and Interstitium as Sites for Effects of Lead

M. W. B. BRADBURY AND R. DEANE

Department of Physiology
King's College
London WC2R 2LS, England

INTRODUCTION

The ability of lead to cause damage to the central nervous system, particularly in young children, is well recognized. Lead encephalopathy occurs after specific exposure at concentrations in blood[1] of 100-800 μg·dl^{-1}. It is still uncertain whether general environmental contamination of lead giving blood levels of 30-50 μg·dl^{-1} in children can cause disturbance of brain development or function. There is certainly evidence of such that gives cause for concern.[2] Since lead encephalopathy, both in children and the young rat, is associated with capillary damage and since it might be surmised that biological membranes, including the blood-brain barrier, are fairly impermeable to large metallic cations, it was considered that lead might exert its initial effects on brain by interfering with a transport or regulatory function of the cerebral endothelium. Bradbury and Deane[3] briefly reviewed the largely negative evidence for an influence of lead on the transport of either neurotransmitter precursors or of the cations, calcium and magnesium *in vivo*. In the same paper, experimental results were presented that showed an unexpectedly high permeability of the blood-brain barrier to lead, particularly in relation to the supposed very low concentration of lead, [Pb^{2+}], in blood plasma. Further results are now discussed that both confirm that transport into brain depends on [Pb^{2+}] or something closely linked to it and that allow a more precise estimate of this concentration and hence of the permeability-surface area product (PS product) of the blood-brain barrier to this cation. Since lead readily enters brain, a direct influence of this metal on structures or processes on the brain side of the cerebral endothelium cannot be disregarded. Measurements of lead binding to three glycosoaminoglycans are described. The possibility is discussed that the high affinity of lead to certain of these compounds might be a mechanism whereby low concentrations of lead could influence neuronal development.

^{203}PB UPTAKE INTO BRAIN AFTER INTRAVENOUS INFUSION

When a constant concentration of ^{203}Pb was maintained in the blood plasma of rats for up to 4 hours, uptake of the radiotracer into three regions of brain was linear with time.[3] In the absence of added carrier, the uptake into the three regions in relation to the plasma concentration of radiotracer was between 0.065 and 0.085 ml·g^{-1}·hr^{-1}. The large suppression of uptake in the presence of raised total lead in plasma suggested to us, but did not prove, that [Pb^{2+}] determines uptake and that the proportion of this to total lead in plasma is much reduced at raised levels. Uptake of ^{203}Pb into red cells, skeletal muscle and kidney was similarly reduced at raised plasma levels, but uptake into liver and spleen was much less affected. It was suggested that a mechanism or mechanisms transporting Pb^{2+} controls entry of lead into brain, red cells, skeletal muscle and kidney, whereas uptake into liver and spleen is more dependent on a nonspecific process such as vesicular transport.

INFLUENCE OF FERRIC AND ALUMINUM HYDROXIDES IN BLOOD ON ^{203}Pb UPTAKE INTO BRAIN

In the course of the experiments, involving intravenous infusion of ^{203}Pb, a batch of the radioisotope was used that gave markedly lower uptakes into the tissues where entry was thought to depend on [Pb^{2+}], than had previously been observed. This batch of isotope was heavily contaminated with ferric iron. Removal of the contaminating iron and repeat of the uptake experiment restored the results to normal. Final evidence for the anomalous results being due to the iron was obtained by infusing ^{203}Pb with ferric chloride to produce the same effect.

Initially it was thought that ^{203}Pb and ferric ions might be competing for a common transport site at the blood-brain barrier, particularly since there is evidence for interaction of iron and lead during intestinal absorption[4,5] and lead administration may reduce the iron content of brain.[6,7] However, a colleague, Dr. T. J. B. Simons, suggested that the concentration of ferric ion in plasma cannot be raised above an insignificant concentration, probably about 10^{-18} M, because of the very low solubility product of ferric hydroxide.[8] Further, the powerful adsorbent properties of colloidal ferric hydroxide,[9] formed in plasma after infusion of FeCl$_3$, might well cause binding of lead ions and hence reduced [Pb^{2+}] in plasma and lower uptake into brain, for example. Experiments were designed to test this hypothesis and also to investigate whether colloidal hydroxides of another tervalent metal, aluminum, would manifest the same behavior. FeCl$_3$ and AlCl$_3$ were infused in acid solution in separate groups of rats, together with ^{203}Pb. The uptake into three regions of brain and other soft tissues was estimated at 1 hour by the method of Bradbury and Deane.[3] Infusion of FeCl$_3$ or AlCl$_3$, to give a final plasma concentration in the case of iron of 1.1 mM, markedly and significantly suppressed the uptake of ^{203}Pb by 70-90% for brain, choroid plexus, cerebrospinal fluid, red blood cells, skeletal muscle and kidney[10] (Figs. 1 and 2). In contrast uptake into liver and spleen was not significantly lowered by either FeCl$_3$ or AlCl$_3$ infusion. Indeed the mean ^{203}Pb uptake spaces of liver and spleen were all greater after iron or aluminum infusion than in the corresponding control tissues except in the case of liver with aluminum.

Final plasma was subjected to ultracentrifugation at 100,000 g for 60 min and ultrafiltration. Little ^{203}Pb could be sedimented by ultracentrifugation in the absence of iron and aluminum. After infusion of FeCl$_3$ or AlCl$_3$, 70% of ^{203}Pb was in the pellet (this value might have been greater, if it had not been necessary to add 2 U·ml^{-1} heparin to the plasma; see below) (TABLE 1). Ultrafilterable ^{203}Pb was reduced by more than 90% after infusion of either chloride (TABLE 1). Finally, by the use of a lead electrode, it was shown that ferric hydroxide prepared *in vitro* indeed potently binds lead ions. It was considered that the infusion results could only be explained by adsorption of ^{203}Pb to colloidal and aluminum hydroxides formed *in vivo*. Such hydroxides are likely to be stabilized in solution in plasma by the presence of protein. The results also added some weight to the hypothesis that lead entry into the brain depends on $[Pb^{2+}]$ in plasma.

PERMEABILITY OF BBB TO ^{203}Pb MEASURED BY CAROTID INFUSION

The high uptake of ^{203}Pb into brain after intravenous infusion without added carrier in relation to an estimated control value of $[Pb^{2+}]$ of $< 0.5 \times 10^{-8}$ M suggested to us that measurable uptake of ^{203}Pb might well occur during short saline perfusion

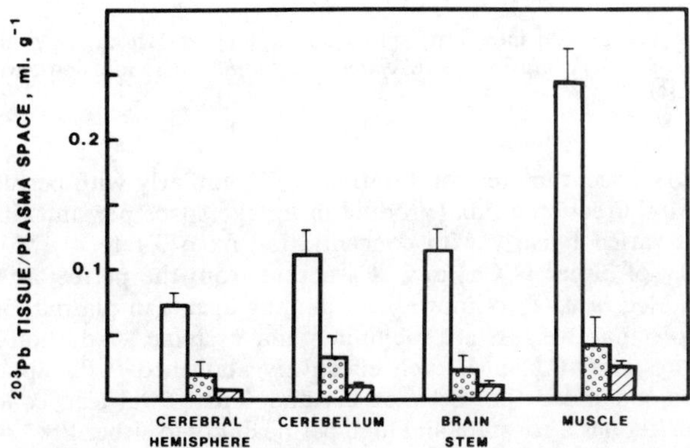

FIGURE 1. Uptake of ^{203}Pb into three regions of brain and skeletal muscle at 1 hour. *Open column:* control; *circle-filled column:* FeCl$_3$ infusions; *cross-hatched column:* AlCl$_3$ infusions. Mean $\pm$ SEM; n = average 6.

of the carotid circulation by the method of Takasato *et al.*[11] If the perfusion fluid contains no substances that complex lead, the concentration of ionized lead can be set at a much higher level than can occur in plasma. Perfusions of the right cerebral hemisphere of the rat were carried out by the technique, as described by its originators, except that the right common carotid artery was cannulated and infused rather than the external carotid and the short test infusion was followed by a 20-sec wash at the same rate with 1.0 mM EDTA in 154 mM NaCl to remove residual ^{203}Pb in or adherent to the cerebral vessels. The rat was then immediately decapitated and the brain dissected into regions for weighing and counting. The test perfusion fluid was isotonic and contained physiological concentrations of sodium, potassium, calcium and magnesium, but not bicarbonate. It was buffered to pH 7.4 with 5 mM HEPES. In addition to ^{203}Pb, nonradioactive lead nitrate was added to make the total lead concentration of 0.1, 1 or 4 μM.

After 60-sec perfusions at 1 μM, the mean uptake of ^{203}Pb into parietal cortex was 0.105 ml·g^{-1}. At this concentration of lead, uptake into cortex varied linearly

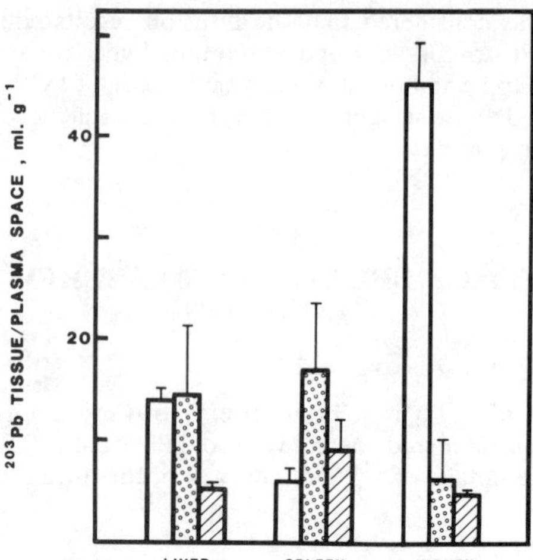

FIGURE 2. Uptake of ^{203}Pb into liver, spleen and kidney at 1 hour. *Open column:* control; *circle-filled column:* FeCl$_3$ infusions; *cross-hatched column:* AlCl$_3$ infusions. Mean ± SEM; n = average 6.

with time up to 60 sec, the greatest duration used. Similarly with perfusions of 1 min, the calculated unidirectional flux (product of uptake space per unit time and infused concentration) varied linearly with concentration from 0.1 to 4 μM. There was no detectable entry of either [^{45}Ca] or [^{14}C]sucrose from the perfusion solution, when either was included in it. Two known lead binding agents in plasma, one of high and one of low molecular weight, are albumin[12] and cysteine.[8] Addition of albumin, 5 g·dl^{-1}, or of cysteine, 0.2 mM, each effectively abolished ^{203}Pb uptake into brain. Thus, it can be concluded that between 0.1 and 4 μM total lead concentration, the blood-brain barrier has a constant and high permeability to either Pb^{2+} or to something that varies directly with it, e.g., PbOH$^+$.

These experiments with the Takasato method allow a better approximate estimation of the [Pb^{2+}] in blood plasma of control rats. Since unidirectional influx of lead into cortex can be calculated as the product of uptake space per unit time from either intravenous infusion or Takasato experiments and the concentration of total lead in either plasma or perfusion fluid, respectively, and since the [Pb^{2+}] is known, [Pb^{2+}] can be calculated from

$$\frac{[\text{Pb}^{2+}] \text{ in plasma}}{[\text{Pb}^{2+}] \text{ in perfusate}} = \frac{\text{Influx into brain (intravenous)}}{\text{Influx into brain (Takasato)}}$$

The concentration of total lead in perfusion fluid of 1 μM was multiplied by 0.070 to give [Pb^{2+}]. This factor corrects for the ratio of [Pb^{2+}]/[PbOH$^+$] at pH 7.4 and for the lower activity of lead, measured with an electrode in the presence of 140 mM chloride rather than in 140 mM perchlorate. [12] Since the activity coefficient of lead will be less than unity even in the presence of perchlorate, [Pb^{2+}] will still to some extent be overestimated. On this basis, the final estimate is approximately 10^{-10} M at a normal concentration of total lead in plasma of control rats of 0.18 μM. The PS product of the blood-brain barrier calculated in relation to $[Pb^{2+}]$ is 2 ml$\cdot$g$^{-1}\cdot$min^{-1}, an extraordinarily high value. If $[Ca^{2+}]$ is assumed to be 50% of total plasma calcium, the corresponding PS product for calcium ion[13] is about 0.0015 ml$\cdot$g$^{-1}\cdot$min^{-1}, or the permeability of the blood-brain barrier to lead is more than 1000 times greater than that to calcium! There is a comparably high permeability of the red cell[12] and other cell[3] membranes to lead.

BINDING OF LEAD TO GLYCOSO-AMINOGLYCANS

Observations of high ultrafiltration of lead from plasma in the presence of heparin[3,14] led us to study the binding characteristics of lead to various glycoso-aminoglycans. The presence of these molecules in interstitial fluid might profoundly influence distribution of lead, and more speculatively effects of lead on cellular development might be mediated through lead binding to them. The experiments were made with a solid-state lead electrode (Orion) immersed at room temperature in a well-stirred solution of 50 mg of the glycoso-aminoglycan in 50 ml of 5 mM HEPES buffer and 10 mM or 140 mM $NaClO_4$ at pH 6.4 or 7.4. The electrode was calibrated against lead perchlorate standards in the same concentration of perchlorate as the test molecule before and after each test run. Successive small volumes of 0.1 M lead were then added and the free (ionized) and bound lead calculated from the voltage and the known amount of added lead.

At pH 6.4 and 10 mM perchlorate, there was substantial binding of lead to heparin, chondroitin sulfate and hyaluronate. This was most marked in the case of heparin (FIG. 3). Initially, it was thought that the binding curves could be satisfactorily fitted by a two-site model, each site having a separate affinity constant, K_b, and binding capacity, B_{max}. While good computer fits could be obtained with such a model, unidirectional variation of the computed constants with number of points used suggested that such an approach is oversimple. If the binding is best represented by a multi-site model, a good index of the binding potential at low ligand concentration can be obtained from the initial slope of bound against free concentrations. This slope is equivalent to B_{max}/K_b. At 10 mM $NaClO_4$, the slope for heparin has a value of about 1000 μol$\cdot$g^{-1} per μM. It is suppressed to 35, when sodium is raised to 140 mM (FIG. 4 and TABLE 2), but is unaffected by pH.

Binding of cations to anionic groups is subject to "masking" by high concentrations of a monovalent cation such as sodium. The lack of effect of pH is compatible with

TABLE 1. Ultracentrifugation and Ultrafiltration of Final Plasma from Rats Infused with ^{203}Pb Alone or with $FeCl_3$ or $AlCl_3$

	Percentage of ^{203}Pb	
	Pellet	Ultrafiltrate
Control	4.4	5.5 $\pm$ 1.0
$FeCl_3$ infusion	70.6 $\pm$ 11.2	0.3
$AlCl_3$ infusion	67.9 $\pm$ 2.5	0.37 $\pm$ 0.14

NOTE: Ultracentrifugation was at 100,000 g for 60 min, and ultrafiltration through YMT membrane (Amicon) with cut-off at 25,000 to 30,000 daltons.

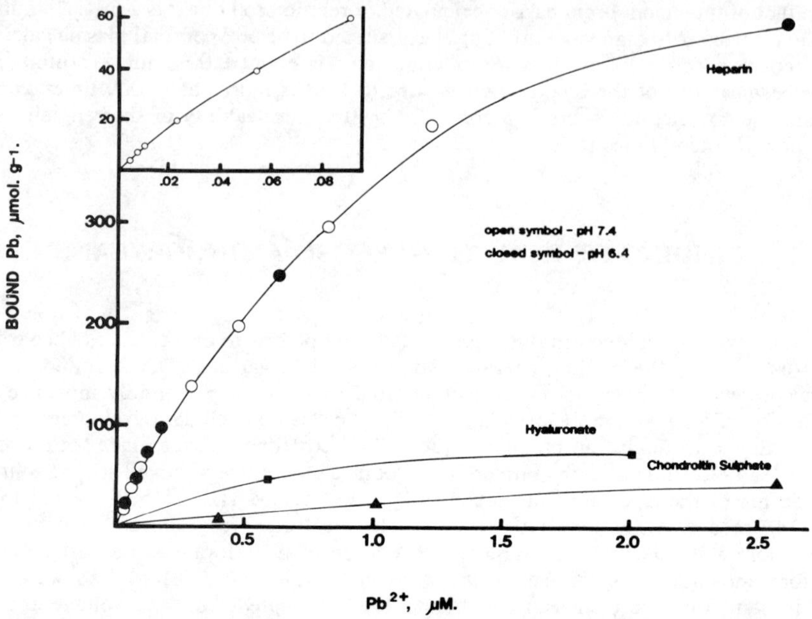

FIGURE 3. Plot of lead bound to three separate glycoso-aminoglycans against ionized lead [Pb^{2+}] in presence of 10 mM $NaClO_4$ and 5 mM HEPES.

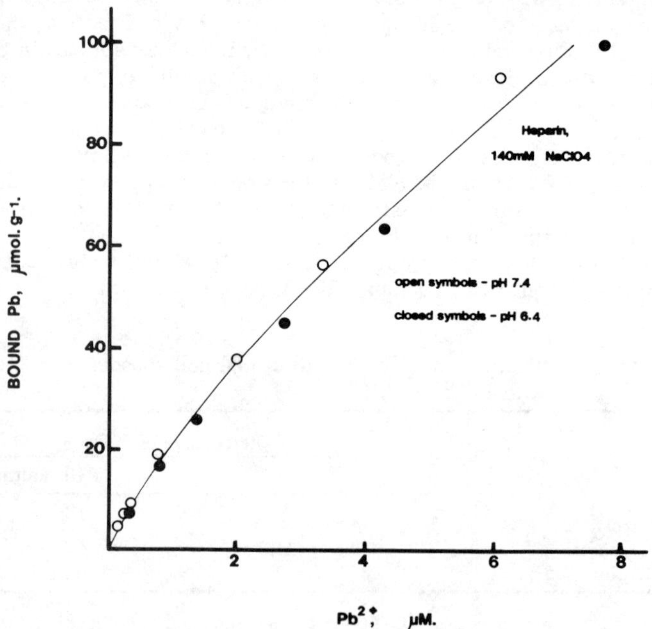

FIGURE 4. Plot of lead bound to heparin against [Pb^{2+}] in presence of 140 mM $NaClO_4$ and 5 mM HEPES.

electrostatic binding of lead to sulfate, sulfamino or carboxyl anionic groups, all of which have low pKs. No doubt binding affinity and capacity for lead depends on an appropriate separation and conformation of two or more anionic groups. Maximum binding capacity of heparin appears to correspond to about one atom of lead per two disaccharide units.

Electrostatic binding occurs between certain proteins and most glycoso-aminoglycans. Thus, Gelman and Blackman have published a series of papers following the interaction of synthetic basic polypeptides with tissue glycoso-aminoglycans of which heparin had the highest affinity; their work was reviewed by Comper and Laurent.[15] The extracellular structural proteins, fibronectin and laminin, both contain heparin-binding domains.[16] Reinnervation of denervated synaptic sites depends on guidance and adhesion of ingrowing axons to a region rich in fibronectin, a heparan sulfate proteoglycan and an extracellular glycoprotein.[17] The neuronal cell adhesion molecule, N-CAM, has a heparan-binding domain.[18] Adhesion of retinal probe cells to retinal cell monolayers appears to depend on reciprocal interactions between a surface heparan sulfate on one cell and N-CAM on the other.[19,20] A heparan sulfate proteoglycan on the surface of dorsal root ganglion neurones has been shown to be the mitogen that stimulates Schwann cell proliferation.[21] Thus, if proteoglycans are important in cell

TABLE 2. Multi-Site Binding of Lead and Calcium to Glycoso-Aminoglycans

	$\mu\text{mol}\cdot\text{g}^{-1}$ per μM	
	Pb^{2+}	Ca^{2+}
Heparin, 140 mM NaClO$_4$	35	—
Heparin, 10 mM NaClO$_4$	1000	30
Hyaluronate	200	—
Chondroitin sulfate	30	—

NOTE: Values are of slope of bound against free $[Pb^{2+}]$ at low lead or calcium concentration.

sprouting, adhesion and proliferation, it is not impossible that lead might influence such processes by its high-affinity binding to these molecules.

CONCLUSIONS

The results from a series of *in vivo* experiments on the passage of ^{203}Pb across the capillary endothelium into brain have been reviewed. These experiments included intravenous infusion of ^{203}Pb to maintain a constant plasma level, similar infusions in the presence of ferric and aluminum hydroxides in blood, and short perfusion, using the method of Takasato, of the carotid circulation with saline solution containing ^{203}Pb. The results led to the conclusion that entry of lead into brain depends on $[Pb^{2+}]$ or on something that varies directly with it, and that the permeability of the blood–brain barrier to the free ion is unexpectedly high (> 1000 times greater than that to calcium). This high permeability to lead does not reflect an isolated character of the blood–brain barrier, but a general property of cell membranes in relation to this ion. Since lead readily enters brain from blood, a direct action on processes on the brain side of the barrier is possible and the need to postulate an effect on transport or regulatory mechanisms at the cerebral endothelium is less demanding. Studies of the characteristics of lead binding to certain glycoso-aminoglycans lead us to the tentative proposal that lead may influence developmental processes in brain through this interaction.

REFERENCES

1. WINDER, C., L. L. GARTEN & P. D. LEWIS. 1983. Neuropath. Appl. Neurobiol. **9:** 87–108.
2. NEEDLEMAN, H. L., C. GUNNOE, A. LEVITON, H. PERESIE, C. MAHER & P. BARRETT. 1979. New Engl. J. Med. **300:** 689–695.
3. BRADBURY, M. W. B. & R. DEANE. 1986. Ann. N.Y. Acad. Sci. **481:** 142–160.
4. MAHAFFEY SIX, K. & R. A. GOYER. 1972. J. Lab. Clin. Med. **79:** 128–135.
5. BARTON, J. C., M. E. CONRAD, S. NUBY & L. HARRISON. 1978. J. Lab. Clin. Med. **92:** 536–547.
6. FICK, K. R., C. B. AMMERMAN, S. M. MILLER, C. F. SIMPSON & P. E. LOGGINS. 1976. J. Anim. Sci. **42:** 515–523.
7. MILLER, G. D., T. F. MASSARO & E. KOPEREK. 1984. Biol. Trace Element Res. **6:** 519–530.
8. MAY, P. M., P. W. LINDER & D. R. WILLIAMS. 1977. J. Chem. Soc. Dalton Trans.: 588–595.
9. MELLOR, J. W. 1934. A Comprehensive Treatise on Inorganic and Theoretical Chemistry. Vol. 13 (part 2),: 831–905. London, New York & Toronto.
10. DEANE, R., S. H. PARK & M. W. B. BRADBURY. 1986/7. J. Pharmacol. Exp. Ther. In press.
11. TAKASATO, Y., S. I. RAPOPORT & Q. R. SMITH. 1984. Am. J. Physiol. **247:** H484–H493.
12. SIMONS, T. J. B. 1986. J. Physiol. **378:** 267–286.
13. WONG, R. P. K. & M. W. B. BRADBURY. 1975. Brain Res. **84:** 361–364.
14. VANDER, A. J., D. L. TAYLOR, K. KALLITIS, D. R. MOUW & W. VICTERY. 1977. Am. J. Physiol. **233:** F532–F538.
15. COMPER, W. D. & T. C. LAURENT. 1978. Physiol. Rev. **58:** 255–315.
16. SANES, J. R., M. SCHACHNER & J. COUVALT. 1986. J. Cell Biol. **102:** 420.
17. YAMADA, K. M. 1983. Annu. Rev. Biochem. **52:** 761–799.
18. COLE, G. J., D. SCHUBERT & L. GLASER. 1985. J. Cell. Biol. **100:** 1192–1199.
19. COLE, G. J. & L. GLASER. 1986. J. Cell. Biol. **102:** 403–412.
20. COLE, G. J., A. LOEWY & L. GLASER. 1986. Nature **320:** 445–447.
21. RATNER, N., R. P. BUNGE & L. GLASER. 1985. J. Cell. Biol. **101:** 744–754.

Role of Secretion and Bulk Flow of Brain Interstitial Fluid in Brain Volume Regulation

HELEN F. CSERR

Section of Physiology and Biophysics
Brown University
Providence, Rhode Island 02912

In systemic tissues, net fluid shifts across the capillary wall and drainage of interstitial fluid (ISF) into the lymphatics are important mechanisms of ISF volume regulation. The situation in the central nervous system is more complex in that (1) cerebral ISF is separated from blood by the blood-brain barrier, (2) there are no lymphatics, and (3) the ependymal and pial surfaces of the brain are in contact with another extracellular fluid, the cerebrospinal fluid (CSF).

In this paper evidence is presented supporting a model of brain ISF volume regulation proposed by Bradbury.[1,2] The model, illustrated in FIGURE 1, is based on ISF secretion at the blood-brain barrier coupled with shifts of extracellular fluid between brain and CSF. Interstitial fluid is separated from blood by the highly impermeable blood-brain barrier, whereas it is connected to CSF by a series of patent extracellular channels. Interstitial fluid volume will be influenced by net exchanges across both of these interfaces. According to the model, ISF is produced at the blood-brain barrier by the process of secretion, whereas volume shifts between brain and the surrounding CSF are via bulk flow. The direction and rate of bulk flow will be a function of the hydrostatic pressure gradient between brain ISF and CSF ($P_{csf} - P_{isf}$). In normal brain, it is proposed that ISF volume is maintained constant despite continual secretion of ISF through drainage into CSF. Although not confirmed experimentally, it is assumed that ISF drainage is driven by an appropriate hydrostatic pressure gradient, i.e., $P_{isf} > P_{csf}$, generated by the secretion of ISF. Reversal of this gradient, through an increase in P_{csf} and/or a decrease in P_{isf}, will lead to a reversal of flow. In normal brain, retrograde flow of CSF into brain will expand ISF, causing interstitial edema, whereas in situations where ISF volume is reduced, retrograde flow provides a mechanism for restoring ISF volume to normal.

In this paper we summarize evidence from this laboratory in favor of the proposed model of ISF volume regulation. Specifically, this evidence indicates that: (1) ISF drains continuously from the parenchyma in normal brain and the rate of drainage provides an estimate of the rate of ISF secretion at the blood-brain barrier; (2) elevation of plasma osmolality, which shrinks the brain and lowers brain ISF pressure,[3] leads to a reversal of ISF flow with bulk flow of CSF into brain; and (3) brain volume regulation during this hyperosmotic stress can be explained quantitatively on the basis of ISF secretion and retrograde flow of CSF into brain (i.e., on the basis of the proposed model). In addition, new evidence is presented, indicating a shift of fluid within the osmotically dehydrated brain from ISF into brain cells.

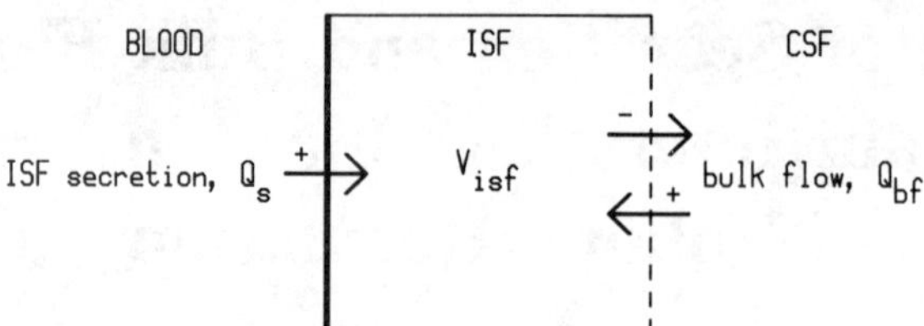

FIGURE 1. Model of brain ISF volume regulation based on ISF secretion at the blood-brain barrier coupled with bulk flow of extracellular fluid between brain and CSF.

DRAINAGE OF ISF FROM NORMAL BRAIN

If there are no lymphatics in the central nervous system, how is ISF cleared from the tissue? In our laboratory we have studied bulk flow and drainage of brain ISF using a tissue clearance technique. In this method, extracellular tracers are microinjected directly into the brain tissue (in a volume of 0.5-1.0 μl) and the characteristics of fluid drainage evaluated by following the pathways and kinetics of tracer clearance from the tissue injection site. Anatomic tracers are used to outline the pathways of flow and radiolabeled extracellular tracers to study the dynamics of exchange. Results obtained using this tracer clearance technique, summarized below, indicate that in the central nervous system, as in other tissues, there is a system of channels for the flow of ISF; furthermore, they provide a quantitative estimate of the rate of ISF drainage from normal brain tissue.

In our initial experiments, we used blue dextran[4] and horseradish peroxidase[5] as markers of ISF flow. In agreement with earlier workers,[6,7] we found that anatomic tracers move through the tissue along preferential channels, yielding a pattern of distribution consistent with bulk flow rather than diffusion. Channels of flow outlined in these experiments include the Virchow Robin, or perivascular, spaces that follow both arteries and veins from the subarachnoid space into the brain, as far as the arterioles and venules but not to the capillaries[8]; spaces between fiber tracts in white matter; and spaces within the subependymal layer of the ventricles. Subsequent analysis of the kinetics of removal from the tissue of three radiolabeled tracers of markedly different diffusion coefficient provided further evidence for bulk flow of ISF. The three test compounds (two polyethylene glycols [900 and 4000 daltons] and albumin [69,000 daltons]), which differ in diffusion coefficient by up to a factor of 5, were cleared from brain according to a single rate constant.[9] Removal according to a single rate constant cannot be by diffusion alone; it is, however, consistent with removal by convection from a well-mixed compartment. If it is assumed that clearance of these large molecules is due entirely to drainage of ISF from the tissue, this flow may be estimated as between 0.18 and 0.29 μl·g brain^{-1}·min^{-1} for different regions of rat brain[10] and as 0.11 μl·g brain^{-1}·min^{-1} for the caudate nucleus of the rabbit.[11]

Although our studies do not relate to the mechanism of ISF production, considerable evidence suggests that ISF is secreted by the capillary endothelium. The blood-brain barrier consists of two membranes, the capillary endothelium and surrounding layer of glial cells. As reviewed by Crone,[12] modern studies have localized barrier function to the capillary endothelium and shown that this membrane has permeability and transport characteristics similar to those of a "tight" epithelial membrane (such as the bladder or distal renal tubule). Evidence that the brain capillary endothelium secretes ISF is indirect and includes a high endothelial density of mitochondria[13] and high concentration of Na, K-ATPase in the abluminal endothelial membrane.[14] Thus, modern evidence is consistent with the view that the turnover of ISF in the brain differs from that in tissues generally both with respect to the mechanism of production (secretion rather than filtration) and the channels of flow (perivascular spaces and spaces within white matter and the subependymal zone rather than lymphatics). Further, our measurements of ISF drainage provide an estimate of the rate of ISF secretion. For the rat, using a mean rate of ISF drainage of 0.22 μl·g brain^{-1}·min^{-1}, the rate of ISF secretion may be estimated at 1.03 μl ISF·g dry wt^{-1}·min^{-1} (0.22 μl ISF·g brain^{-1}·min^{-1} × 4.68 g wet wt·g dry wt^{-1}).

BRAIN VOLUME REGULATION DURING HYPEROSMOTIC STRESS

The stability of normal brain volume in the face of a continual turnover of ISF is consistent with the proposed model of ISF volume regulation (FIG. 1), the rate of ISF secretion being matched in the steady state by the rate of drainage from the tissue. To test this model further, we studied the volume-regulatory response to brain shrinkage induced by elevation of plasma osmolality. We first evaluated the magnitude and kinetics of the volume-regulatory response and then evaluated the separate contributions of ISF secretion and of bulk flow of extracellular fluid between brain and CSF to the observed degree of brain volume regulation.

In our hyperosmotic model, plasma osmolality of anesthetized rats is elevated by a single i.p. injection or by i.v. infusion of NaCl or mannitol. The i.v. infusion schedule is designed to produce a step increase in plasma osmolality. Terminal plasma osmolality varies from 290 - 385 mosM and the experimental duration from 15 - 120 min.

Results obtained during hypernatremia (FIG. 2) show that brain water content

FIGURE 2. Osmotic water loss from brain in hypernatremia. Brain water content given as a function of plasma osmolality 15, 30 and 120 min after a single i.p. injection of hypertonic NaCl. Comparison of 30- and 120-min experimental values with those predicted if brain behaves as a perfect osmometer (*interrupted line*) shows that osmotic water loss is approximately one-third of that predicted.

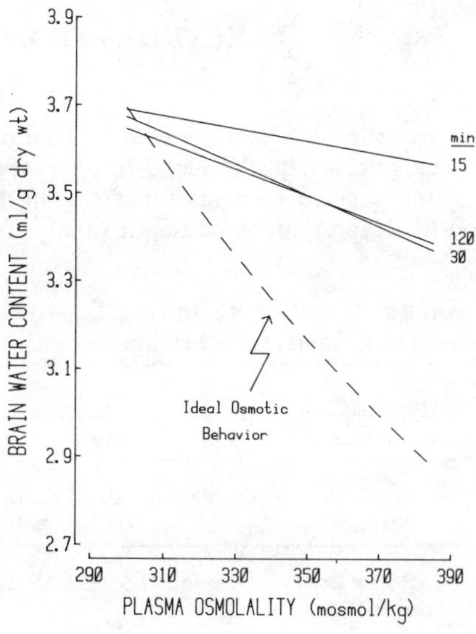

(ml·kg dry wt^{-1}) decreases in response to hyperosmotic stress and that the reduction in volume is proportional to the increase in plasma osmolality.[15] The decrease is less than predicted, however, if the brain behaves as a perfect osmometer. Fifteen minutes after i.p. administration of NaCl the loss was only 13% of the predicted value. At the longer time points — 30 or 120 min — the loss of water plateaued at roughly one-third of the theoretical decrease, values for the two time points being 36% and 35% of predicted values, respectively. The slow time course of the response (greater than 15 min) is consistent with the low hydraulic conductivity of the blood-brain barrier[16] and the "nonideal" osmotic behavior with the presence of volume-regulatory mechanisms. Since a stable brain volume was achieved within 30 min of the hyperosmotic stress, studies of volume-regulatory mechanisms were concentrated over this time period.

As reviewed previously,[1,2,17] the volume-regulatory response to hyperosmotic stress depends on the accumulation of osmotically active solutes within brain tissue. This prevents the full osmotic loss of water from brain. Analysis of the electrolyte content of brain tissue (mEq·kg dry wt^{-1}) 30 min after the induction of hyperosmotic stress showed that the "nonideal" osmotic behavior of brain tissue in the current series of experiments[15] is based on shifts of roughly equimolar amounts of Na, Cl and K into the tissue (TABLE 1). Tissue Na, Cl and K content increased whether plasma osmolality was elevated with NaCl or with mannitol, although the magnitude of the increase was greater with hypernatremia.

What is the mechanism of the volume-regulatory uptake of electrolytes? Theoretically, the accumulation of Na, Cl and K could occur through uptake from plasma, across the blood-brain barrier, or through bulk flow of CSF into brain. However, since CSF concentrations of Na and Cl are much higher than CSF K concentration, it can be anticipated that CSF will contribute more to the tissue uptake of Na and Cl than of K. For the purposes of this paper, which include the evaluation of bulk flow of CSF into brain, we will focus on the mechanism of the volume-regulatory gain of Na by brain tissue during acute hyperosmotic stress.

ELECTROLYTE INFLUX FROM PLASMA

In order to assess the contribution of influx across the blood-brain barrier to the tissue gain of Na, the blood-to-brain transfer constant, K_1, for Na was measured in isosmotic and hyperosmotic rats using the technique of Ohno, Pettigrew and Rapoport.[18] Hypertonicity caused a small increase in K_1 and the increase in the unidirec-

TABLE 1. Slopes of Regression Equations Relating Increase in Brain Electrolyte Content to Increase in Plasma Osmolality 30 min after Injection of Hypertonic Solute

Hyperosmotic Solute	n	Slope[a]		
		Na	Cl	K
NaCl	50	0.442 ± 0.039	0.450 ± 0.042	0.302 ± 0.062
Mannitol	22	0.295 ± 0.055	0.320 ± 0.045	0.359 ± 0.085

[a] mEq·kg dry wt^{-1}·mosml^{-1}·kg. In all cases the slope was significantly different from zero ($p < 0.001$).

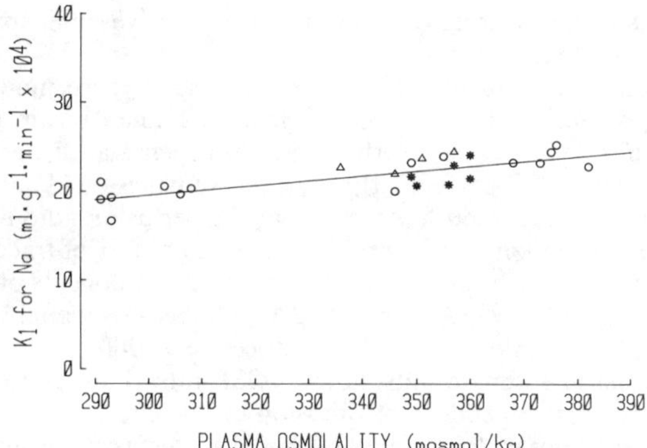

FIGURE 3. Change in the blood-to-brain transfer constant, K_1, for sodium during hyperosmotic stress induced by i.p. injection of different osmolytes. (O) NaCl; (*) sucrose plus NaCl; (△) mannitol plus NaCl. Least-squares line relating K_1 ($ml \cdot g^{-1} \cdot min^{-1} \times 10^4$) and C_{osm} ($mosml \cdot kg^{-1}$) derived by regression analysis of the results. $K_1 = 0.054 \, C_{osm} + 3.4$ ($r = 0.80$; $p < .001$; $n = 25$).

tional rate constant was the same whether plasma osmolality was elevated with NaCl or with mannitol (FIG. 3).[19] With hypernatremia, there will be an inwardly directed concentration gradient across the blood-brain barrier (i.e., $[Na]_p > [Na]_{isf}$). Both this gradient and the increase in K_1 with hypertonicity favor Na influx into brain during hypernatremia. However, even with extreme assumptions concerning the magnitude of the Na gradient across the blood-brain barrier, we estimate that Na influx across the blood-brain barrier can account for only one-third of the volume regulatory influx of Na into rat brain during acute hypernatremia. This is consistent with a large influx of electrolyte from CSF. With mannitol induced hypertonicity, there will be an outwardly directed gradient for Na ($[Na]_{isf} > [Na]_p$). The fact that the brain loads NaCl (TABLE 1) during mannitol-induced hypertonicity — in the presence of outwardly directed concentration gradients — provides further indirect evidence for a CSF source of electrolyte.

BULK FLOW OF CSF INTO BRAIN IN RESPONSE TO ACUTE HYPEROSMOTIC STRESS

Bulk flow of CSF into brain can be demonstrated by adding tracers to the CSF and following the kinetics and pathways of tracer penetration into the underlying nervous tissue. Numerous studies employing this approach have provided qualitative evidence for bulk flow of CSF into brain under appropriate pressure conditions. For example, retrograde flow of anatomic tracers from CSF into brain tissue has been demonstrated in response to an elevation of CSF pressure[7,20] and also in response to osmotic dehydration of the interstitium,[21] a condition associated with a decrease in ISF pressure.[3] There are, however, few quantitative estimates of the magnitude of the flow involved[22–24] and no previous studies of this flow as a mechanism of brain volume regulation. The channels of retrograde flow include the perivascular spaces (as reviewed by Davson[25]) and white matter.[26]

In order to study bulk flow of CSF into brain in our hyperosmotic rats, we added two radiolabeled extracellular tracers — ^{125}I-labeled human serum albumin (RISA; 69,000 daltons) and [57]Co-labeled diethylenetriaminepenta-acetic acid (DTPA; 393 daltons) — to artificial CSF perfused through the ventricles and subarachnoid space and measured the clearance of both tracers from the perfusion fluid into brain.[27] Since both diffusion and convection may contribute to the total flux of tracer into the tissue, it was necessary to distinguish diffusive and convective components of efflux. Diffusion coefficients ($cm^2 \cdot sec^{-1}$) for RISA[28] and DTPA (Blasberg, personal communication), which were 0.85×10^{-6} and 6.4×10^{-6}, respectively, differ by a factor of 7. This difference allowed us to estimate bulk flow of CSF into brain as that component of tracer clearance that is independent of diffusion coefficient.

Artificial CSF containing RISA and DTPA was perfused through the ventricles and/or subarachnoid space of anesthetized rats according to one of three perfusion protocols; from one lateral ventricle to the cisterna magna, from the supracallosal cistern to the cisterna magna, or simultaneously from both the ventricle and subarachnoid space to the cisterna magna. Tracer clearance from CSF to brain was measured as a function of plasma osmolality. Since 30 min was required to reach a steady-state distribution of tracer within the CSF system, osmolality was elevated after 30 min of CSF perfusion in the hyperosmotic animals; the perfusion was then continued for an additional 30 min in order to include the period of maximal electrolyte uptake by brain tissue. Tracer clearance C (μl CSF $\cdot$ g dry wt brain^{-1}) was estimated according to the relationship

$$C = n_b \, / \, c_{CSF}$$

where n_b is the concentration of tracer in brain at the end of the 60-min experimental period (dpm $\cdot$ g dry wt^{-1}) and c_{CSF} the mean concentration of tracer in CSF (dpm $\cdot \mu$l CSF^{-1}) where the latter is taken as the arithmetic mean of the concentrations in the inflow fluid and the steady-state concentration in the outflow fluid.

Analysis of DTPA and RISA clearance as a function of plasma osmolality revealed that tracer influx into brain consists of two components: one that is a function of diffusion coefficient and one that is independent of diffusion coefficient. FIGURE 4 gives the results for combined ventricular and subarachnoid perfusion. In isosmotic animals (plasma osmolality; 290-308 mosml $\cdot$ kg^{-1}), the ratio of clearances, DTPA/RISA, was 2.94 ± 0.22 (mean $\pm$ SE, $n = 19$); this is close to the theoretical value of 2.7 predicted for free diffusion of the two tracers into an infinite region (where the latter is calculated as the square root of diffusion coefficients, DTPA/RISA, i.e., the square root of the quantity, 6.4/0.85). Similarity between the observed and predicted ratio of clearances implies that in this study, as in previous studies of the kinetics of tracer penetration from CSF into brain,[29,30] diffusion is the principal mechanism of tracer penetration in isosmotic animals. With elevation of plasma osmolality (over the range 290-370 mosml $\cdot$ kg^{-1}), the clearances of DTPA and RISA both increased. The ratio of slopes, DTPA/RISA, of the regression equations relating tracer clearance and plasma osmolality (1.47 ± 0.38; mean $\pm$ SE, $n = 19$) was not significantly different from 1.0 ($p > 0.1$) (i.e., the increase was the same for DTPA and RISA, consistent with bulk flow), whereas it was significantly different ($p < 0.01$) from the ratio of 2.7 predicted for diffusion.

Results for ventriculo-cisternal perfusion and for subarachnoid perfusion were similar to those for combined ventricular and subarachnoid perfusion (FIG. 4) in that tracer clearance could be divided into two components — one mediated by diffusion (in isosmotic animals) and one by bulk flow (stimulated by plasma hypertonicity). The magnitude of the fluxes was greater, however, with combined ventricular and subarachnoid perfusion than with either of the more limited perfusion protocols.

The conclusion that the osmotically stimulated increase in tracer clearance is mediated by bulk flow of CSF into brain indicates that this increase can be used to estimate the volume shift of CSF into the tissue. Although there was no significant difference between the slopes of the regression equations relating tracer clearance to plasma osmolality for RISA and DTPA, values for DTPA were consistently greater than for RISA, for all three perfusion protocols. Therefore, as a conservative estimate of flow we have used the value for RISA.[27] Using the RISA slope for combined

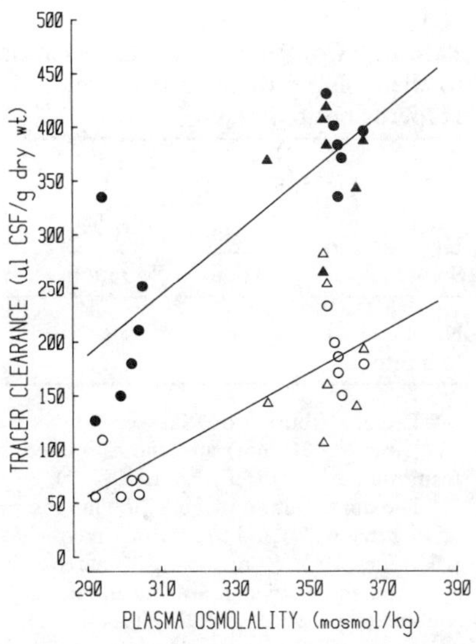

FIGURE 4. Clearances of RISA (*open symbols*) and DTPA (*solid symbols*) from CSF to brain increase during hyperosmotic stress induced by i.p. injection of NaCl ($\bigcirc$) or mannitol ($\triangle$). Rates of increase for RISA and DTPA are similar despite a seven-fold difference in diffusion coefficient, consistent with osmotic stimulation of bulk flow of CSF into brain. There is no difference between results for hypernatremia and mannitol-induced hypertonicity.

ventricular and subarachnoid perfusion, the volume flow of CSF into brain in response to a 60 mosml$\cdot$kg^{-1} increase in osmolality can be estimated as 114 μl CSF$\cdot$g dry wt^{-1} (1.90 μl CSF$\cdot$g dry wt$^{-1}\cdot$mosml$^{-1}\cdot$kg $\times$ 60 mosmol$\cdot$kg^{-1}). For a typical 1.8-g rat brain, this yields an estimated value of 44 μl CSF as the total volume which enters brain (1.8 g wet wt $\times$ 0.214 g dry wt$\cdot$g wet wt^{-1} $\times$ 114 μl CSF$\cdot$g dry wt^{-1}). Comparison of this value with the total volume of CSF in an adult rat of 250 μl[31] and with the estimated rate of CSF production over the 30-min period of hyperosmotic stress of 90 μl (3 μl CSF$\cdot$min^{-1} $\times$ 30 min; Cserr and DePasquale, unpublished material) clearly indicates that the volume of CSF available (290 + 90 = 340 μl) is more than adequate to account for the estimated shift into brain tissue (44 μl).

MECHANISM OF BRAIN VOLUME REGULATION DURING ACUTE HYPEROSMOTIC STRESS

Results summarized in TABLE 2 evaluate the contributions of bulk flow of CSF into brain and of ISF secretion to tissue Na gain during 30 min of plasma hypertonicity. During hypernatremia, ISF secretion and bulk flow together account for 93% of the observed gain of Na by brain tissue and during mannitol-induced hypertonicity for 142%. The fact that the estimated recovery is close to 100% for both experimental situations supports the model of ISF volume regulation illustrated in FIGURE 1, i.e., brain volume regulation in response to osmotic dehydration of the interstitium can be explained on the basis of ISF secretion plus bulk flow of CSF into brain. The difference between the two recoveries — 93% after i.p. injection of NaCl and 142% for mannitol administration — may be related to differences in the direction of net

TABLE 2. Contribution of ISF Secretion[a] and of Bulk Flow of CSF into Brain[b] to the Volume-Regulatory Uptake of Na by Brain Tissue[c] during Acute Hyperosmotic Stress

Hyperosmotic Solute	Na uptake (μEq·g dry wt^{-1})			Contribution to Total Na Uptake (%)		
	Total	Secretion	Bulk Flow	Secretion	Bulk Flow	Secretion + Bulk Flow
NaCl	26	5.1	19	20	73	93
Mannitol	17	5.1	19	30	112	142

[a] The contribution of ISF secretion as the product of the volume secreted (1.03 μl ISF·g dry wt^{-1}·min^{-1}×30 min) and the estimated Na concentration of nascent ISF during acute hyperosmotic stress (0.165 μEq·μl ISF^{-1}).

[b] The contribution of bulk flow as the product of the volume shift of CSF into brain (114 μl CSF·g dry wt^{-1}) and the mean Na concentration of CSF perfusion fluid (0.165 μEq·μl CSF^{-1}). All values for a 60 mosm·kg^{-1} increase in plasma osmolality of 30-min duration.

[c] Total increase in brain Na uptake calculated using regression equations relating brain Na content to plasma osmolality (TABLE 1).

diffusional exchange of Na across the blood-brain barrier between the two experimental protocols. As indicated previously, there will be an inwardly directed gradient for Na across the blood-brain barrier during hypernatremia, i.e., $[Na]_p > [Na]_{isf}$, whereas this gradient will be outwardly directed during mannitol-induced hypertonicity. Net diffusional exchange across the blood-brain barrier cannot be estimated precisely, without further information on the magnitude and time course of the Na gradient; however, it is clear that correction for a small diffusional uptake across the blood-brain barrier during hypernatremia and for a comparable loss during mannitol-induced hypertonicity will bring the estimated values for tissue Na gain (93% and 142%, respectively) closer together.

Analysis of the contributions of bulk flow and secretion to tissue Cl gain during acute hyperosmotic stress, comparable to the analysis for Na (TABLE 2), further supports the proposed model of brain volume regulation. In contrast, the model only accounts for a small fraction of K uptake, given the low K concentrations of plasma and CSF, and analysis of the uptake of this ion[19] suggests that a major component of uptake is mediated by a selective, osmotically stimulated increase in barrier permeability to K.

Flow of CSF into brain tissue in response to plasma hypertonicity implies a net hydrostatic pressure gradient favoring flow into the interstitium, i.e., $P_{csf} > _{isf}$, although this has not been confirmed experimentally. Both P_{isf} and P_{csf} decrease in response to the shrinkage induced by elevation of plasma osmolality.[3] However, a gradient favoring flow of CSF into brain may develop with time, given the more rapid replacement of CSF through secretion by the choroid plexus ($t_{1/2}$ for turnover $= 42$ minutes; estimated as the product of 250 μl total CSF volume $\times \frac{1}{2} \times \frac{1}{3}$ min$\cdot\mu$l CSF^{-1}, using values given above) as compared to the much slower turnover of ISF[10] ($t_{1/2} = 6$-10 hours).

CELL VOLUME REGULATION DURING HYPERNATREMIA

Total tissue water is partitioned between the cells and extracellular space. Previous attempts to evaluate the separate contributions of cellular and extracellular mechanisms to brain volume regulation during osmotic disturbances have estimated ISF volume on the basis of a tissue Cl space. This estimation assumes that Cl is primarily extracellular and that intracellular Cl concentration is a constant fraction of the concentration in tissue ISF.[32] On the basis of response of the tissue Cl space to plasma hypertonicity, it has been suggested that ISF volume is conserved during acute hyperosmotic stress both in the rat[33] and in an elasmobranch,[34] the little skate. According to this view osmotic water loss from the brain during hyperosmotic stress is primarily from the intracellular compartment and extracellular volume is regulated.

Many cells regulate their volume in response to osmotic shrinkage based on the cellular uptake of osmolytes. The osmolytes involved vary but in some cells these include Na and Cl, as reviewed by Grinstein et al.[35] This raises the possibility that hyperosmotic stress may lead to a shift of NaCl plus an isosmotic equivalent of water from the interstitium into brain cells. In order to investigate this possibility we have measured the dynamic response of tissue ISF volume to acute plasma hypertonicity (Cserr, Nicholson and Rice, unpublished material) using ion-selective micropipettes and ionophoretic point sources as described by Nicholson and colleagues.[36,37] Measurements were made in the cerebral cortex of anesthetized rats at a depth of 800 μm before and after i.p. injection of hypertonic NaCl. Results (FIG. 5) show a biphasic response to plasma hypertonicity. For the first 33 minutes (2000 seconds) of hypernatremia, ISF volume remained relatively stable, followed by a marked decrease over the following 1.7 hours (6000 seconds). If ISF volume responded passively to the hyperosmotic stress (i.e., as a perfect osmometer), it would be expected to decrease by 17% for a 60 mosml$\cdot$kg^{-1} increase in plasma osmolality. Unfortunately, the accuracy of the volume measurement is not sufficient to determine whether the volume decreased by this amount or was regulated during the initial 33-minute period. The marked reduction in ISF volume at longer time periods, when total tissue Na and Cl are still increasing,[15] indicates a shift of ISF into brain cells. We interpret this shift as regulation of brain cell volume at the expense of extracellular volume, suggesting that mechanisms of cell volume regulation take precedence over those of ISF volume regulation. Furthermore, these results provide an example of the fallibility of the tissue Cl space as an estimate of ISF volume and indicate the need for direct measurement of this value.

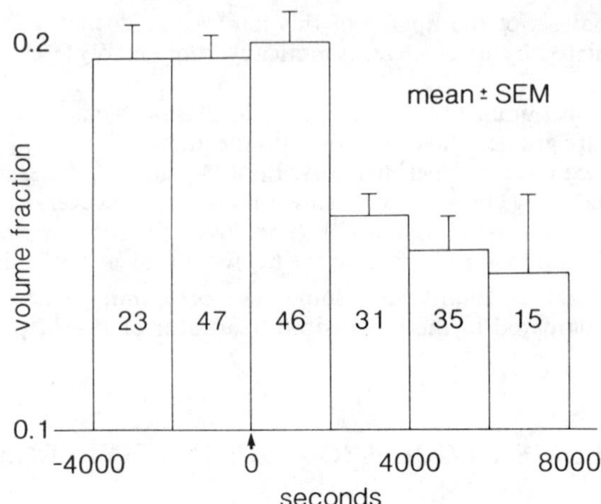

FIGURE 5. Dynamic response of ISF volume (volume fraction) of rat cerebral cortex to hyperosmotic stress. Hyperosmotic NaCl injected i.p. at t = 0. Data from ten experiments are given as means ± SE and numbers in the blocks refer to the number of measurements (Cserr, Nicholson & Rice, unpublished material).

REFERENCES

1. BRADBURY, M. W. B. 1976. Physiopathology of the blood-brain barrier. *In* Transport Phenomena in the Nervous System. G. Levi, L. Battistin & A. Lajtha, Eds.: 507-516. Plenum. New York.
2. BRADBURY, M. W. B. 1979. The Concept of a Blood-Brain Barrier. Wiley. Chichester.
3. WIIG, H. & R. K. REED. 1983. Rat brain interstitial fluid pressure measured with micro-pipettes. Am. J. Physiol. **244:** H239-H246.
4. CSERR, H. F. & L. H. OSTRACH. 1974. Bulk flow of interstitial fluid after intracranial injection of Blue Dextran 2000. Exp. Neurol. **45:** 50-60.
5. CSERR, H. F., D. N. COOPER & T. H. MILHORAT. 1977. Flow of cerebral interstitial fluid as indicated by the removal of extracellular markers from rat caudate nucleus. Exp. Eye Res. Suppl. **25:** 461-473.
6. HIS, W. 1865. Uber ein perivasculares Canal system in den nervosen Central organen und uber dessen Beziehungen zum Lymphsystem. Z. Wiss. Zool. **15:** 127-141.
7. WEED, L. H. 1914. The dual source of cerebro-spinal fluid. J. Med. Res. **26:** 93-113.
8. RENNELS, M. L., T. F. GREGORY, O. R. BLAUMANIS, K. FUJIMOTO & P. A. GRADY. 1985. Evidence for a "paravascular" fluid circulation in the mammalian central nervous system, provided by the rapid distribution of tracer protein throughout the brain from the subarachnoid space. Brain Res. **326:** 47-63.
9. CSERR, H. F., D. N. COOPER, P. K. SURI & C. S. PATLAK. 1981. Efflux of radiolabeled polyethylene glycols and albumin from rat brain. Am. J. Physiol. **240:** F319-F328.
10. SZENTISTVANYI, I., C. S. PATLAK, R. A. ELLIS & H. F. CSERR. 1984. Drainage of interstitial fluid from different regions of rat brain. Am. J. Physiol. **246:** F835-F844.
11. BRADBURY, M. W. B., H. F. CSERR & R. J. WESTROP. 1981. Drainage of cerebral interstitial fluid into deep cervical lymph of the rabbit. Am. J. Physiol. **240:** F329-F336.
12. CRONE, C. 1986. The blood-brain barrier: A modified tight epithelium. *In* The Blood-Brain Barrier in Health and Disease. A. J. Suckling, M. G. Rumsby & M. W. B. Bradbury, Eds.: 17-40. Ellis Horwood. Chichester.

13. BETZ, A. L., J. A. FIRTH & G. W. GOLDSTEIN. 1980. Polarity of the blood-brain barrier: Distribution of enzymes between the luminal and antiluminal membranes of brain capillary endothelial cells. Brain Res. **192:** 17-28.

14. OLDENDORF, W. H. & W. J. BROWN. 1975. Greater number of capillary endothelial cell mitochondria in brain than in muscle. Proc. Soc. Exp. Biol. Med. **149:** 736-738.

15. CSERR, H. F., M. DEPASQUALE & C. S. PATLAK. 1987. Volume regulatory gain of electrolytes by rat brain during acute hyperosmolality in rats. Am. J. Physiol. **253:** F522-F529.

16. FENSTERMACHER, J. D. & J. JOHNSON. 1966. Filtration and reflection coefficients of the rabbit blood-brain barrier. Am. J. Physiol. **211:** 341-346.

17. POLLOCK, A. S. & A. I. ARIEFF. 1980. Abnormalities of cell volume regulation and their functional consequences. Am. J. Physiol. **239:** F195-F205.

18. OHNO, K., K. D. PETTIGREW & S. I. RAPOPORT. 1978. Lower limits of cerebrovascular permeability to nonelectrolytes in the conscious rat. Am. J. Physiol. **235:** H299-H307.

19. CSERR, H. F., M. DEPASQUALE & C. S. PATLAK. 1987. Volume regulatory influx of electrolytes from plasma to brain during acute hyperosmolality. Am. J. Physiol. **253:** F530-F537.

20. MILHORAT, T. H., R. G. CLARK, M. K. HAMMOCK & P. P. MCGRATH. 1970. Structural, ultrastructural, and permeability changes in the ependyma and surrounding brain favoring equilibration in progressive hydrocephalus. Arch. Neurol. **22:** 397-407.

21. WEED, L. H. & P. S. MCKIBBEN. 1919. Experimental alteration of brain bulk. Am. J. Physiol. **48:** 531-555.

22. SAHAR, A., G. M. HOCHWALD & J. RANSOHOFF. 1969. Alternate pathway for cerebrospinal fluid absorption in animals with experimental obstructive hydrocephalus. Exp. Neurol. **25:** 200-206.

23. ROSENBERG, G. A., L. I. WOLFSON & R. KATZMAN. 1978. Pressure-dependent bulk flow of cerebrospinal fluid into brain. Exp. Neurol. **60:** 267-276.

24. ROSENBERG, G. A., W. T. KYNER & E. ESTRADA. 1980. Bulk flow brain interstitial fluid under normal and hyperosmolar conditions. Am. J. Physiol. **238:** F42-F49.

25. DAVSON, H. 1956. Physiology of the Ocular and Cerebrospinal Fluids.: 50. Churchill. London.

26. ROSENBERG, G. A., W. T. KYNER & E. ESTRADA. 1982. The effect of increased CSF pressure on interstitial fluid flow during ventriculocisternal perfusion in the cat. Brain Res. **232:** 141-150.

27. PULLEN, R. G. L., M. DEPASQUALE & H. F. CSERR. 1987. Bulk flow of cerebrospinal fluid into brain in response to acute hyperosmolality. Am. J. Physiol. **253:** F538-F545.

28. LANDIS, E. M. & J. R. PAPPENHEIMER. 1963. Exchange of substances through the capillary walls. *In* Handbook of Physiology. Circulation. W. F. Hamilton & P. Dow, Eds. Sect. 2, Vol. **11:** 961-1034. American Physiological Society. Washington, DC.

29. RALL, D. P., W. W. OPPELT & C. S. PATLAK. 1962. Extracellular space of brain as determined by diffusion of inulin from the ventriclar system. Life Sci. **2:** 43-48.

30. FENSTERMACHER, J. D. & C. S. PATLAK. 1975. The exchange of material between cerebrospinal fluid and brain *In* Fluid Environment of the Brain. H. F. Cserr, J. D. Fenstermacher & V. Fencl, Eds.: 201-214. Academic Press. New York.

31. BASS, N. H. & P. LUNDBERG. 1973. Postnatal development of bulk flow in the cerebrospinal fluid system of the albino rat: Clearance of [carboxyl-^{14}C] inulin following intrathecal infusion. Brain Res. **52:** 323-332.

32. VAN HARREVELD, A. 1966. Brain Tissue Electrolytes.: 34-35. Butterworths. Washington, DC.

33. HOLLIDAY, M. A., M. N. KALAYCI & J. HARRAH. 1968. Factors that limit brain volume changes in response to acute and sustained hyper- and hyponatremia. J. Clin. Invest. **47:** 1916-1928.

34. CSERR, H. F., M. W. B. BRADBURY, K. MACKIE & E. J. MOODY. 1983. Control of extracellular ions in skate brain during osmotic disturbances. Am. J. Physiol. **245:** R853-R859.

35. GRINSTEIN, S., A. ROTHSTEIN, B. SARKADI & E. W. GELFAND. 1984. Responses of lymphocytes to anisotonic media: volume-regulating behavior. Am. J. Physiol. **246:** C204-C215.

36. NICHOLSON, C., J. M. PHILLIPS & A. R. GARDNER-MEDWIN. 1979. Diffusion from an iontophoretic point source in the brain: Role of tortuosity and volume fraction. Brain Res. **169:** 580-584.
37. NICHOLSON, C. & J. M. PHILLIPS. 1981. Ion diffusion modified by tortuosity and volume fraction in the extracellular microenvironment of the rat cerebellum. J. Physiol. (London) **321:** 225-257.

Structural and Functional Variations in Capillary Systems within the Brain[a]

JOSEPH FENSTERMACHER, PAUL GROSS, NADINE SPOSITO, VIRGIL ACUFF, SUSAN PETTERSEN, AND KURT GRUBER

Department of Neurological Surgery
State University of New York
Stony Brook, New York 11794-8122

INTRODUCTION

The cerebrovascular system supports the activities of brain cells by delivering essential substrates such as oxygen and glucose to brain tissue and removing the byproducts of cerebral metabolism such as carbon dioxide and hydrogen ions. Cerebral blood vessels also carry blood-borne hormones and homeostatic information (e.g., plasma osmolality) to the brain and distribute neurally synthesized humoral agents from brain to systemic sites. In this way, the cerebrovascular system assists in the integration of the central nervous and endocrine systems.

The endothelial cell lining of cerebral blood vessels, which is only one cell layer thick, forms the interface between the flowing blood and the interstitial fluid of the surrounding tissue and sets, in combination with the blood flow, the rates of solute and water exchange between these two fluids. Most cerebral blood vessels have a very tight endothelial lining, which greatly restricts the blood-to-brain transfer of many substances, including a number that are neurotoxic or untowardly neuroactive, and forms a "blood-brain barrier," the term commonly used to describe both the endothelium and this transport-restricting function.

As indicated above, the two major "vascular" components of blood-brain distribution are the flow of blood up to, through, and away from the capillaries and the exchange of material across the microvascular endothelium. In many instances, blood delivery greatly exceeds transendothelial exchange, and tissue uptake or loss is viewed as being "permeability limited." Both flow and permeability factors must, however, be considered if we are to understand adequately blood-brain transfer. For example, two recent reviews of the pharmacology of the blood-brain barrier[1,2] emphasize that many drugs are sufficiently lipid-soluble to pass readily across the cerebral endothelium, but their apparent rates of uptake are curiously slow because of processes such as rapid clearance from the blood by other tissues and extensive plasma protein binding,

[a] This work was supported by Grants NS21157 and HL35791 from the National Institutes of Health.

both of which functionally reduce the amount of exchangeable drug delivered by the blood to the capillaries. Accordingly, cerebral blood flow and capillary permeability will be integrated in this treatment of the blood-brain barrier.

Within the central nervous system there are broad differences in local neural activity, cellular organization, blood flow, neural transmitters, endocrine activity, glucose utilization, and other metabolic processes. In view of these variations in brain microphysiology, we hypothesize that matching or compensatory differences exist in the structure and function of local capillary beds. At the physiological level, this proposition was tested by measuring local blood flow, volume of perfused microvessels, mean tissue transit time of albumin, transcapillary influx rate, and capillary permeability-surface area (PS) product. At the morphologic level, it was tested by quantitating capillary volume and surface area as well as frequency of endothelial cell fenestrae, vesicular profiles, and mitochondria.

METHODS

Physiological Studies

Male Sprague-Dawley rats weighing 280-350 g were lightly anesthetized with a mixture of 1.0-1.5% halothane, 30% oxygen, and 70% nitrous oxide. Polyethylene catheters were inserted into one femoral vein and in each femoral artery, the wounds infiltrated with local anesthetic, and the hindquarters of each rat loosely wrapped in a plaster bandage. The anesthetic gas was removed and the rats were allowed to recover under a heating lamp for a least two hours with the bandaged hindquarters taped to a lead brick to restrain movement. During the recovery period, arterial blood pressure, blood gases and pH, hematocrit, and rectal temperature were measured to assure a normal physiological state before beginning the experiments.

Each physiological determination was made in a separate set of animals using radioactive tracers; the techniques and key references are listed below. Because each method required measurement of radioactivity in arterial blood during the experiments, blood samples were drawn from a femoral artery at prescribed times throughout each study and their levels of radioactivity were assessed by scintillation or gamma counting.

At the conclusion of all experiments, the brain was removed, frozen, and cut in a cryostat at $-17°C$ into 20-μ-thick coronal sections. Adjacent sections were collected and stained with thionin for histologic identification of individual brain structures. After drying on coverslips, the sections for autoradiography were placed in cassettes with film for 5-21 days, the length of time depending on the tracer used. All analyses of tissue radioactivity were conducted with quantitative autoradiography and image processing.[3-5] The following measurements were made within a number of brain structures: (1) rate of *capillary blood flow* using [^{14}C]iodoantipyrine[6]; (2) rate of *blood-to-brain influx* of a neutral and nonmetabolizable amino acid, [^{14}C]alpha-aminoisobutyric acid (AIB)[3,7]; (3) *tissue microvascular plasma volume,* using [^{125}I]albumin[7,8]; and (4) *tissue microvascular erythrocyte volume,* using [^{51}Cr]erythrocytes.[7,8] The experimental data were combined to derive the following indices of microvascular physiology[3,7,8,9]: *tissue microvascular blood volume* (sum of microvascular plasma and erythrocyte volumes); *capillary plasma flow* (blood flow $\times$ [tissue plasma volume divided by tissue blood volume]); *mean albumin transit time* (tissue plasma volume divided by capillary plasma flow); an *extraction fraction* (E) for AIB, i.e., the percentage of AIB retained in the tissue per amount delivered (rate of AIB influx divided by plasma flow); and a *permeability* $\times$ *surface area (PS) product* for AIB (PS $=$ $-$plasma flow $\times$ ln [1 $-$ E]).

Morphometric Studies

Male Sprague-Dawley rats weighing 250–300 g were anesthetized with sodium pentobarbital (40 mg/kg, i.p.) and were initially perfused from a cannula in the left ventricle with heparinized Ringer's solution. A second perfusion through the same cannula was made with a solution of 2% paraformaldehyde and 2% glutaraldehyde in 0.1 M phosphate buffer at pH 7.4 and 37°C. The brain was removed and stored for at least 24 hr in cold 2% glutaraldehyde in 0.1 M buffer.

Blocks (200-μ-thick sections) containing specific brain structures were cut the following day on a vibratome. The blocks were placed in a 0.1 M phosphate buffer and were postfixed in buffered 1% osmium tetroxide. These sections were then dehydrated through a graded series of alcohols and propylene oxide, and embedded in Embed 812. From each tissue specimen, 2-μ-thick sections were cut with glass knives and stained with 1% toluidine blue.

For morphometric analysis at the light microscopic level, photographs of the sections were taken with a Nikon Optiphot at a magnification of 50$\times$ and enlarged 8$\times$. An area for analysis was outlined on the micrographs for individual structures and all cross-sectioned microvessels within that area having profiles with diameters less than 7.5 μ were counted and analyzed (hereafter, we refer to these vessels as "capillaries"). A Nikon Microplan II was used for the quantitative analysis of total tissue area, the number and area of capillary profiles within the measured area, and capillary diameter. From these measurements, the capillary density, volume fraction, and the surface area of capillaries were calculated within the region of interest according to stereological theory and formulae.[10,11]

Morphometric analyses at the ultrastructural level were conducted with the same tissue blocks as used for light microscopy. Ultrathin silver sections (about 75 nm thick) were cut with a diamond knife and stained with uranyl acetate and lead citrate. A JEOL 100 CX transmission electron microscope was used to obtain micrographs of endothelial cells at 10,000$\times$ magnification. These micrographs were subsequently enlarged to 25,000$\times$ for morphometric analysis on the Microplan II. The area of endothelial cells was measured and the number of cell junctions, mitochondria, fenestrations, and vesicular profiles within the cells were counted in all of the structures of interest.

Brain Structures

For the physiological or quantitative autoradiographic studies, the following brain structures were analyzed: inferior colliculus (IC), sensorimotor cortex (SMC), ventromedial nucleus of the hypothalamus (VMH), supraoptic nucleus of the hypothalamus (SON), corpus callosum (CC), pineal gland (PG), subfornical organ (SFO), median eminence (ME), and neural lobe of the pituitary (NL). For the morphometric studies, all of the same brain structures except for the pineal gland and median eminence were analyzed. Since these studies are still in progress, only the mean ± the standard error (SE) or the mean ± the standard deviation (SD) are presented.

RESULTS AND DISCUSSION

Most brain structures have tight, blood-brain barrier (BBB) capillaries. Included in this group of structures and the present study are the sensorimotor cortex (telencephalic gray matter), the corpus callosum (telencephalic white matter), the inferior colliculus (midbrain gray matter), the supraoptic nucleus of the hypothalamus, and the ventromedial nucleus of the hypothalamus (both diencephalic gray matter). Within the brain, however, there is a series of small brain structures that border the ventricular system and have relatively permeable or open capillaries. These structures are collectively referred to as circumventricular organs (CVOs) and have secretory or endocrine functions. In the present study, the capillary systems of four CVOs (the subfornical organ, the neural lobe of the pituitary gland, the pineal gland, and the median eminence) were examined.

The rate of capillary blood flow varies broadly among CNS structures.[6,9] Gray matter structures such as the inferior colliculus have relatively high rates of blood flow, which can be modified by various stimuli, whereas white matter structures such as the corpus callosum have fairly low, nearly constant rates of blood flow. The two most likely causes of these blood flow variations are differences in the linear velocity of flow within the capillaries and the number or amount of perfused capillaries.[9,12,13] If the cause of interstructural differences in blood flow is variations in linear velocity, then the capillary beds of most cerebral structures will have similar surface areas of endothelium for blood-brain exchange. On the other hand, if the cause of flow dif-

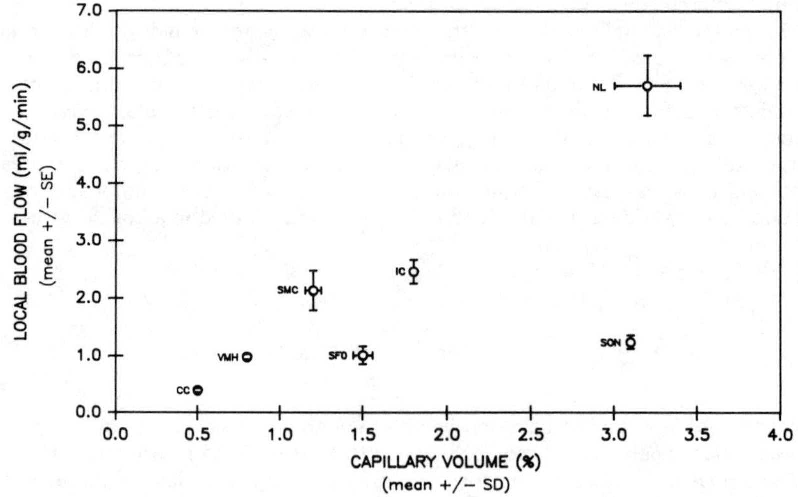

FIGURE 1. Local blood flow versus capillary volume fraction (%).

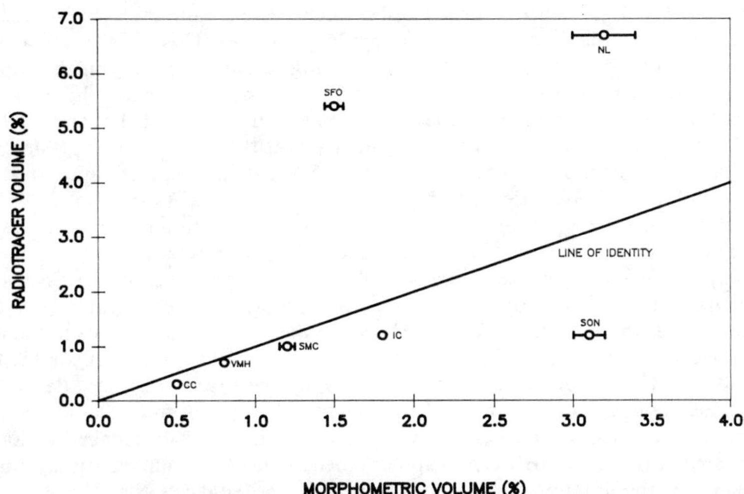

FIGURE 2. "Capillary" blood volume as assessed by radiotracer distribution and morphometry.

ferences is the number of perfused capillaries, then the surface area of endothelium will vary among CNS structures. Although other aspects of blood-tissue exchange are affected by the two blood flow cases (e.g., mean capillary concentration), the difference in surface area amply illustrates the importance of considering both blood flow and capillary transfer in an analysis of the BBB.

To test the proposition that flow varies as a function of the number of perfused capillaries (the second model or case), blood flow, as measured by iodoantipyrine, has been plotted against morphometrically determined capillary volume fraction, a parameter that takes into account both capillary size and capillary density (FIG. 1). Although there is a trend toward higher flow with increasing capillary volume fraction (e.g., compare SMC and NL), VMH, SFO, and SON have similar flow rates but a four-fold range in capillary volume fraction. This general lack of corroboration implies that either the linear velocity of flow varies to some extent (the first model) or the percentage of *perfused* capillaries differs appreciably among some structures, namely, the capillary volume fraction is not a good indicator of the perfused capillary volume (the second model).

The latter possibility can be examined since the radiation arising from blood contained in large vessels can be eliminated by image analysis of autoradiograms and "microvascular" radioactivity and volume ascertained. By presuming that all microvessels perfused during the experimental period (2-5 min) still contain radiotracer and combining the distribution volumes of [^{125}I]albumin- and ^{51}Cr-labeled red cells, an estimate of the "perfused microvessel volume" can be made, and the results compared with the morphometrically measured capillary volume fraction (FIG. 2).

For structures with BBB capillary beds, the volume of radiolabeled microvessels is less than the capillary volume fraction. Moreover, the radiotracer volume: morphometric volume ratios vary from VMH and SMC (88% and 83%, respectively) to IC and CC (67% and 60%, respectively) to SON (39%). The values of these ratios suggest that only a proportion of the capillary beds was perfused during the experimental period and that the perfused fraction varies among the capillary beds of non-CVO structures.

For the CVO structures, the perfused microvessel (radiotracer) volume greatly exceeds the anatomic (morphometric) capillary volume. This indicates that radiolabeled albumin (RISA) and/or red cells are being retained not only in the capillaries, but also in one or more noncapillary tissue compartments. Because the capillaries of CVO structures are permeable to macromolecules,[14] it is possible that some RISA has passed from blood into tissue and consequently that the *intravascular* plasma volume is overestimated in these structures with the RISA distribution technique. Indeed, this "extravasation of RISA" seems to be more sizable in the SFO than in the NL[14] and can account for the discrepancy between the radiotracer and morphometric estimates of capillary volume for the SFO (unpublished observations).

Using the labeled microvessel blood volume (radiotracer volume in FIGURE 2) as an approximation of the volume of perfused capillaries, we can graphically test the hypothesis that the differences in blood flow are the result of differences in the number of perfused capillaries (FIG. 3). For four of the non-CVO capillary beds (SMC, IC, CC, and VMH), the relationship between flow and perfused microvascular volume is fairly linear and the deviation of the SON to the right is much less than that in FIGURE 1. This observation demonstrates that much of the difference in blood flow among structures with BBB-type capillary beds can be accounted for by matching differences in the amount of perfused capillaries and suggests that the local rate of cerebral blood flow strongly depends on the number of perfused capillaries (the second model). As indicated above, the microvessel blood volume of the SFO and NL sizably exceeds the total capillary volume and comparison of this parameter to blood flow is pointless.

The mean transit time of albumin, which is inversely proportional to the linear velocity of plasma flow, should be fairly constant if the main determinant of microvascular blood flow for non-CVO capillary systems is the number of perfused capillaries. The albumin transit times were found to be 0.3 sec (i.e., the fastest linear velocity of flow) for SMC and IC, 0.4 sec for VMH, 0.45 sec for CC, and 0.6 sec (i.e., the slowest linear velocity) for SON capillary beds. This finding is consistent with that in FIGURE 3, where flow and microvessel volume are highest for SMC and

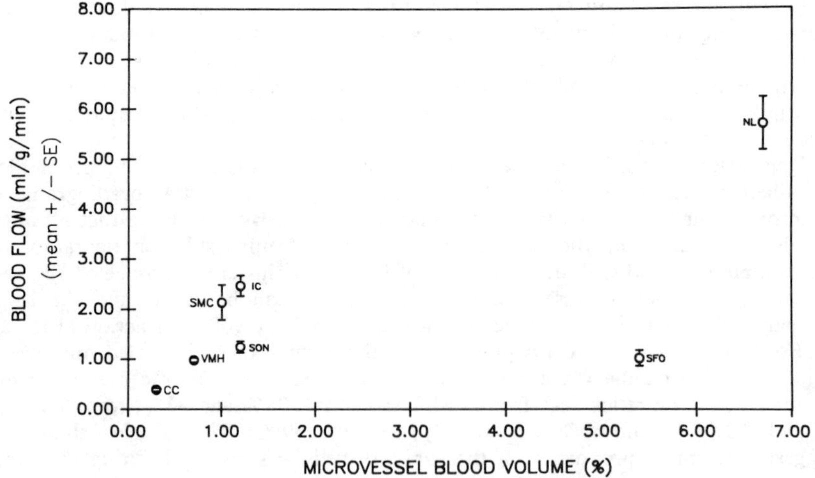

FIGURE 3. Local blood flow rate versus microvessel blood volume.

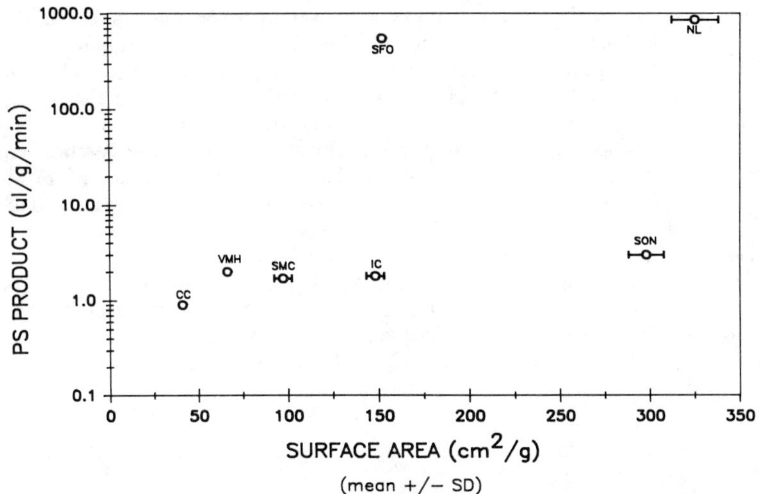

FIGURE 4. PS product for AIB versus capillary surface area.

IC and lowest for SON, and implies that dissimilarities in linear flow velocity account for some of the difference in local rates of blood flow (the first model) as previously reported by Phelps *et al.*[12]

Although the local PS products for AIB of the examined capillary beds varied, the relative range of these variations greatly exceeded that of blood flow; moreover the ranking or order of PS products (e.g., from lowest to highest values) differed somewhat from that of blood flow (FIGS. 1 and 4). In addition, these PS products seemingly have little dependence on capillary surface area (S) as measured morphometrically (FIG. 4). The PS products of the SFO and NL capillary systems for AIB are about 600 and 870 μl/g per min, respectively, and may be underestimations of the real PS products of these capillary beds because of the conditions employed in the experiments and the assumptions used in the calculations (see Refs. 7 and 14 for further details). The PS product for the IC, which has a capillary surface area similar to that of the SFO, is about 2 μl/g per min and for the SON, which has a capillary surface area similar to that of the NL, is about 3 μl/g per min. These results clearly show the several-hundred-fold difference in permeability per se between CVO and non-CVO capillary systems.

FIGURE 4 also indicates that the PS products of VMH, SMC, IC, and SON capillary beds for AIB are quite similar (less than a two-fold difference), but that the surface areas of these capillary systems range from 60 (VMH) to 300 (SON) cm^2/g tissue. This implies that the permeability of these capillaries to AIB per unit surface area (i.e., the P part of the PS product) varies several-fold among these structures. If, however, not all capillaries are perfused in these areas under the conditions of these experiments as previously suggested, then the proper surface area to use for comparison to PS is that of the perfused capillaries. Since the mean diameter of the capillaries is nearly the same for the CC, VMH, SMC, and IC (about 5μ), capillary surface area is proportional to capillary volume fraction among this set of structures, and their perfused capillary surface areas would be expected to be proportional to their microvessel blood volumes (FIG. 3). Using this approach to approximate the "perfused" capillary surface area (S), a plot of PS product versus S was made and a fairly linear relationship was found (graph not shown). This suggests that a major cause of the variation in the PS products of AIB among brain capillary systems is the difference in S. The deviation of this plot from linearity hinted, however, that the permeability of capillaries to AIB *per se* also varied somewhat among brain areas.

Among the various quantitated ultrastructural features of the cerebral capillary endothelium, frequency of fenestrations correlated best with capillary PS product for AIB (FIG. 5). Non-CVO capillaries were unfenestrated, whereas CVO capillaries had numerous fenestrae. Accordingly, fenestrae may function as a pathway for rapid exchange of solutes (e.g., AIB) across CVO capillaries.

The microvesicles of endothelial cells, which are also referred to as pinocytotic vesicles, are often considered to be "ferry boats," which transport various substances and fluid across capillary walls. The frequency of vesicular profiles per capillary profile was found to be about 8 for CC and IC, 25 for SON, 33 for NL, 57 for SFO. The

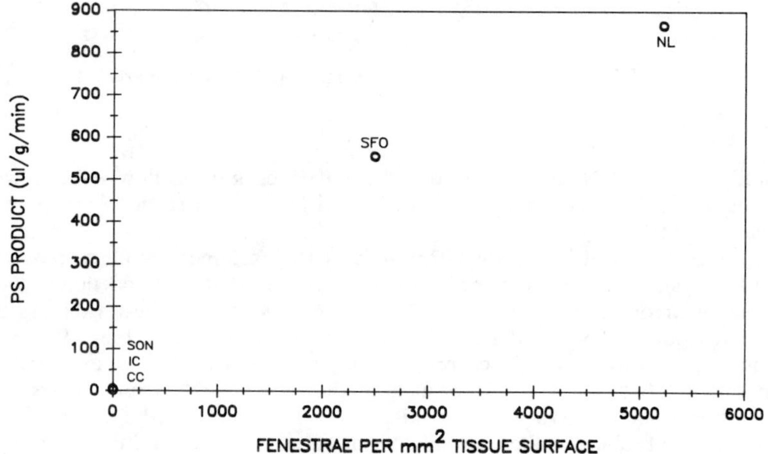

FIGURE 5. PS product for AIB versus frequency of endothelial fenestrations.

number of endothelial-associated vesicular profiles per unit tissue surface or volume were nearly the same for the NL and SFO and only 20% less for the SON. The PS products of the capillary beds of these structures are 3 (SON), 600 (SFO), and 870 (NL) μl/g per min, and thus the PS product for AIB seems to be independent of vesicular frequency. In other words, it appears that few, if any of these vesicles, are "ferrying" AIB across capillary walls in any of these brain structures under these conditions and the frequency of endothelial cell vesicles does not demarcate capillary permeability for this solute.

About a decade ago, Oldendorf and coworkers reported that brain capillaries have more mitochondria than other capillaries and suggested that much of the energy produced by cerebral endothelial mitochondria could be used to transfer materials such as glucose, phenylalanine, and potassium ions across the blood-brain barrier.[15] Our findings indicate that the number of mitochondria per capillary profile in BBB capillary systems are 4.6 (IC), 4.7 (CC), and 5.8 (SON). In contrast, the frequencies of mitochondria per capillary profile in CVO capillary beds were 1.7 (SFO) and 1.0 (NL). These differences imply that CVO capillaries are much less active than BBB capillaries and that the flow of many essential solutes and nutrients across CVO capillaries may not be driven by energy derived from the endothelial cell. That is, most transfer across these capillaries occurs by simple diffusion through aqueous channels (pores) or fenestrae and is not facilitated by carrier systems in the endothelial cell membranes.

SUMMARY

The major hypothesis of this study is that there are differences among brain areas in capillary bed structure and function. Three general differences between circum-ventricular organ and non-CVO capillary beds were found. First, the PS products for AIB were about 300 times greater in CVO capillaries than in non-CVO (blood-brain barrier) capillaries. Second, the frequency of endothelial cell fenestrations was much greater in CVO capillaries than in non-CVO capillaries and the fenestrae may be structural modifications of endothelial cells that permit ready passage of solutes such as AIB. Third, the frequency of mitochondria was greater in BBB capillaries than in CVO capillaries; this high metabolic potential of BBB capillaries may be associated, in part, with "carrier-mediated" transport of various solutes between plasma and cerebral interstitial fluid.

Capillary bed differences among all (i.e., both CVO and non-CVO) brain structures were also observed. Among these differences are: rate of blood flow, mean transit time of albumin, capillary volume and surface area, perfused microvessel blood volume, apparent percentage of perfused capillaries, PS products for AIB, and frequency within the endothelium of vesicular profiles.

ACKNOWLEDGMENTS

The technical assistance of Lisa Rybacki, Grace Richardson, Hiroyuki Nakata, and Atsushi Tajima and the secretarial help of Patricia Milligan are gratefully acknowledged. Thanks are also due to Dr. Thomas Kaye and Paul Stang for critically reading the manuscript.

REFERENCES

1. FENSTERMACHER, J. D. 1983. Drug transfer across the blood-brain barrier. *In* Topics in Pharmaceutical Sciences. D. D. Breimer & P. Speiser, Eds.: 143-154. Elsevier. Amsterdam.
2. FENSTERMACHER, J. D. The pharmacology of the blood-brain barrier. *In* Clinical Impact of the Blood-Brain Barrier and Its Manipulation. E. Neuwelt, Ed. Plenum. New York.

3. BLASBERG, R., P. MOLNAR, D. GROOTHUIS, C. PATLAK, E. OWENS & J. FENSTERMACHER. 1984. Concurrent measurements of blood flow and transcapillary transport in avian sarcoma virus-induced experimental brain tumors: Implications for chemotherapy. J. Pharmacol. Exp. Ther. **231:** 724-735.

4. GOOCHEE, C., W. RASBAND & L. SOKOLOFF. 1983. Application of computer-assisted image processing to autoradiographic methods for studying brain functions. Trends Neurosci. **6:** 257-260.

5. HAWKINS, R. A., A. M. MANS & J. F. BIEBUYCK. 1982. Amino acid supply to individual cerebral structures in awake and anesthetized rats. Am. J. Physiol. **242:** E1-E11.

6. SAKURADA, O., C. KENNEDY, J. JEHLE, J. D. BROWN, G. CARBIN & L. SOKOLOFF. 1978. Measurement of local cerebral blood flow with iodo [^{14}C] antipyrine. Am. J. Physiol. **234:** H59-H66.

7. BLASBERG, R. G., J. FENSTERMACHER & C. PATLAK. 1983. Transport of alpha-amino-isobutyric acid across brain capillary and cellular membranes. J. Cereb. Blood Flow Metab. **3:** 8-32.

8. CREMER, J. E., V. J. CUNNINGHAM & M. P. SEVILLE. 1983. Relationships between extraction and metabolism of glucose, blood flow, and tissue blood volume in regions of rat brain. J. Cereb. Blood Flow Metab. **3:** 291-302.

9. FENSTERMACHER, J. D. & S. I. RAPOPORT. 1984. Blood-brain barrier. *In* Handbook of Physiology — The Cardiovascular System. E. M. Renkin & C. C. Michel, Eds. Vol. **IV:** 969-1000. Williams & Wilkins. Baltimore, MD.

10. BAR, T. 1980. The Vascular System of the Cerebral Cortex. Springer Verlag. Berlin.

11. WEIBEL, E. W., G. S. KISTLER & W. F. SCHERLE. 1966. Practical stereological methods for morphometric cytology. J. Cell. Biol. **30:** 23-38.

12. PHELPS, M. E., S. C. HUANG, E. J. HOFFMAN, C. SELIN & D. E. KUHL. 1981. Cerebral extraction of N-13 ammonia: Its dependence on cerebral blood flow and capillary permeability-surface area product. Stroke **12:** 607-619.

13. FENSTERMACHER, J. D. 1985. Current models of blood-brain transfer. Trends Neurosci. **8:** 449-453.

14. GROSS, P. M., R. G. BLASBERG, J. D. FENSTERMACHER & C. S. PATLAK. 1987. The microcirculation of rat circumventricular organs and pituitary gland. Brain Res. Bull. **18:** 73-85.

15. OLDENDORF, W. H., M. E. CORNFORD & W. J. BRAUN. 1977. The large apparent work capability of the blood-brain barrier: A study of the mitochondrial content of capillary endothelial cells in brain and other tissues of the rat. Ann. Neurol. **1:** 409-417.

Endothelial Cell–Astrocyte Interactions

A Cellular Model of the Blood–Brain Barrier

GARY W. GOLDSTEIN

Departments of Pediatrics and Neurology
University of Michigan
Ann Arbor, Michigan 48109-0570

Recent advances in cell biology open the way for studying aspects of blood-brain barrier (BBB) formation and function that have puzzled investigators since Ehrlich and Goldmann first described the barrier. Although the endothelial cells in the microvessels of all tissues appear to arise from common mesodermal precursors, considerable organ-to-organ variability exists in their ultrastructure and permeability. The most extreme example of restricted microvascular permeability is found in brain, where the endothelial cells are sealed together by continuous complex tight junctions to form a polarized "epithelium" that regulates the passage of small organic molecules and ions between blood and brain.[1]

A constant feature of brain microvessels with barrier properties is an almost total investment with processes from astrocytes.[2] This contact is so extensive that the astrocytes were once thought to physically create the BBB. We now know from the classic tracer studies of Reese, Karnovsky, and Brightman that the anatomic basis of the BBB is provided by the unfenestrated endothelial cells with their continuous and complex tight junctions.[3] Although astrocytic processes encircle the capillaries and attach to a basement membrane shared with the endothelial cells, the foot processes are not sealed to each other and small gaps between the astrocytes allow passage of proteins and other polar molecules in the interstitial fluid up to the abluminal surface of the endothelial cells (FIG. 1).

Even though the astrocytes do not create a physical barrier, their close contact with brain microvessels suggests that they have an important role in the function of the BBB.[4] Evidence for a direct influence of brain cells upon microvascular function is provided by the elegant chimeric experiments of Stewart and Wiley.[5] These investigators implanted embryonic avascular quail brain into the abdominal cavity of a chick embryo. Because of the marked difference in nuclear structure between the quail and the chicken, Stewart and Wily could prove that the blood vessels that formed in the transplanted brain arose from the abdominal vasculature of the chicken (FIG. 2). Despite their systemic origin, the chick microvessels growing into the quail brain exhibited the restricted permeability, special structure, and enzymes characteristic of the BBB. In contrast, microvessels growing into embryonic quail muscle implanted into chick brain were highly permeable and lacked BBB enzymes even though they were derived from chick brain. These studies demonstrate a direct influence of brain upon endothelial cell differentiation.

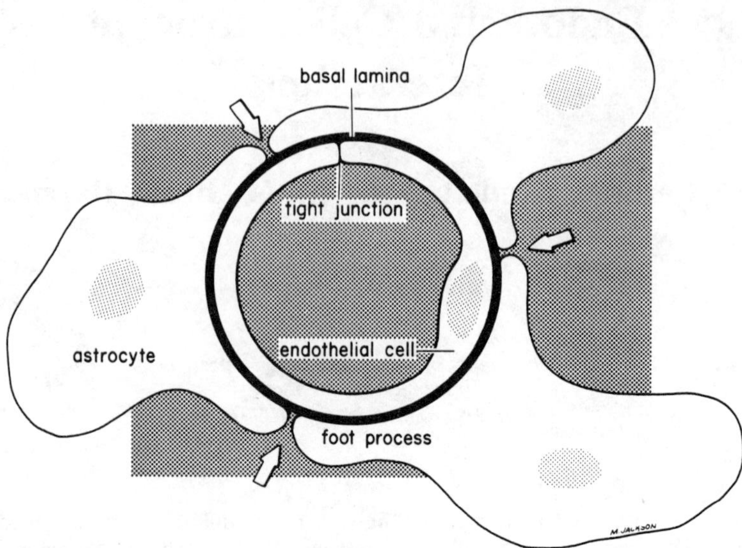

FIGURE 1. Diagram illustrating the close association of endothelial cells in brain capillaries with foot processes extending from astrocytes. The blood-brain barrier is produced by the continuous endothelium. The astrocytes encircle the microvessels, but are not sealed together and interstitial fluid has access (*arrows*) to the basement membrane and abluminal surface of the endothelial cell.

The astrocyte seems a likely candidate for mediating this effect of brain tissue upon capillary structure and function. Ultrastructural studies of regions of the brain without a barrier, astrocytic brain tumors, and microvessels in the retina support this hypothesis. Several small regions of the brain stem and hypothalmus are involved in neuroendocrine feedback and are more permeable than the rest of the brain.[6] Microvascular endothelial cells in these areas are fenestrated and lack the special features found elsewhere in the brain. The characteristic close apposition of astrocytic processes to the endothelial cell is also absent in these regions, leaving open spaces between the basement membrane of the capillary and the nearby astrocytes (FIG. 3). Similarly, the highly permeable microvessels within astrocytic brain tumors also lack close contact with the transformed astrocytes.[7] In contrast, microvascular permeability in the retina, like the brain, is markedly restricted and contributes to the formation of the blood-retinal barrier.[8] Consistent with the proposed role of the astrocyte in brain capillary differentiation, the retinal microvessels are also surrounded by foot processes from astrocytes[9] (FIG. 4).

The constant close association between astrocytes and endothelial cells in areas of the brain and retina with a barrier, and the lack of these contacts in tumors and brain regions without a barrier together with the transplantation studies, provide the *in vivo* evidence for attributing formation of the BBB to an interaction between astrocytes and endothelial cells. The development of methods to isolate endothelial cells and astrocytes from brain and to grow the two cell types separately in culture provides a new approach for studying features of their interaction that may be important to formation of the BBB.

Highly purified suspensions of microvessels can be isolated from brain tissue. These preparations have been particularly useful in characterizing the transporters involved in moving solutes into and out of the endothelial cells. Since the endothelial cell is an obligate part of the pathway for passage of polar molecules across the BBB, the ability to isolate microvessels provided the first direct approach to study the cells controlling exchange of molecules across the BBB. During the process of isolation, the astrocytes are stripped away from the capillary walls. A recent review describing the use of microvessels isolated from brain to study cellular characteristics of the BBB is available elsewhere.[10]

When isolated brain microvessels are treated with collagenase or other preparative enzymes, it is possible to separate and grow the endothelial cells in tissue culture. Work in our laboratory by Bowman *et al.*[11] and Dorovini-Zis *et al.*[12] established that primary cultures of brain endothelial cells retain some barrier features. Using either rat-tail collagen or fibronectin as an attachment substrate, brain microvascular endothelial cells form a continuous monolayer. The cells are connected by tight junctions that restrict the intercellular passage of horseradish peroxidase, a protein similar in size to plasma albumin (FIG. 5). The endothelial cells are not fenestrated and contain few pinocytotic vesicles. The tight junctions can be reversibly separated by exposing the monolayers to hypertonic solutions of arabinose and show a dose-response similar to that which opens the BBB *in vivo* after infusion of the same hypertonic solutions

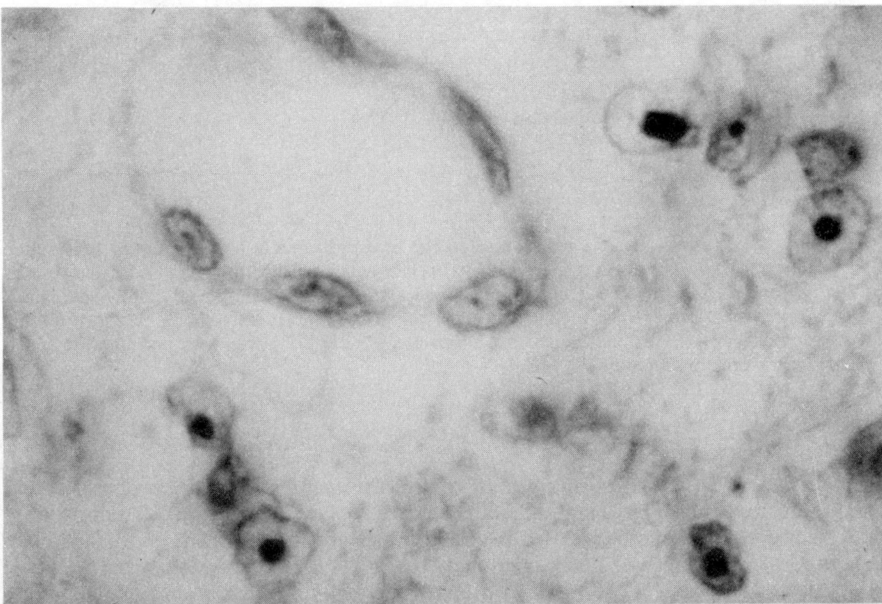

FIGURE 2. Chick endothelial cells vascularizing a graft of quail brain. The nuclei in astrocytes surrounding the microvessel have a distinctive large central nucleolus consistent with their quail origin, whereas the endothelial nuclei have the typical appearance of chick host. Despite the origin of the endothelial cells from the abdominal vasculature, they exhibited the permeability and enzymatic characteristics of the blood-brain barrier.[5] (This photomicrograph was provided by Drs. Stewart and Wiley and is printed with their permission.)

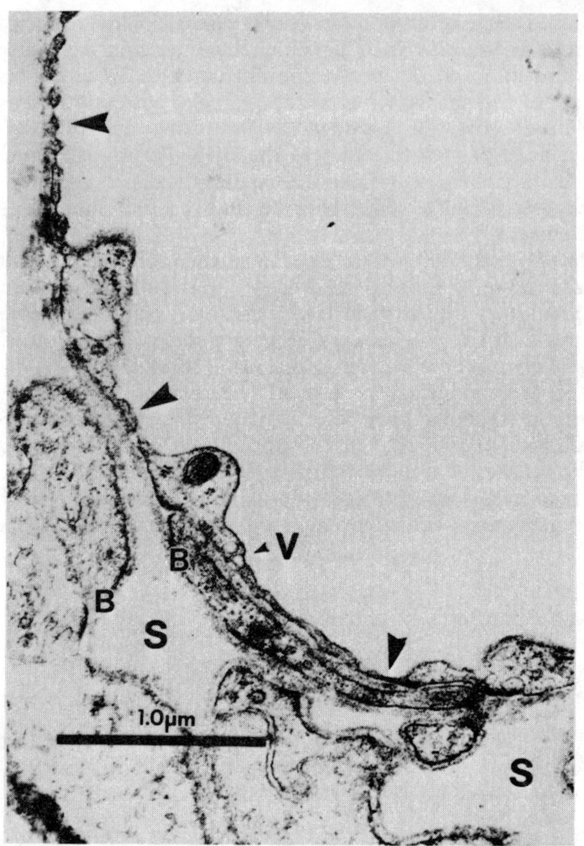

FIGURE 3. Ultrastructure of a microvessel in the area postrema of a mouse brain. This region lacks a blood-brain barrier and the endothelial cell is fenestrated (*arrows*) and contains vesicles (V). The astrocyte in the *lower left corner* is separated from the endothelial cell by spaces (S); the basement membrane (B) on the astrocyte is distinct from the basement membrane surrounding the endothelial cell. (This photomicrograph was provided by Drs. Coomber and Stewart and is printed with their permission.)

into the carotid artery.[13] Using a similar preparation, Audus and Borchardt found saturable transport of large neutral amino acids across the endothelial cell monolayer.[14]

Astrocytes can also be isolated from brain and grown in culture. Some studies of endothelial cell-astrocyte interaction have used a cell line of rat astrocytes (C-6, American Cell Tissue Type) derived from a chemically induced brain tumor. Despite the fact that microvessels growing into implants of C-6 cell tumors *in vivo* are highly permeable and lack BBB characteristics, the C-6 astrocytes *in vitro* were shown to stimulate passaged endothelial cells to exhibit many features typical of normal brain capillaries. In these studies DeBault, Cancilla, and subsequently Beck, and their coworkers isolated an endothelial cell line from mouse brain. These passaged cells when grown as a single cell type did not exhibit barrier features. However, when co-cultured with C-6 glioma cells, many properties associated with the intact barrier could be identified. These include appearance of γ-glutamyl transpepidase activity in the endothelial cell,[15] evidence for polarity of small neutral amino acid transport,[16] and by cytochemical analysis, induction of sodium-potassium ATPase and alkaline phosphatase.[17] Despite the tumor origin of the glial cells, these studies demonstrate a positive interaction between endothelial cells and astrocytes.

Methods are now available to isolate and culture astrocytes from normal newborn rat brain.[18] These nontransformed cells can be propagated for at least several passages and contain glial fibrillary acidic protein, which provides a useful marker of their astrocytic origin. Using astrocytes from neonatal brain, Tao-Cheng, Nagy, and Brightman examined the effect of co-culture upon the freeze-fracture appearance of tight junctions between brain endothelial cells.[19] They found that endothelial cells prepared from bovine brain no longer formed tight junctions after several cell passages, even though they retained other markers of their endothelial cell origin. However, when the passaged endothelial cells were co-cultured with astrocytes, tight junctions appeared that resembled the complex junctions seen *in situ*. These *in vitro* studies demonstrate the ability of C-6 glioma and neonatal astrocytes to influence the morphology and metabolic activity of cultured brain endothelial cells. The biochemical basis for this interaction is not yet understood.

Brain tissue contains several growth factors as well as insoluble matrix material that may act as signals for the endothelial cell-astrocyte interactions important for formation and maintenance of the BBB. For example, extracts of both immature and mature rat brain have a potent mitogenic effect upon brain endothelial cells in culture.[20]

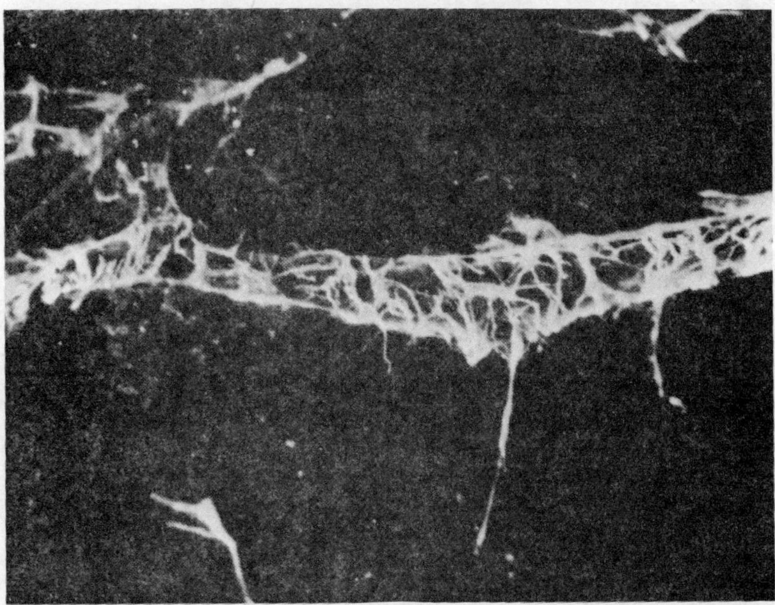

FIGURE 4. Retinal microvessel surrounded by glial processes. Consistent with their barrier features, endothelial cells in the retina have close contact with astrocytes. In the micrograph the astrocytic processes are stained with antibody against glial fibrillary acidic protein. (From Bjorklund and Dahl.[9] Reprinted by permission from the *Journal of Neuroimmunology.*)

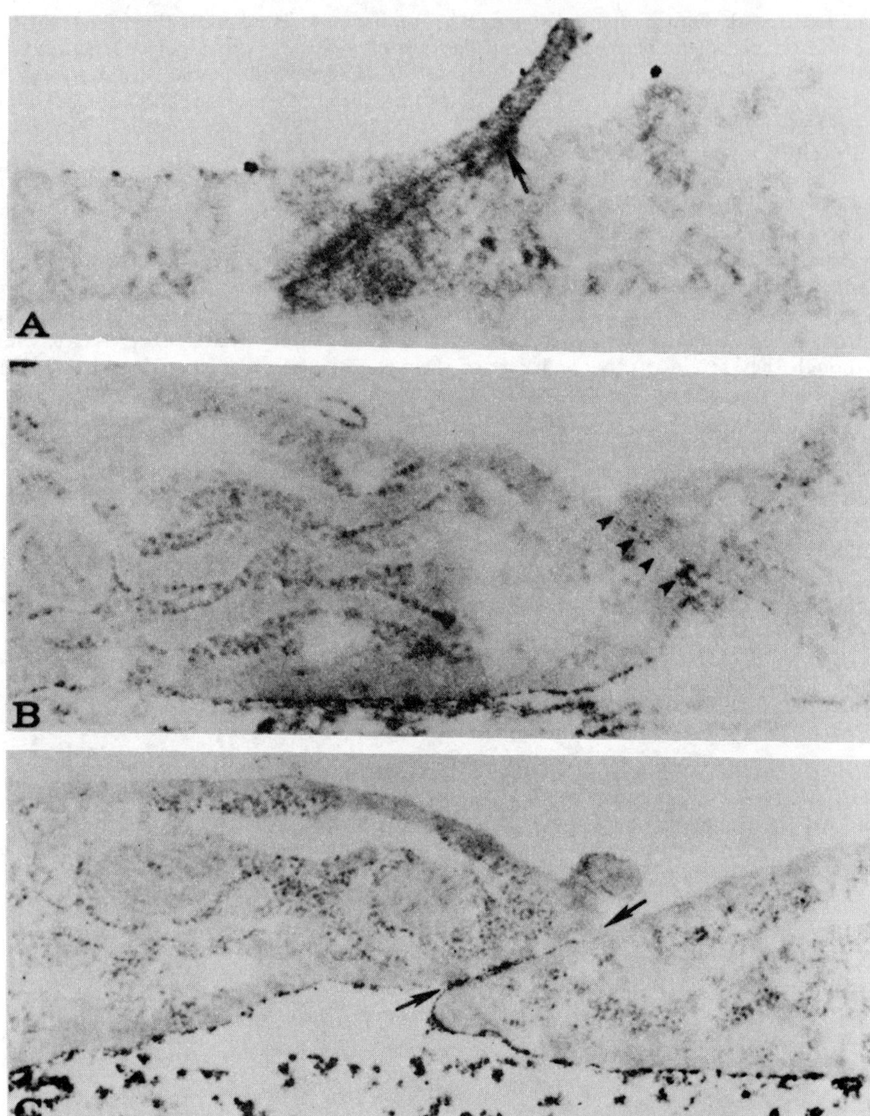

FIGURE 5. Formation of an *in vitro* barrier by brain endothelial cells: effect of hypertonic arabinose. Transcellular movement of the protein tracer horseradish peroxidase (HRP) was monitored by appearance of its reaction product. (**A:**) Control; 5 minutes after exposure to HRP, the cleft between two adjacent endothelial cells contains no tracer. The HRP forms small discrete patches on the surface of the cells, but does not penetrate beyond the first tight junction (*arrow*). The basal surface is free of HRP. ×82,500. (**B:**) Two-minute exposure to 1.6 M arabinose. The extracellular spaces between successive tight junctions (*arrowheads*) contain deposits of HRP. The basal surfaces are labeled with small amounts of the tracer. ×66,000. (**C:**) Five-minute exposure to 1.6 M arabinose HRP penetrates throughout the entire length of an interendothelial cleft (*between arrows*) and forms dense deposits on the basal surface of the cells. ×39,000. (From Dorovini-Zis *et al.*[12] Reprinted by permission from *Brain Research.*)

It is possible to purify this activity from brain, and on the basis of affinity for heparin, molecular weight, and isoelectric constants, it has homology with fibroblast growth factor[21] and endothelial cell growth factor.[22] In addition to having a mitogenic effect upon brain endothelial cells, these factors can induce differentiation in cultured astrocytes, producing a transition from a spherical cell to one with multiple extended processes.[23]

The cellular origins of the various brain factors important in signaling endothelial cell and astrocytic growth, differentiation, and interaction are just now being investigated. It seems likely that signals are operating in both directions. For example, endothelial cells can produce platelet-derived growth factor,[24] for which receptors linked to proliferation exist on the astrocyte[25] but not on the endothelial cell, whereas

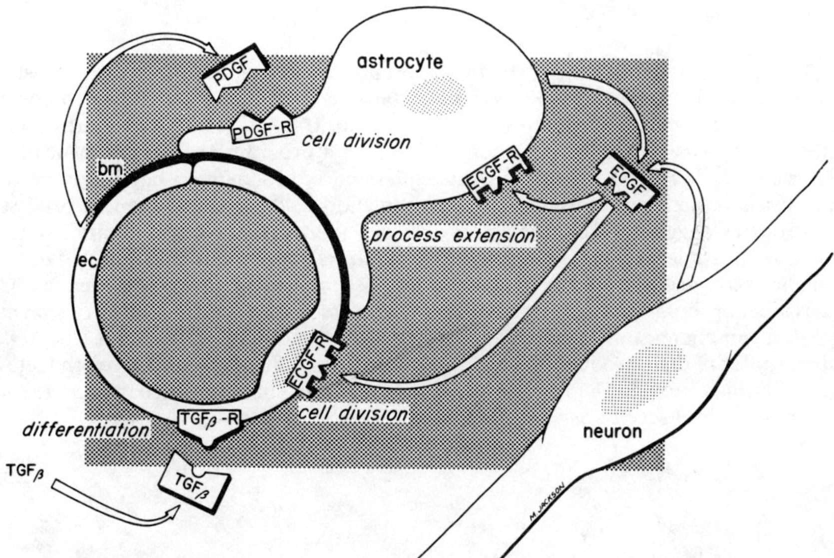

FIGURE 6. A diagram illustrating possible interactions between brain endothelial cells (ec) and astrocytes. Receptors for growth factors important in regulation of proliferation and differentiation are found when these cells are maintained in culture. Platelet-derived growth factor (PDGF) can be synthesized by endothelial cells and has a mitogenic effect upon glial cells. Endothelial cell growth factors (ECGF) are synthesized by neurons or astrocytes and stimulate endothelial cells to proliferate and glial cells to differentiate. Transforming growth factor β (TGFβ) blocks the mitogenic effect of other growth factors and may cause endothelial cells to differentiate. Finally, basement membrane (bm) formed by the endothelial cells and astrocytes may influence the behavior of both cell types.

astrocytes may produce factors such as endothelial cell growth factor, for which receptors exist on both the endothelial cell (linked to proliferation) and the astrocyte itself (linked to differentiation).[23] Transforming growth factor-β, a newly discovered peptide mediator found in many tissues,[26] is able to counter the mitogenic effect of endothelial cell growth factor, presumably by causing differentiation of the endothelial cell and loss of its responsiveness to endothelial cell growth factor.[27] Extracellular matrix (i.e. basement membrane) also appears important in cell interactions, either as a potential signal itself or as a permissive factor for response to the soluble peptides. FIGURE 6 is a hypothetical model showing how some of these signals might coordinate the interaction of endothelial cells and astrocytes to form the BBB.

Considerable information is available concerning the many special anatomic, physiologic, and biochemical features that produce the BBB. The vulnerability of this complex system to injury during different stages of development is also well characterized. Investigations now in progress at a cell and molecular level should extend our understanding of the BBB and encourage development of new techniques to manipulate the barrier for treatment of the neurologic diseases that either damage brain microvessels or require selective bypass to deliver drugs to specific brain regions.

SUMMARY

Microvascular endothelial cells in the brain have a number of special properties that underlie formation of the blood-brain barrier (BBB) and contribute to control of the neuronal microenvironment. Evidence from transplantation experiments indicates that signals arising within brain rather than a programmed commitment of the endothelial cells are responsible for the expression of blood-brain barrier properties. The close anatomic relationship between brain endothelial cells and the foot processes of astrocytes suggests a role for astrocytes as a source of the differentiation signals. It is now possible to isolate and separately culture populations of brain-derived endothelial cells and astrocytes. When the two cell types are grown together, a characteristic morphologic organization occurs that is associated with induction of enzymes and tight junctions similar to those found *in vivo*. Endothelial cells and astrocytes in culture differ in their production of and response to specific polypeptide growth factors. These findings provide the basis for a model of endothelial cell-astrocyte interaction that may explain several aspects of BBB behavior.

REFERENCES

1. GOLDSTEIN, G.W. & A. L. BETZ. 1986. The blood-brain barrier. Sci. Amer. **254:** 74-83.
2. WOLFF, J. R. 1970. The astrocyte as link between capillary and nerve cell. Triangle **9:** 153-164.
3. BRIGHTMAN, M.W. 1977. Morphology of the blood-brain barrier. Exp. Eye Res. **Suppl:** 1-25.
4. BRADBURY, W.B. 1984. The structure and function of the blood-brain barrier. Fed. Proc. **43:** 186-190.
5. STEWART, P.A. & M.J. WILEY. 1981. Developing nervous tissue induces formation of blood-brain barrier characteristics in invading endothelial cells: A study using quail-chick transplantation chimeras. Dev. Biol. **84:** 184-192.
6. COOMBER, B.L. & P. A. STEWART. 1985. Morphometric analysis of CNS microvascular endothelium. Microvas. Res. **30:** 99-115.
7. LONG, D. M. 1970. Capillary ultrastructure of the blood-brain barrier in human malignant brain tumors. J. Neurosurg. **32:** 127-144.
8. RAVIOLA, G. 1977. The structural basis of the blood-ocular barrier. Exp. Eye Res. Suppl. **25:** 27-63.
9. BJORKLUND, H. & D. DAHL. 1985. Glial fibrillary acidic protein (GFAP)-like immunoreactivity in the rodent eye. J. Neuroimmunol. **8:** 331-345.
10. BETZ, A. & G. W. GOLDSTEIN. 1984. Brain Capillaries: Structure and function. *In* Handbook of Neurochemistry. A. Lajtha, Ed. Vol. **7:** 465-484. Plenum Press. New York.

11. BOWMAN, P.D., S. R. ENNIS, K. D. RAREY, A. L. BETZ & G. W. GOLDSTEIN. 1983. Brain microvessel endothelial cells in tissue culture: A model for study of blood-brain barrier permeability. Ann. Neurol. **14:** 396-402.

12. DOROVINI-ZIS, K., P. D. BOWMAN, A. L. BETZ & G. W. GOLDSTEIN. 1984. Hyperosmotic arabinose solutions open the tight junctions between brain capillary endothelial cells in tissue culture. Brain Res. **302:** 383-386.

13. RAPOPORT, S. I. & H. M. KLATZOL. 1972. Testing of a hypothesis for osmotic opening of the blood-brain barrier. Am. J. Physiol. **223:** 323-331.

14. AUDUS, K. & R. T. BORCHARDT. 1986. Characteristics of the large neutral amino acid transport system of bovine brain microvessel endothelial cell monolayers. J. Neurochem. **47:** 484-488.

15. DEBAULT, L. E. & P. A. CANCILLA. 1980. γ-Glutamyl transpeptidase in isolated brain endothelial cells: induction by glial cells in vitro. Science **207:** 653-655.

16. BECK, D. W., H. V. VINTERS, M. N. HART & P. A. CANCILLA. 1984. Glial cells influence polarity of the blood-brain barrier. J. Neuropathol. Exp. Neurol. **43:** 219-224.

17. BECK, D. W., R. L. ROBERTS & J. J. OLSON. 1986. Glial cells influence membrane-associated enzyme activity at the blood-brain barrier. Brain Res. **381:** 131-137.

18. FRANGAKIS, M. V. & H. K. KIMELBERG. 1984. Dissociation of neonatal rat brain by dispase for preparation of primary astrocyte cultures. Neurochem. Res. **9:** 1689-1698.

19. TAO-CHENG, J.-H., Z. NAGY & M. W. BRIGHTMAN. 1987. Tight junctions of cerebral endothelium in vitro are enhanced in the company of glia. J. Neurosci. **7:** 3293.

20. ROBERTSON, P. L., M. DU BOIS, P. D. BOWMAN & G. W. GOLDSTEIN. 1985. Angiogenesis in developing rat brain: an in vivo and in vitro study. Develop. Brain Res. **23:** 219-223.

21. ABRAHAM, J. A., A. MERGIA, J. L. WHANG, A. TUMOLO, J. FRIEDMAN, K. A. JHERRILD, D. GOSPODAROWICZ & J. C. FIDDES. 1986. Nucleotide sequence of a bovine clone encoding the angiogenic protein, basic fibroblast growth factor. Science. **233:** 545-548.

22. JAYE, M., R. HOWK, W. BURGESS, G. A. RICCA, I.-M. CHIU, M. W. RAVERA, S. J. O'BRIEN, W. S. MODI, T. MACIAG & W. N. DROHAN. 1986. Human endothelial cell growth factor: Cloning, nucleotide sequence, and chromosome localization. Science. **233:** 541-545.

23. MORRISON, R. S., J. DE VELLIS, Y. L. LEE, R. A. BRADSHAW & L. F. ENG. 1985. Hormones and growth factors induce the synthesis of glial fibrillary acidic protein in rat brain astrocytes. J. Neurosci. **14:** 167-176.

24. COLLINS, T., D. GINSBURG, J. M. BOSS, S. H. ORKIN & J. S. POBER. 1985. Cultured human endothelial cells express platelet-derived growth factor B chain: cDNA cloning and' structural analysis. Nature. **316:** 748-750.

25. HELDIN, C. H., B. WESTERMARK & A. WATESON. 1981. Special receptors for PDGF on cells from connective tissue and glia. Proc. Natl. Acad. Sci. USA **78:** 3664-3668.

26. SPORN, M. B., A. B. ROBERTS, L. M. WAKEFIELD & R. K. ASSOIAN. 1986. Transforming growth factor-β: biological function and chemical structure. Science. **233:** 532-534.

27. SCHRODER, M., G. MULLER, W. BIRCHMEIER & P. BOLEN. 1986. Transforming growth factor-beta inhibited endothelial cell proliferation. Biochem. Biophys. Res. Commun. **37:** 295-302.

Regional Transport of Some Essential Nutrients across the Blood-Brain Barrier in Normal and Diseased States[a]

RICHARD A. HAWKINS,[b,c,d] ANKE M. MANS,[c]
LYNDON S. HIBBARD,[c,e] DONALD W. DAVIS,[c] AND
JULIEN F. BIEBUYCK[c]

Departments of [c]Anesthesia, [d]Physiology, and [e]Radiology
Hershey Medical Center
The Pennsylvania State University
College of Medicine
Hershey, Pennsylvania 17033

INTRODUCTION

The brain requires a continuous supply of essential nutrients in order to maintain normal cerebral function: glucose and ketone bodies for energy metabolism, amino acids for protein and neurotransmitter synthesis, choline for acetylcholine synthesis, purines for nucleic acid synthesis, and so forth. The availability of these nutrients is determined by the blood-brain barrier as well as their plasma concentrations. The movement of nutrients into the brain is mediated by specific transport mechanisms in the capillary endothelial cells, which constitute the blood-brain barrier. Several transport systems have now been described and the presence of these systems can be taken as evidence of the brain's requirement for the substrates carried.[1] In normal animals, the blood-brain barrier mediates a delicate balance between supply and need, but in metabolic disorders such as hepatic encephalopathy, starvation or diabetes the permeability characteristics of several transport systems can be altered and this may be of etiologic significance to the development of cerebral dysfunction.

[a]This work was supported by Grants NS16389 and NS16737 from the National Institutes of Health, Grant BNS8506479 from the National Science Foundation and by the American Diabetes Foundation.

[b]Address for correspondence: Richard A. Hawkins, Department of Anesthesia, Hershey Medical Center, P.O. Box 850, Hershey, Pennsylvania 17033.

STUDY OF THE BLOOD-BRAIN BARRIER AT THE LEVEL OF INDIVIDUAL BRAIN STRUCTURES

Our interest has been in determining the substrate movement into individual brain structures as it occurs *in vivo* in normal and pathologic situations. Therefore, we developed a method for use in laboratory rats based on quantitative autoradiography.[f] Briefly, after arterial and venous catheters are placed, a tracer quantity of ^{14}C-labeled substrate is infused intravenously in such a way as to establish and maintain a steady plasma level of ^{14}C. Initially the accumulation of label in the brain is linear with time. The radioactivity in the brain relative to the radioactivity and substrate concentration in the blood is used to calculate either the permeability-to-surface-area product (PA) or the substrate flux into any cerebral structure with an intact blood-brain barrier.[2] Thus PA is calculated:

$$PA = {}^{14}C \text{ in brain/integral of plasma } {}^{14}C$$

while PA is a function of the transport system and substrate concentration:

$$PA = T_{max}/[S + K_t(1 + \Sigma I/K_i)] + K_d$$

where T_{max} = maximal transport capacity, S = substrate concentration, I = concentration of competing substrate, K_t = affinity of the carrier for substrate, K_i = affinity of carrier for competing substrate, K_d = carrier-independent clearance. Accordingly it can be seen that PA will be a function of the substrate concentration and the concentration of competing substrates. The units of PA are the same as flow or clearance (μl.min^{-1}.g^{-1}) and in fact the method may be thought of as determining the initial rate of clearance. The influx of a particular substrate is given by:

$$\text{Influx} = (PA) \times (S).$$

The amount of radioactive tracer taken up by the brain is measured by quantitative autoradiography. After the experiment, the brain is removed and placed in a beaker containing a two-phase mixture of a heavy (1-bromobutane) and a light (2-methyl-butane) organic liquid maintained at about $-30°$ centigrade. The brain floats gently at the interface of the liquids and is frozen in about 2 minutes without apparent distortion. When the brain is hard, it is removed and sectioned. The sections are placed on glass slides, dried and incubated with photographic film along with a set of calibrated methylmethacrylate standards which allow the conversion of optical densities to carbon dpm/g.

The optical densities of individual brain structures can be measured manually with an ordinary photodensitometer. This procedure is quite satisfactory for obtaining a limited amount of information and it is usually the first step in determining if and where changes in transport are occurring. A more complete, but more labor-intensive

---

[f]The characteristics of the various transport systems have been studied mostly by the "Old-endorf technique," a convenient dual-label method whereby the uptake of a substance from a defined medium is measured following a single pass through the cerebral capillary bed.[1] The solution is injected into one of the carotid arteries and uptake is measured in a portion of the ipsilateral cerebral hemisphere, and the technique is used to determine relative permeability primarily in the cerebral cortex. The method was not designed for determining influx of the substrate as affected by factors such as plasma competitors, which are normally present in circulation.

procedure involves the use of computers and scanning microdensitometry to reconstruct the data set.[3]

In experiments where three-dimensional maps of cerebral nutrition are made, every fourth brain section is digitized using a scanning microdensitometer at a resolution of 100 μm. Each image comprises 10,000 to 15,000 individual measurements of optical density. Each of the optical density measurements are converted to values of PA or influx using appropriate algorithms. The sections are then placed in register with each other by a set of programs that reorients the images on the basis of their intrinsic mathematical properties. Once the data have been assembled in a three-dimensional matrix they can be viewed in a variety of ways. FIGURE 1 shows transverse sections of the apparent PA of phenylalanine in normal and portacaval-shunted rats. It is also possible to combine several whole brains together thereby forming an average brain where each point has a mean and standard deviation. This enables the comparison of experimental and control brains for global patterns of statistically significant changes. Such analyses are possible because of accurate and objective image registration obtained using digital image processing techniques.

The infusion-autoradiographic method is expected to slightly underestimate the actual value of plasma-to-brain flux *in vivo* because no correction is made for efflux (although it should be kept in mind that experiments are designed so that efflux is minimal). Nevertheless the results compare well with those predicted from kinetic constants determined by the Oldendorf technique. In fact, in many instances the values obtained with the infusion-autoradiographic method are somewhat higher than predicted. An important limitation of the infusion-autoradiographic method is that it is difficult, although not impossible, to determine kinetic constants. (Kinetic constants have been determined in whole brain by others, using sophisticated infusion programs to establish and maintain different circulating concentrations *in vivo*.[1,4])

By means of the infusion-autoradiographic method the transport of several substrates has been studied in normal and metabolic disease states of rats[2,5–9] (TABLES 1-3). These studies, summarized below, have shown that transport varies considerably between the different brain structures and they have made it possible to identify the location and magnitude of some permeability changes that may have important consequences for cerebral function.

RESULTS FROM SELECTED STUDIES

Glucose Transport

Glucose is carried across the blood-brain barrier by a transport system that obeys saturation kinetics.[1] It is well known that regional cerebral glucose utilization and regional cerebral blood flow are closely coupled. The autoradiographic method showed that glucose influx was likewise closely related to cerebral glucose utilization and blood flow.[6] Thus, glucose influx, blood flow and glucose utilization are closely coordinated processes. Glucose influx, which is about 50% greater than utilization, is governed by the number of carriers available, the kinetic constants of the carrier, and plasma glucose concentrations. Cremer and her associates found that glucose transport is a function of glucose metabolism; when metabolism is increased, glucose transport follows suit.[10] The mechanism by which glucose transport is modulated, however,

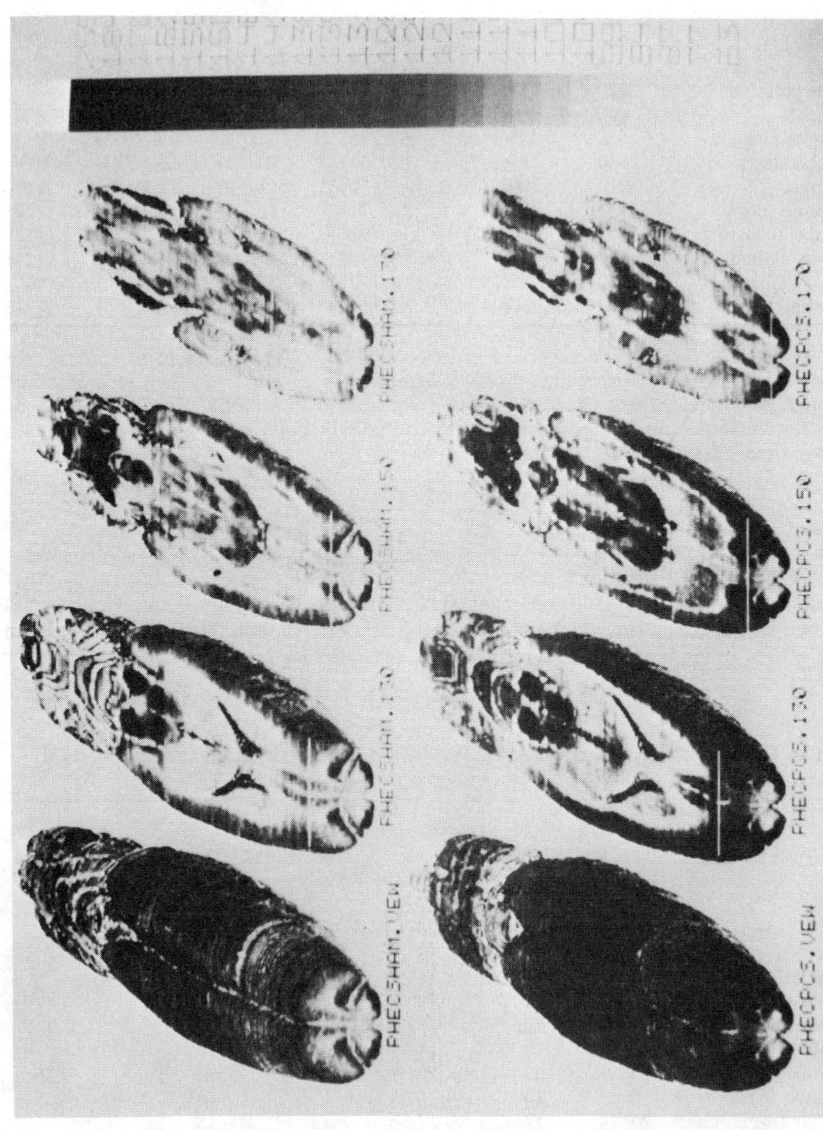

FIGURE 1. Autoradiographs of coronal sections were automatically aligned with respect to a fixed coordinate system and put together by computer.[3] The views shown are stereograms of the apparent phenylalanine PA in a normal rat (*top row*) and a portacaval-shunted rat (*bottom row*). The adjacent key relates gray levels to PA expressed in $\mu l \cdot min^{-1} \cdot g^{-1}$.

TABLE 1. Influx and Apparent Permeability-to-Surface-Area Product of Several Essential Substrates in Selected Brain Structures of Normal Rats

	Glucose Utilization	Glucose	Phenylalanine	Tryptophan	Leucine	Lysine
Frontal cortex	940	1,400(233)	8.3(127)	4.2(23)	15.4 (98)	9.3(26)
Caudate nucleus	930	1,150(191)	6.7(102)	3.1(17)	11.0 (70)	5.8(17)
Globus pallidus	510	948(158)	5.4 (82)	2.3(13)	9.0 (58)	3.6(10)
Amygdala	680	966(161)	6.6(101)	3.0(17)	11.2 (72)	7.3(21)
Hippocampus	660	936(156)	5.8 (88)	2.6(14)	9.7 (64)	5.0(13)
Hypothalamus	640	1,180(197)	7.9(121)	3.7(21)	13.4 (88)	6.2(18)
Thalamus	1,010	1,390(231)	8.8(130)	4.6(25)	15.1 (98)	9.5(27)
Substantia nigra	730	1,270(211)	7.4(112)	4.2(23)	13.7 (87)	5.9(17)
Reticular formation	710	1,070(178)	5.8 (98)	2.6(14)	11.2 (73)	6.0(17)
Superior colliculus	1,030	1,930(321)	7.9(120)	4.4(25)	14.6 (95)	9.7(27)
Inferior colliculus	1,760	2,110(352)	11.3(172)	7.2(40)	21.0(136)	14.0(40)
Cerebellar gray	970	1,460(243)	7.8(119)	4.0(22)	15.9(105)	7.5(21)

NOTE: All influx values are expressed as $nmol \cdot min^{-1} \cdot g^{-1}$. The values in parentheses are apparent permeability-to-surface-area products expressed in $\mu l \cdot min^{-1} \cdot g^{-1}$. All measurements were made on normal fed rats. Glucose utilization was measured with [6-^{14}C]glucose in unstressed rats.[20] Glucose values are calculated for a plasma concentration of 6 $\mu mol/ml$. Values for amino acids are from Mans *et al.*[7]

remains open. There are at least three possibilities. First more or less capillaries are available for blood flow and glucose transport, depending on the metabolic activity. Second, the number of capillaries that are open may be constant, but transport is regulated by adjusting the number of transport carriers present. Third, transport from either side is influenced by the concentration of substrate on the other side.

TABLE 2. Permeability-to-Surface-Area Product of 3-Hydroxybutyrate in Normal and Metabolic Disease States

	Fed	Starved		Diabetic		Portacaval-Shunted for
		2 days	4 days	6 days	4 weeks	7–8 weeks
Frontal cortex	16	28	26	29	30	5
Caudate nucleus	9	17	17	20	17	4
Globus pallidus	6	12	11	13	11	4
Amygdala	14	21	19	26	22	4
Hippocampus	11	19	19	19	18	5
Hypothalamus	8	14	14	14	18	6
Thalamus	12	20	20	24	22	4
Substantia nigra	9	14	15	15	14	4
Reticular formation	7	12	12	13	13	2
Superior colliculus	15	21	22	17	15	6
Inferior colliculus	18	23	23	27	30	5
Cerebellar gray	11	15	17	18	17	4

NOTE: All values are expressed as $\mu l \cdot min^{-1} \cdot g^{-1}$. For more details see Hawkins *et al.*[8]

The close relationship between glucose transport and metabolism is borne out by the observation that the glucose content of various brain regions is almost identical whereas glucose utilization differs markedly throughout the brain. Homogeneous glucose content could only occur if glucose transport and glucose use were perfectly matched. A conundrum exists in hepatic encephalopathy. In this disease glucose utilization is decreased by about 20-30%.[11] It may be expected that brain glucose content would rise during this disease as a result of decreased utilization, but instead brain glucose content is decreased by half.[12,13] The simplest explanation for this observation would be that glucose influx is decreased, yet no evidence of decreased influx could be found by the Oldendorf method.[14] We studied this subject more closely by quantitative autoradiography and found that glucose influx was decreased but only to that extent expected by decreased glucose consumption.[15] In other words, the decrease in transport was not enough to account for the very marked decrease in brain glucose content. The observation that brain glucose content is decreased in

TABLE 3. Changes in the Apparent Permeability-to-Surface-Area Product of Some Amino Acids in Hepatic Encephalopathy

	Phenylalanine	Tyrosine	Tryptophan	Leucine	Lysine
Frontal cortex	+80%	+63%	+204%	+33%	−63%
Caudate nucleus	+79	+78	+229	+44	−78
Globus pallidus	+72	+82	+192	+29	−75
Amygdala	+84	+94	+253	+38	−69
Hippocampus	+106	+86	+257	+53	−66
Hypothalamus	+40	+41	+133	+9	−78
Thalamus	+86	+59	+196	+31	−80
Substantia nigra	+65	+61	+135	+11	−80
Superior colliculus	+83	+63	+208	+37	−63
Inferior colliculus	+72	+60	+145	+22	−80
Cerebellar gray	+88	+87	+281	+62	−77

NOTE: All values are expressed as % change compared to control. Permeability-to-surface-area products were measured 7-8 weeks after a portacaval shunt operation. For other details see Mans *et al.*[7]

hepatic encephalopathy remains without explanation, but it could be explained by an asymmetry in the transport system.

Transportation of Ketone Bodies in Starvation, Diabetes, and Hepatic Encephalopathy.

Ketone bodies are an alternate fuel of brain energy metabolism—the only other fuel besides glucose that brain can rely on to any appreciable extent. Ketone bodies are metabolized as fast as they enter brain; the blood-brain barrier is the rate-limiting step.[1,8,16] Autoradiographic experiments have shown that the ketone-body-to-glucose consumption ratio is not constant throughout brain.[17] That is, ketone bodies can provide more nourishment relative to glucose in some areas than in others.

Evidence is mounting that the transport of both glucose and ketone bodies across the blood-brain barrier is modulated, and that significant changes can occur in pathologic circumstances including diabetes, starvation and hepatic encephalopathy.[1] Although the enzyme activities of ketone body metabolism do not change during starvation or diabetes the ability of brain to use ketone bodies under these conditions is improved by an increase in the permeability of the blood-brain barrier. The infusion-autoradiographic method has demonstrated that this permeability increase, about 50-60%, occurs throughout the brain and is near maximal after 2 days starvation (TABLE 2). Four days' starvation, 6 days of diabetes or 28 days of diabetes does not cause a further increase over that caused by 2 days' starvation.[8]

In hepatic encephalopathy ketone body transport is reduced by about 60%.[5] Thus, ketone bodies could not provide a significant source of fuel in hepatic encephalopathy even if they were present in high circulating concentrations. The reduced capacity of the blood-brain barrier to transport ketone bodies and the decreased glucose content means that the brain may be compromised easily if the oxidative demand rises; there is little margin for error.

Amino Acid Transport in Hepatic Encephalopathy and Diabetes

Transport across the blood-brain barrier is believed to be the rate-limiting step for the penetration of amino acids into brain cells since the maximal velocities of neuronal membrane transport systems are much greater.[1] Studies of amino acid transport *in vivo* showed that most amino acids could be assigned to one of three carriers: one for acidic amino acids, one for basic amino acids, and one for neutral amino acids. The neutral amino acid system carries nine large neutral amino acids at similar rates. Because the affinity of the carrier for each amino acid is similar to its plasma concentration, the rate of entry of any particular amino acid is influenced by the presence of the competing amino acids.[1]

In hepatic encephalopathy or portal systemic shunting there are characteristic alterations in the concentrations of many amino acids in both brain and plasma. In the plasma, significant decreases are present in branched-chain amino acids (leucine, isoleucine and valine), threonine, and lysine. On the other hand, there are increased concentrations of aromatic amino acids (tyrosine, phenylalanine, histidine and unbound tryptophan). In the brain, tryptophan, phenylalanine, tyrosine, methionine, histidine, and glutamine are increased, whereas the other neutral amino acids are not significantly affected. Basic amino acid content is generally decreased although the changes are usually not statistically significant.

The amino acid transport systems at the level of individual structures were studied in hepatic encephalopathy caused by portacaval shunting to determine whether and where changes in blood-brain barrier transport took place.[7] The transport of all the neutral amino acids measured was increased throughout the brain (TABLE 3). It was found that the changes were most pronounced in several limbic structures and the reticular formation, whereas the hypothalamus was least affected. The PA of tryptophan was increased by about 200%, phenylalanine and tyrosine by about 80%, and leucine by 30% (TABLE 3). This differential increase was unexpected because all four amino acids are believed to be transported by the same mechanism. The changes in the permeability index suggested that the kinetic characteristics of the blood-brain barrier were changed in such a way that the increase in transport was greater for tryptophan than for phenylalanine and tyrosine, and greater for phenylalanine and tyrosine than for leucine. It is possible that more than one carrier mechanism exists for the neutral amino acids. Several studies have shown the presence of an apparently nonsaturable component of neutral amino acid transport.[17–19] Perhaps the separate components of the multicomponent system are affected differently by portacaval shunting. Another explanation may be related to the observation that an "A" amino acid transport system exists on the antiluminal membrane which actively removes amino acids from the brain. Such a system acting selectively on leucine, phenylalanine and tyrosine in opposition to the passive "L" system, would have the effect of retarding net influx to the brain of these amino acids. Regardless of the explanation of the increased transport capacity the combination of elevated circulating concentrations of aromatic amino acids and the increased transport capacity causes the brain to be flooded with precursors of important neurotransmitters and this may be important in the disease process of hepatic encephalopathy.[7]

After 7-8 weeks of portacaval shunting the blood-brain barrier became almost impermeable to lysine and presumably other basic amino acids transported by the same carrier (TABLE 3).[7] This finding raises the possibility that the availability of basic amino acids group may be limiting to the extent that it interferes with protein synthesis. The rates of transport of the basic amino acids into brain are very similar to the rates of protein synthesis. Thus, a significant reduction in entry of one or more of these amino acids could conceivably affect the rate of protein synthesis, depending on the efficiency of reutilization of amino acids from protein breakdown within the brain (TABLE 4).

These studies in portacaval-shunted rates showed that amino acid transport systems are selectively altered during experimental liver failure; neutral amino acid transport is increased whereas basic amino acid transport is decreased. The plasma neutral amino acid pattern shows a characteristic abnormality with increased aromatic amino acids and decreased branched-chain amino acids and threonine. The mechanism of the changes at the blood-brain barrier are unknown, but the question arises whether chronic exposure to an abnormal distribution of amino acids, either in the brain or plasma, might have an effect on the transport mechanisms. It was, therefore, interesting to study another metabolic disease state, characterized by abnormal amino acid levels. In diabetes, the branched-chain amino acids leucine, isoleucine and valine are markedly elevated in the plasma, in contrast to hepatic encephalopathy. If the neutral amino

TABLE 4. Amino Acid Influx and Protein Synthesis

Amino Acid	Normal Requirement for Protein Synthesis	Influx	
		Control	Portacaval Shunt
Phenylalanine	4.2	7.8	22.3
Tyrosine	2.5	4.3	12.7
Tryptophan	1.0	4.1	11.3
Leucine	8.9	14.8	16.9
Lysine	7.1	8.8	2.0

NOTE: All values are expressed as $nmol \cdot min^{-1} \cdot g^{-1}$. The normal requirement for protein synthesis was calculated from the molar content of amino acids in brain protein,[21] taking the fractional rate of turnover to be $1.03 \cdot 10^{-4}/min$.[22] The influx values are the unweighted arithmetic means of 25 individual brain structures.[7]

TABLE 5. Changes in the Apparent Permeability-to-Surface Area Product of Phenylalanine in Diabetes

Frontal cortex	−40%
Caudate nucleus	−46
Globus pallidus	−48
Amygdala	−42
Hippocampus	−41
Hypothalamus	−48
Thalamus	−43
Substantia nigra	−43
Superior colliculus	−40
Inferior colliculus	−40
Cerebellar gray	−38

NOTE: All values are expressed as % change compared to the control permeability-to-surface-area products were measured 4 weeks after diabetes was caused by streptozotocin. For other details see Mans et al.[9]

acid transport system were found to be altered, this would lend support to the idea that the chronic presence of abnormal plasma branched-chain amino acid concentrations could stimulate changes in the transport carrier. Therefore transport of phenylalanine and lysine into the brain was measured in a 4-week streptozotocin-diabetic rat model to address the question of whether chronic exposure to a changed amino acid pattern is involved in altering the kinetic properties of the blood-brain barrier.[9] It was found that the PA of phenylalanine was significantly depressed throughout the brain (TABLE 5). However, the observed alterations in plasma concentrations of the neutral amino acids seemed to be almost entirely responsible for this finding; there was no evidence of a change in the activity of the transport system. It appears therefore that, in diabetes, chronic presence of an abnormal distribution of substrates in the plasma does not cause changes in the transport carriers involved.[g]

In conclusion, the infusion-autoradiographic method has proven to be useful in determining nutrient influx *in vivo*. The method has revealed marked changes in blood-brain barrier transport systems in hepatic encephalopathy, diabetes and starvation. Whether changes in blood-brain barrier permeability occur in other metabolic diseases and whether these changes are of etiologic significance remains to be investigated.

ACKNOWLEDGMENT

We thank Ms. J. Heisey for editorial assistance.

[g] Although the activity of the transport systems was unaltered by diabetes, the change in the plasma amino acid spectrum did have an impact on the content of amino acids in brain; tryptophan, phenylalanine, tyrosine, methionine and lysine were markedly decreased. Whether these changes alter neurotransmitter content and cerebral function remains to be clarified.

REFERENCES

1. PARDRIDGE, W. M. 1983. Brain metabolism: A perspective from the blood-brain barrier. Physiol. Rev. **63:** 1481-1535.
2. HAWKINS, R. A., A. M. MANS & J. F. BIEBUYCK. 1982. Amino acid supply to individual cerebral structures in awake and anesthetized rats. Am. J. Physiol. **242:** E1-E11.
3. HIBBARD, L. S. & R. A. HAWKINS. 1984. The three-dimensional reconstruction of metabolic data from quantitative autoradiography of rat brain. Am. J. Physiol. **247:** E412-E419.
4. PRATT, O. E. 1979. Kinetics of tryptophan transport across the blood-brain barrier. J. Neural. Trans. Suppl. **15:** 29-42.
5. HAWKINS, R. A., A. M. MANS & J. F. BIEBUYCK. 1981. Regional blood-brain barrier permeability in hepatic encephalopathy. J. Cereb. Blood Flow Metabol. (Suppl. 1) **1:** 385-386.
6. HAWKINS, R. A., A. M. MANS, D. W. DAVIS, L. S. HIBBARD & D. M. LU. 1983. Glucose availability to individual cerebral structures is correlated to glucose metabolism. J. Neurochem. **40:** 1013-1018.
7. MANS, A. M., J. F. BIEBUYCK, K. SHELLEY & R. A. HAWKINS. 1982. Regional blood-brain barrier permeability to amino acids after portacaval anastomosis. J. Neurochem. **38:** 705-717.
8. HAWKINS, R. A., A. M. MANS & D. W. DAVIS. 1986. Regional ketone body utilization by rat brain in starvation and diabetes. Am. J. Physiol. **250:** E169-E178.
9. MANS, A. M., M. R. DEJOSEPH, D. W. DAVIS & R. A. HAWKINS. 1987. Regional amino acid transport into brain during diabetes: Effect of plasma amino acids. Am. J. Physiol. **253:** E575-E583.
10. CREMER, J. R., D. E. RAY, G. S. SARNA & V. J. CUNNINGHAM. 1981. A study of the kinetic behavior of glucose based on simultaneous estimates of influx and phosphorylation in brain regions of rats in different physiological states. Brain Res. **221:** 331-342.
11. MANS, A. M., J. F. BIEBUYCK, D. W. DAVIS, R. M. BRYAN & R. A. HAWKINS. 1983. Regional cerebral glucose utilization in rats with portacaval anastomosis. J. Neurochem. **40:** 986-991.
12. HOLMIN, T & B. K. SIESJO. 1974. The effect of portacaval anastomosis upon the energy state and upon acid-base parameters of the rat brain. J. Neurochem. **22:** 403-412.
13. MANS, A. M., J. F. BIEBUYCK, D. W. DAVIS & R. A. HAWKINS. 1984. Portacaval anastomosis: Brain and plasma metabolite abnormalities and the effect of nutritional therapy. J. Neurochem. **43:** 697-705.
14. SARNA, G. S., M. W. BRADBURY, J. E. CREMER, J. C. LAI & H. M. TEAL. 1979. Brain metabolism and specific transport at the blood-brain barrier after portocaval anastomosis in the rat. Brain Res. **160:** 69-83.
15. MANS, A. M., D. W. DAVIS & R. A. HAWKINS. 1986. Regional blood-brain barrier transport of glucose after portacaval anastomosis. Metab. Brain Dis. **1:** 119-128.
16. HAWKINS, R. A. & J. F. BIEBUYCK. 1979. Ketone bodies are selectively used by individual brain regions. Science **205:** 325-327.
17. MANS, A. M., J. F. BEIBUYCK, S. J. SAUNDERS, R. E. KIRSCH & R. A. HAWKINS. 1979. Tryptophan transport across the blood-brain barrier during acute hepatic failure. J. Neurochem. **33:** 409-418.
18. BRENDER, J., P. E. ANDERSEN & O. J. RAFAELSEN. 1975. Blood-brain barrier transfer of D-glucose, L-leucine, and L-tryptophan in the rat. Acta Physiol. Scand. **93:** 490-499.
19. ETIENNE, P., S. N. YOUNG & T. C. SOURKES. 1976. Inhibition by albumin of tryptophan uptake by rat brain. Nature (London) **262:** 144-145.
20. HAWKINS, R. A., A. M. MANS, D. W. DAVIS, J. R. VINA & L. S. HIBBARD. 1985. Cerebral glucose use measured with [14C]glucose labeled in the 1, 2 or 6 positions. Am. J. Physiol. **248:** C170-C176.
21. DUNLOP, D. S. 1978. Measuring protein synthesis and degradation in CNS tissue. *In* Research Methods in Neurochemistry, Vol. 4. N. Marks & R. Rodnight, Eds.: 91-141. Plenum. New York.
22. DUNLOP, D. S., W. VAN ELDER & A. LAJTHA. 1975. A method for measuring protein synthesis in young and adult rats. J. Neurochem. **24:** 337-344.

New Directions in Blood-Brain Barrier Research[a]

Studies with Isolated Human Brain Capillaries

WILLIAM M. PARDRIDGE

Department of Medicine and
Brain Research Institute
UCLA School of Medicine
Los Angeles, California 90024

INTRODUCTION

The transfer of circulating substances through the brain capillary wall, i.e., the blood-brain barrier (BBB), has been traditionally approached from a physiological perspective using a variety of classical *in vivo* transport techniques in laboratory animals. However, it is desirable to characterize transport physiology at the biochemical level and also to ultimately characterize the biochemistry of the human BBB. This latter goal may now be realized since brain capillaries can be isolated from human brain in high yield and in relatively pure form.[1] Human brain capillaries are enriched in BBB-specific proteins such as factor VIII antigen,[1] gamma-glutamyl transpeptidase,[1] and a 46-K protein[2] that may be associated with the BBB tight junctions.[3] In addition, the transport systems mediating the transport of nutrients and peptides at the human BBB may also be analyzed using isolated human brain capillaries. This review will discuss recent studies in the two areas, nutrient and peptide transport, and clinical analogues of these areas will also be discussed. Aspartame, a BBB phenylalanine transport problem,[4] will be discussed in the context of the kinetics of phenylalanine uptake by isolated human brain capillaries. Additionally, Alzheimer's disease, a BBB peptide transport problem,[5] will be discussed in the context of identification of human BBB peptide receptor transport system using the isolated human brain capillary.

PHENYLALANINE UPTAKE BY ISOLATED HUMAN BRAIN CAPILLARIES

Owing to the presence of epithelial-like high-resistance tight junctions in the capillaries of brain of virtually all vertebrates,[6] circulating nutrients gain access to brain

[a] This work was supported by NIH Grant NS-24429, the California Department of Health Services, and the Juvenile Diabetes Foundation.

cells only via carrier-mediated transport through the BBB.[4] Eight independent transport systems have been identified for different classes of circulating nutrients (FIG. 1). The neutral amino acid carrier transports approximately fourteen neutral amino acids, including phenylalanine, and this system has a preference for the large neutral amino acids.[7] This system is analogous to the leucine-preferring or L-system, and is sodium-independent and insulin-insensitive.[4] There is no alanine-preferring system (A-system) localized on the luminal surface of the BBB, although the A-system may be situated on the brain side of the BBB.[8]

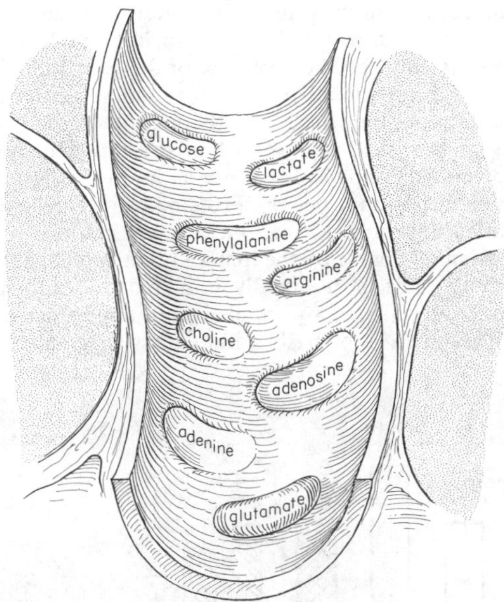

FIGURE 1. Eight major blood-brain barrier nutrient transport systems have been identified using *in vivo* techniques such as the carotid injection technique. The eight transport systems mediate the brain uptake of different classes of nutrients in the blood and only a representative member of the eight different classes of nutrients is shown in the figure. The eight different classes of nutrients include: hexoses (such as glucose, mannose, galactose); monocarboxylic acids (such as pyruvate, lactate, or ketone bodies); neutral amino acids; basic amino acids; choline; purine bases; nucleosides; and acidic amino acids. (From Pardridge.[4] Reprinted by permission from Raven Press.)

The brain of the laboratory rat is known to be unique among the organs of the body in that the K_m of the neutral amino acid transport system at the BBB is of very high affinity (very low K_m).[4] The K_m of transport of most of the large neutral amino acids is less than or equal to the normal plasma concentration of neutral amino acids, e.g., $\leq 50\text{-}100\ \mu M$.[4] Therefore, the BBB transport system is normally heavily saturated under physiologic conditions. In contrast, the K_m of neutral amino acid transport across cell membranes of nonbrain organs is high, e.g., on the order of 1-10 mM or higher, and these values are at least ten-fold greater than normal plasma amino acid concentrations.[4] Therefore, neutral amino acid transport systems in nonbrain organs are minimally saturated. Thus, when plasma phenylalanine rises, there is essentially no change in the saturation of the transport system in organs other than brain, and the cellular uptake of other neutral amino acids, e.g., tryptophan, which compete with phenylalanine for entry into the cell, is essentially unchanged. However, when the concentration of a given neutral amino acid (e.g., phenylalanine) rises in the rat, the brain uptake of other essential neutral amino acids (e.g., tryptophan or tyrosine) is inhibited owing to increased saturation of the BBB transport system. Thus, competition effects normally do not occur under physiologic conditions in organs other than brain, whereas competition effects normally occur in brain owing to the high-affinity nature of BBB transport of neutral amino acids in the rat.[9]

The above paradigm forms the foundation for nutritional regulation of brain amino acid metabolism, including neurotransmitter synthesis.[10] A major factor in nutritional regulation of brain amino acid metabolism is competition effects at the BBB neutral amino acid transport system. The extrapolation of this paradigm from rats to humans allows one to understand nutritional regulation of amino acid metabolism in human brain; however, this assumes that the K_m of the neutral amino acid uptake system at the human BBB is as low as that observed in the rat. This assumption has been recently tested (and verified) using isolated human brain capillaries.[11] FIGURE 2 shows the uptake (clearance in μL/min per mg protein) of [³H]phenylalanine by isolated human brain capillaries in the presence of 50-μM concentrations of a variety of different amino acids. Phenylalanine uptake by the human brain capillary is markedly inhibited by large neutral amino acids, is minimally inhibited by small neutral amino acids (e.g., alanine), and is uninhibited by iminoglycine amino acids (proline, glycine), acidic amino acids (glutamate), or basic amino acids (arginine). A concentration of

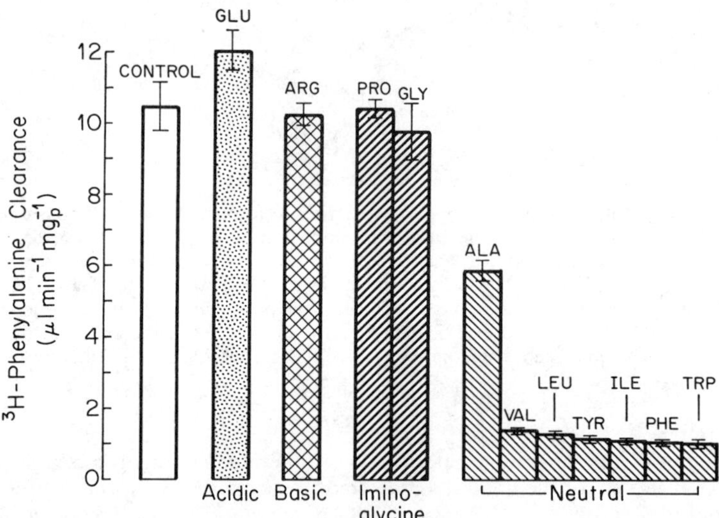

FIGURE 2. Competition for [³H]phenylalanine uptake by isolated human brain capillaries is marked for large neutral amino acids, minimal for small neutral amino acids (alanine), and immeasurably low for iminoglycine (proline, glycine), acidic (glutamate), or basic (arginine) amino acids. Amino acids were present individually at concentrations of 50 μM. ABBREVIATIONS: GLU, glutamic acid; ARG, arginine; PRO, proline; GLY, glycine; ALA, alanine; VAL, valine; LEU, leucine; TYR, tyrosine; ILE, isoleucine; PHE, phenylalanine; TRP, tryptophan. (From Choi and Pardridge.[11] Reprinted by permission from the *Journal of Biological Chemistry.*)

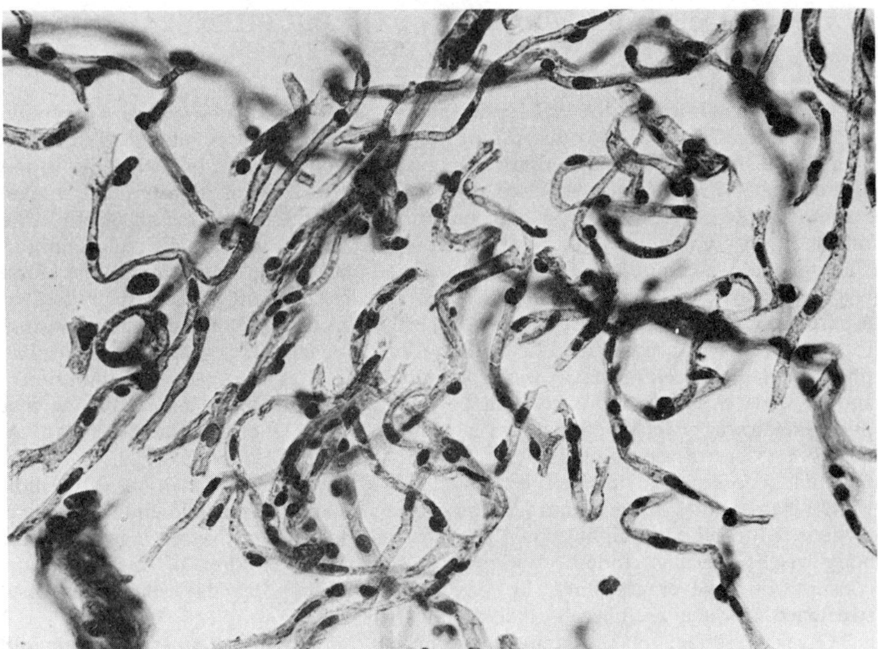

FIGURE 3. Light photomicrograph of isolated human brain microvessels stained with toluidine blue showing that capillaries can be obtained from autopsied human brain intact and free of adjoining brain tissue. (From Choi and Pardridge.[11] Reprinted by permission from the *Journal of Biological Chemistry.*)

50 μM neutral amino acid is well within the normal plasma concentration of amino acid. This concentration causes a marked inhibition of phenylalanine uptake by human brain capillaries, and this is indicative of the very high-affinity nature of the human BBB transport system. This was confirmed by a formal saturation plot which showed that the K_m of phenylalanine uptake by human brain capillaries is low, 22 μM.[11] This value compares closely with the K_m of phenylalanine uptake by isolated rat brain capillaries, 11 μM.[11]

The human brain capillaries shown in FIGURE 3 are isolated from fresh autopsy brain obtained from the pathologist anywhere from 12–40 hours post mortem. To show that a postmortem artifact is not introduced prior to isolation of the human brain capillaries, two different types of studies were performed.[11] First, rabbits were sacrificed and placed in the cold room for either 0 or 42 hours prior to removal of the brain and isolation of capillaries. The K_m of phenylalanine uptake in the rabbit brain capillaries was not significantly different when comparing the results obtained from capillaries isolated from brain that had undergone either 0 or 42 hours of autolysis.[11] Secondly, the proteins making up the rabbit brain capillaries were analyzed by SDS-polyacrylamide gel electrophoresis and Coomassie blue staining. There were few differences between the molecular weights of the major proteins seen on the electrophoretic gel when the capillaries were isolated from brains after either 0 or 42 hours of autolysis.[11] These studies indicate that there is minimal postmortem autolysis of the BBB phenylalanine transport system and that the kinetics of this system at the human BBB may be assessed using isolated human brain capillaries.

ASPARTAME: A BLOOD-BRAIN BARRIER PHENYLALANINE TRANSPORT PROBLEM

Aspartame (L-aspartylphenylalanine methyl ester) is a new non-nutritive dipeptide sweetener that has been introduced into the food supply in large quantities in recent years.[4] USDA figures indicate that the per capita consumption of aspartame in the United States in 1986 has increased to a level nearly 20% of refined sugar intake. Since aspartame is consumed in high quantities, it is reasonable to ask whether the intake of this compound would be expected to introduce any metabolic abnormalities. The dipeptide is composed of aspartic acid, methanol, and phenylalanine. The blood level of aspartic acid and methanol rise minimally, if at all, after a high dose of aspartame intake, so there are likely no ill effects from aspartame consumption due to changes in acidic amino acid or methanol metabolism.[4] However, it is known that phenylalanine concentration in the blood does rise after aspartame consumption.[12] Indeed, it is expected that there will be four-fold rise in blood phenylalanine if a phenylketonuria (PKU) heterozygote consumes 50 mg/kg of aspartame per day.[4] A 50 mg/kg dose of aspartame is about five servings per 50 lb. body weight per day.[4] It should also be noted that it is estimated that there may be as many as 20 million PKU heterozygotes in the country.[13] It is expected that 7-12-year-old children, owing to their reduced body weight, would consume on the order of five servings per 50 lb. body weight per day.[4] Indeed, studies have shown that children of this age group consume, on the average, 50 mg/kg/day and up to 77 mg/kg/day when aspartame-sweetened products are liberally included in their diet.[14]

Given these facts on aspartame consumption, one might ask whether a four-fold increase in blood phenylalanine would be expected to have any deleterious effects on the human brain. It is known, of course, that when blood phenylalanine is raised twenty-fold and in a sustained way, as in the case of PKU, there are significant and deleterious effects on the human brain. If increases in blood phenylalanine and effects on the human brain are related in a linear way (as opposed to a threshold relationship), then it might be expected that a moderate, e.g., four-fold, increase in blood phenylalanine may have subtle but distinct effects on brain function in humans. This idea is supported by studies showing that the I.Q. of babies born of mothers with a four-fold increase in blood phenylalanine is decreased 10.5 points.[4,13] In addition, there is a 10% change in choice-reaction time, a test of higher cognitive function in humans, in association with a 250-μM increment (i.e., a four-fold increase) in blood phenyl-alanine.[15]

One mechanism by which increased blood phenylalanine may cause brain dys-function is via saturation of BBB neutral amino acid transport sites. Indeed, it has been shown that the inhibition of brain protein synthesis associated with phenylalanine administration can be normalized by the administration of other large neutral amino acids that enter brain via the same phenylalanine BBB transport system.[16] Taken together, these considerations provide circumstantial evidence in support of the hy-pothesis that there may be subtle brain dysfunction in humans, particularly PKU heterozygotes, who consume up to 50 mg/kg/day of aspartame, and that this could be reversed and even normalized by the co-administration of other large neutral amino acids that compete with phenylalanine for the same human BBB transport system.

PEPTIDE UPTAKE BY ISOLATED HUMAN BRAIN CAPILLARIES

Like circulating nutrients, some peptides in blood may gain access to brain inter-stitial fluid via specific transport systems localized in the luminal and antiluminal membranes of the brain capillary (FIG. 4). Specific receptors have been identified for insulin, insulin-like growth factor (IGF)-I, IGF-II, and transferrin.[17] The dissociation constant and binding capacity of the human BBB receptors for these peptides are given in TABLE 1. In addition, substances such as cationized albumin have been found to traverse brain capillaries via absorptive endocytosis mechanisms which resemble receptor-mediated transport by being saturable and temperature-dependent.[17] (Cationized albumin is generated when primary amine groups are covalently coupled to native albumin such that the isoelectric point of the plasma protein is raised from 3.9 to 8.5.) As shown in FIGURE 5, the receptor-mediated transport of peptides through

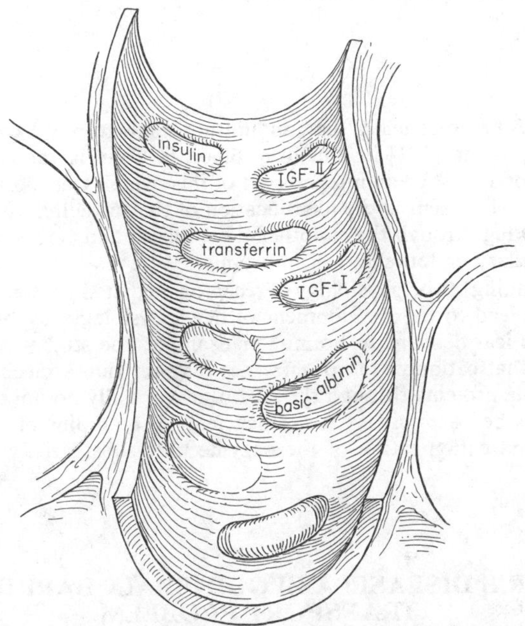

FIGURE 4. At least four blood-brain barrier peptide transport systems have been identified in isolated human brain capillaries for peptides such as insulin, insulin-like growth factor (IGF)-I, IGF-II, and transferrin, and one absorptive endocytosis mechanism has been identified for cationized (basic) albumin. Cationized albumin is formed by the covalent addition of primary amine groups to albumin such that the isoelectric point of the plasma protein is increased from 3.9 to 8.5. (From Pardridge.[17] Reprinted by permission from *Endocrine Reviews*).

the BBB is believed to involve three sequential steps: (*a*) receptor-mediated endocytosis at the luminal or blood side of the BBB, (*b*) diffusion through the endothelial cytoplasm, and (*c*) receptor-mediated exocytosis at the antiluminal or brain side of the BBB.[17]

Receptor-mediated endocytosis is a passive process that does not require energy and can be demonstrated in isolated brain capillaries.[18] However, receptor-mediated exocytosis is an ATP-dependent process.[19] Since isolated capillaries obtained with either a mechanical or an enzymatic homogenization technique are depleted of ATP,[20] receptor-mediated exocytosis cannot be documented with this cell preparation.[18] How-

TABLE 1. Human Brain Capillary Peptide Receptors

Peptide	K_D (nM)	B_{max} (pmol mg_p^{-1})
Insulin	1.2 ± 0.5	0.17 ± 0.08
IGF-I	2.1 ± 0.4	0.17 ± 0.02
IGF-II	1.1 ± 0.1	0.21 ± 0.01
Transferrin	5.6 ± 1.4	0.10 ± 0.02

NOTE: K_D = molar dissociation constant; B_{max} = maximal binding capacity of high-affinity receptor. Determined from Scatchard plots reported previously.[1,28,29]

ever, recent *in vivo* studies using thaw-mount autoradiography have documented net transport of circulating [^{125}I] insulin from blood to brain in rabbits,[21] and receptor-mediated exocytosis must be an integral part of this overall transport process. A major research problem at present is the identification of the subcellular organelles involved in peptide trafficking through the endothelial cytoplasm and across the two membrane poles of the capillary endothelial cell in brain.

An understanding of the physiology of receptor-mediated peptide transport through the BBB should lead to the development of new physiologically based strategies for neuropharmaceutical delivery to brain. For example, the studies with cationized albumin indicate that cationization of enzymes may provide a mechanism for specific delivery of plasma proteins through the BBB that normally do not cross the barrier.[17] Enzymes such as hexosaminidase A are deficient in the brains of patients with Tay-Sachs disease, and cationization of the enzyme may substantially increase its brain uptake.

ALZHEIMER'S DISEASE: A BLOOD-BRAIN BARRIER PEPTIDE TRANSPORT PROBLEM

The cardinal lesions of Alzheimer's disease are neurofibrillary tangles found in the cell body and proximal dendrites, neuritic plaques composed of an extracellular amyloid core surrounded by dystrophic nerve endings, and vascular amyloid found in small arterioles.[22] All three lesions are congophilic, i.e., they bind the Congo red dye.[22] This is a feature typical of β-pleated sheet or amyloid-like proteins. Thus, the amyloid-like fibrils are both intracellular (neurofibrillary tangles) and extracellular (neuritic plaque amyloid and vascular amyloid). In addition, all three lesions have nearly identical amino acid compositions.[5,23–25] These studies indicate that all three lesions may be made up of the same amyloid precursor peptide or protein. Indeed, recent studies have documented the isolation of a 4000-dalton peptide from neurofibrillary tangles,[26] neuritic plaques,[24] and extracerebral or meningeal vascular amyloid in Alzheimer's disease.[27] Since most other forms of amyloid arise from the circulation of a blood-borne peptide precursor,[22] it is possible that the amyloid-like protein making up the three principal lesions of Alzheimer's disease arises from the brain uptake of the 4200-dalton peptide precursor in blood via transport through the BBB.

An important link in the foregoing hypothesis is the demonstration that the vascular amyloid around parenchymal intracerebral vessels in a brain with Alzheimer's disease is composed of the 4200-dalton peptide. Since human brain microvessels can be isolated (FIG. 3), it was possible to document the presence of the congophilic amyloid-like peptide in isolated microvessels from a brain with Alzheimer's disease (FIG. 6). The congophilic material in the Alzheimer's disease microvessels was isolated by solubilization of the capillaries in formic acid followed by separation using sodium dodecylsulfate (SDS)-urea polyacrylamide gel electrophoresis.[5] Staining of the gel revealed a unique 4200-dalton peptide that was found in Alzheimer's disease brain with congophilic angiopathy. This 4200-dalton peptide was subsequently purified by high-performance liquid chromatography, and amino acid composition analysis indicated that this peptide was nearly identical to the composition of the 4000-dalton peptide isolated from neurofibrillary tangles, neuritic plaques, and meningeal vascular amyloid in Alzheimer's disease.[5] In addition, the microvessel 4200-dalton peptide has been partially sequenced and the N-terminal amino acid sequence is as follows:[5]

ASP - ALA - GLU - PHE - ARG - HIS - ASP - SER - GLY - TYR

The major question now is the origin of the 4200-dalton amyloid peptide of Alzheimer's disease, called the A_4 peptide. The peptide may arise from either brain or blood. If the peptide arises from blood, then the uptake by brain of the circulating A_4 peptide is fundamentally a problem of BBB peptide transport. For example, the peptide may be made peripherally and may have access to one of the specific peptide receptor-mediated transport mechanisms in the brain capillary.[17] If the A_4 peptide arises from blood then it should also be possible to develop a radioimmunoassay for the detection of the A_4 peptide or its precursor in the serum of Alzheimer's disease subjects.

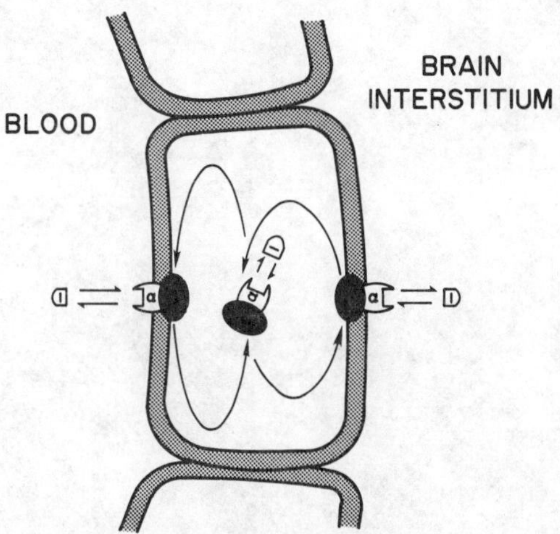

FIGURE 5. Transcytosis of insulin (I) through the blood-brain barrier is viewed as three sequential steps: (*a*) receptor-mediated endocytosis of the peptide at the blood side of the BBB; (*b*) diffusion of the peptide and/or peptide-receptor complex through the endothelial cytoplasm; and (*c*) receptor-mediated exocytosis of the peptide at the brain side of the BBB. (From Pardridge.[17] Reprinted by permission from *Endocrine Reviews.*)

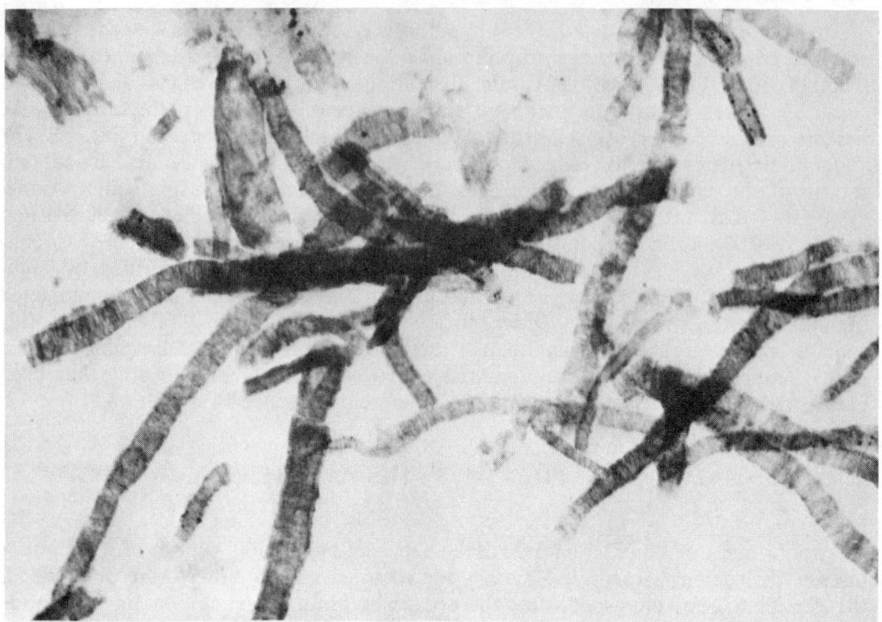

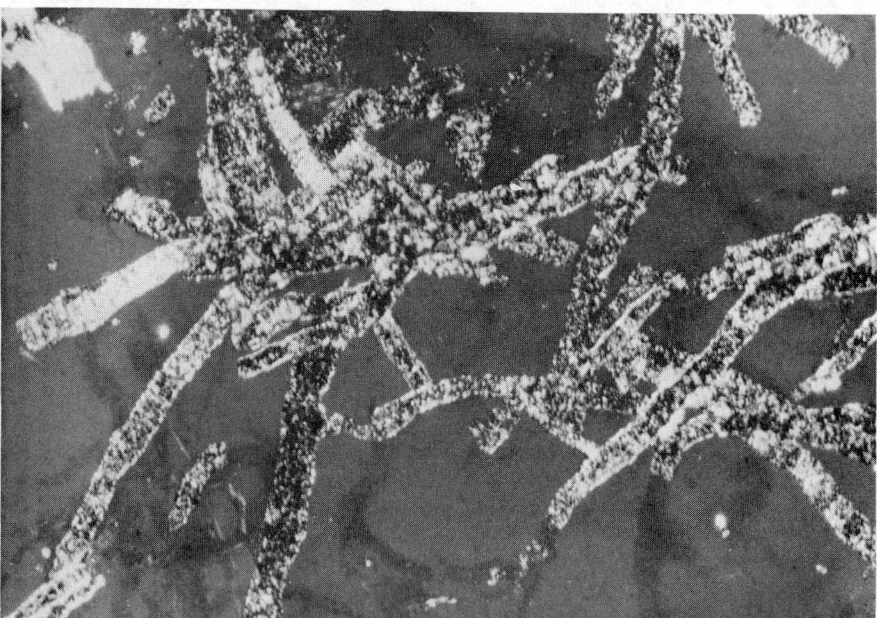

FIGURE 6. Isolated cerebrocortical arterioles from a subject with Alzheimer's disease with amyloid angiopathy. The microvessels were fixed to a glass slide and stained with 0.05% Congo red. (*Top*) nonpolarized light; (*bottom*) polarized light. Birefringence shown at *bottom* was apple-green. (Original magnification ×75; reduced by 30%.)

ACKNOWLEDGMENTS

I am indebted to Jing Yang and Jody Eisenberg for expert technical assistance, to Dawn Brown who skillfully prepared the manuscript, and to Drs. Thomas Choi and Arno Kumagai for many valuable discussions.

REFERENCES

1. PARDRIDGE, W. M. 1985. Human blood-brain barrier insulin receptor. J. Neurochem. **44:** 1771-1778.
2. PARDRIDGE, W. M. 1986. Biochemistry of the human blood-brain barrier: Blood-brain barrier. Interface between internal medicine and the brain. W. M. Pardridge, Moderator. Ann. Int. Med. **105:** 82-95.
3. PARDRIDGE, W. M., J. YANG, J. EISENBERG, & L. J. MIETUS. 1986. Antibodies to blood-brain barrier bind selectively to brain capillary endothelial lateral membranes and to a 46K protein. J. Cereb. Blood Flow Metab. **6:** 203-211.
4. PARDRIDGE, W. M. 1986. Potential effects of the dipeptide sweetener aspartame on the brain. *In* Nutrition and the Brain, Vol. 7. R. J. Wurtman & J. J. Wurtman, Eds.: 199-241. Raven Press. New York.
5. PARDRIDGE, W. M., H. V. VINTERS, J. YANG, J. EISENBERG, T. CHOI, W. W. TOUR-TELLOTTE, V. HUEBNER & J. E. SHIVELY. 1987. Amyloid angiopathy of Alzheimer's disease: Amino acid composition and partial sequence of a 4200 Dalton peptide isolated from cortical microvessels. J. Neurochem. **49:** 1394-1401.
6. BRIGHTMAN, M. W. 1977. Morphology of blood-brain interfaces. Exp. Eye Res. **25**[Suppl.]: 1-25.
7. OLDENDORF, W. H. 1971. Brain uptake of radiolabeled amino acids, amines, and hexoses after arterial injection. Am. J. Physiol. **221:** 1629-1639.
8. BETZ, A. L. & G. W. GOLDSTEIN. 1978. Polarity of the blood-brain barrier. Neutral amino acid transport into isolated brain capillaries. Science **202:** 225-227.
9. FERNSTROM, J. D. & R. J. WURTMAN. 1972. Brain serotonin content: Physiological regulation by plasma neutral amino acids. Science **178:** 414-416
10. WURTMAN, R. J. & J. D. FERNSTROM. 1975. Control of brain monoamine synthesis by diet and plasma amino acids. Am. J. Clin. Nutr. **28:** 538-647.
11. CHOI, T. & W. M. PARDRIDGE. 1986. Phenylalanine transport at the human blood-brain barrier. Studies in isolated human brain capillaries. J. Biol. Chem. **261:** 6536-6541.
12. YOKOGOSHI, H., C. H. ROBERTS, B. CABALLERO & R. J. WURTMAN. 1984. Effects of aspartame and glucose administration on brain and plasma levels of large neutral amino acids and brain 5-hydroxyindoles. Am. J. Clin. Nutr. **40:** 1-7.
13. LEVY, H. L. & S. E. WAISBREN. 1983. Effects of untreated maternal phenylketonuria and hyperphenylalaninemia on the fetus. N. Engl. J. Med. **309:** 1269-1274.
14. FREY, G. H. 1976. Use of aspartame by apparently healthy children and adolescents. J. Toxicol. Environ. Health **2:** 401-415.
15. KRAUSE, W., M. HALMINSKI, L. McDONALD, P. DEMBURE, R. SALVO, D. FREIDES & L. ELSAS. 1985. Biochemical and neuropsychological effects of elevated plasma phenylalanine in patients with treated phenylketonuria. J. Clin. Invest. **75:** 40-48.
16. BINEK-SINGER, P. & T. C. JOHNSTON. 1982. The effects of chronic hyperphenylalaninaemia on mouse brain protein synthesis can be prevented by other amino acids. J. Biochem. **206:** 407-414.
17. PARDRIDGE, W. M. 1986. Receptor-mediated peptide transport through the blood-brain barrier. Endocrine Rev. **7:** 314-330.
18. FRANK, H. J. L., W. M. PARDRIDGE, W. L. MORRIS, R. G. ROSENFELD & T. B. CHOI. 1986. Binding and internalization of insulin and insulin-like growth factors by isolated brain microvessels. Diabetes **35:** 654-661.

19. PAN, B. T. & R. JOHNSTONE. 1984. Selective externalization of the transferrin receptor by sheep reticulocytes in vitro. J. Biol. Chem. **259:** 9776-9782.
20. LASBENNES, F. & J. GAYET. 1984. Capacity for energy metabolism in microvessels isolated from rat brain. Neurochem. Res. **9:** 1-11.
21. DUFFY, K. R. & W. M. PARDRIDGE. 1987. Blood-brain barrier transcytosis of insulin in developing rabbits. Brain Res. **420:** 32-38.
22. GLENNER, G. G. 1980. Amyloid deposits and amyloidosis: The β-fibrilloses. New Engl. J. Med. **302:** 1283-1292 & 1333-1343.
23. ALLSOP, D., M. LANDON & M. KIDD. 1983. The isolation and amino acid composition of senile plaque core protein. Brain Res. **259:** 348-352.
24. MASTERS, C. L., G. SIMMS, N. A. WEINMAN, G. MULTHAUP, B. L. McDONALD & K. BEYREUTHER. 1985. Amyloid plaque core protein in Alzheimer's disease and Down's syndrome. Proc. Natl. Acad. Sci. USA **82:** 4245-4249.
25. SELKOE, D. J., C. R. ABRAHAM, M. B. PODLISNY & L. K. DUFFY. 1986. Isolation of low-molecular-weight proteins from amyloid plaque fibers in Alzheimer's disease. J. Neurochem. **46:** 1820-1834.
26. MASTERS, C. L., G. MULTHAUP, G. SIMMS, J. POTTGIESSER, R. N. MARTINS & K. BEYREUTHER. 1985. Neuronal origin of a cerebral amyloid: Neurofibrillary tangles of Alzheimer's disease contain the same protein as the amyloid of plaque cores and blood vessels. EMBO J. **4:** 2757-2763.
27. GLENNER, G. G. & C. W. WONG. 1984. Alzheimer's disease and Down's syndrome: Sharing of a unique cerebrovascular amyloid fibril protein. Biochem. Biophys. Res. Commun. **122:** 1131-1135.
28. DUFFY, K. R., W. M. PARDRIDGE & R. G. ROSENFELD. 1988. Human blood-brain barrier insulin-like growth factor (IGF) receptor. Metabolism. In press.
29. PARDRIDGE, W. M., J. EISENBERG & J. YANG. 1987. Human blood-brain barrier transferrin receptor. Metabolism **36:** 892-895.

Poster Papers

Analysis of Water and Electrolyte Changes in Vasogenic Brain Edema

Difference between Cortex and White Matter

TAKASHI KONDO [a]

Department of Emergency Medicine
Institute of Clinical Medicine
University of Tsukuba
Tsukuba, Ibakaki, 305 Japan

INTRODUCTION

In order to clarify the pathogenesis of vasogenic brain edema, the changes of the brain water and electrolyte contents in relation to blood-brain barrier (BBB) impairment were studied. Moreover, the pathophysiological difference between cortex and white matter in brain edema was analyzed.

MATERIALS AND METHODS

Twenty-four adult cats were anesthetized with pentobarbital, and intubated intratracheally. Each cat's skull was fixed to a frame, and after the removal of the left eyeball, the left middle cerebral artery (MCA) was exposed by a transorbital approach. Then, the MCA was occluded by aneurysm-clip for 2 hours, and it was carefully recanalized for 2 hours. After sacrifice, the brain was extracted immediately, and cut coronally into slices each 4 mm thick. Brain tissues were separated to a total of 24 pieces of white matter, gray matter and caudate nucleus (CN).

Each piece was approximately 150 to 350 mg, and packed tightly. Two percent Evans-blue (EB) dye was administered at 30 min before sacrifice. The degree of BBB impairment was graded macroscopically: mild (1+), moderate (2+), and severe (3+), according to the blue-staining of EB. Tissue water content was measured by the freeze-vacuum-dry method, and calculated according to the formula of Elliott and Jasper. Each dry tissue was suspended in 10 ml of 0.75 M nitric acid solution for 7 days, and then sodium (Na) and potassium (K) contents were determined in the supernatant by flame-photometry.

[a] Present address: Department of Neurology, School of Medicine, University of Pittsburgh, Pittsburgh, Pennsylvania 15261.

RESULTS

Correlation between the Grade of Extravasation of EB and the Changes of Water Content

All edematous tissues were stained variously with EB. FIGURE 1 shows the changes of water content of white and gray matter and CN in edematous area, compared with the contralateral hemisphere. Brain water contents were observed to increase significantly in proportion to the grade of EB-leakage, by 10% in the white matter, by 5% on the average in cortex.

Changes of Na/K Contents in Edematous Areas

Tissue Na content was also increased in the area of edema, in accordance with the increase in water content (FIG. 2). In comparison with the uninjured side, the gray matter and CN had greater increases in Na than did the white matter. In contrast, tissue K was slightly increased in edematous white matter. However, in the gray matter and CN of the edematous area, tissue K values were significantly decreased. These decreases in tissue K did not correlate with the grading of EB.

Correlation between Brain Water and Na Contents in Edematous Area of White Matter and Cortex

Tissue Na was significantly increased in proportion to the increment of water content in both edematous areas ($p < 0.001$). The regression line for the gray matter and CN had steeper slopes than that for the white matter, and Na accumulation was more prominent than water collection in gray matter and CN.

CONCLUSIONS

Changes in water and electrolyte contents in vasogenic brain edema proved to be different for the cortex and the white matter. It may be suspected that the differences of both blood flow after re-canalization and tissue compliance contribute to these changes of water and Na/K contents.

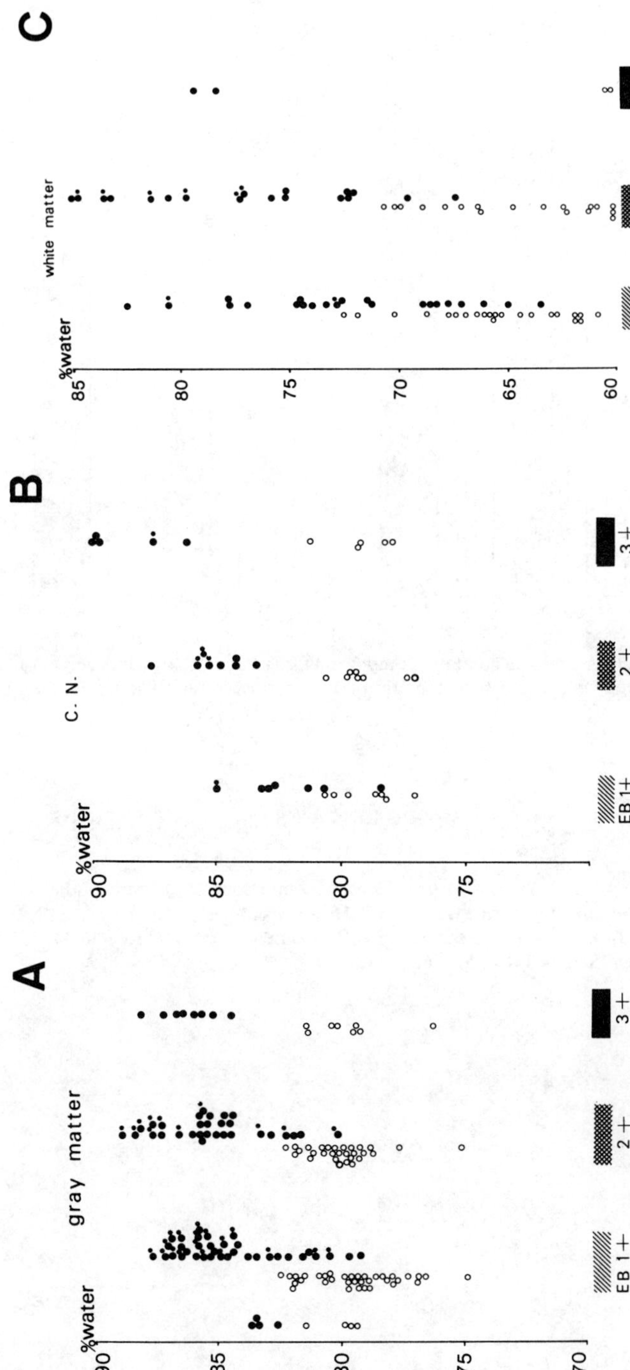

FIGURE 1. Increments of brain water content (%) in each grade of EB-leakage tissues (*solid circles*), compared with the same areas (*open circles*) of contralateral hemispheres. (*smaller solid circles*) show reactive hyperemias.

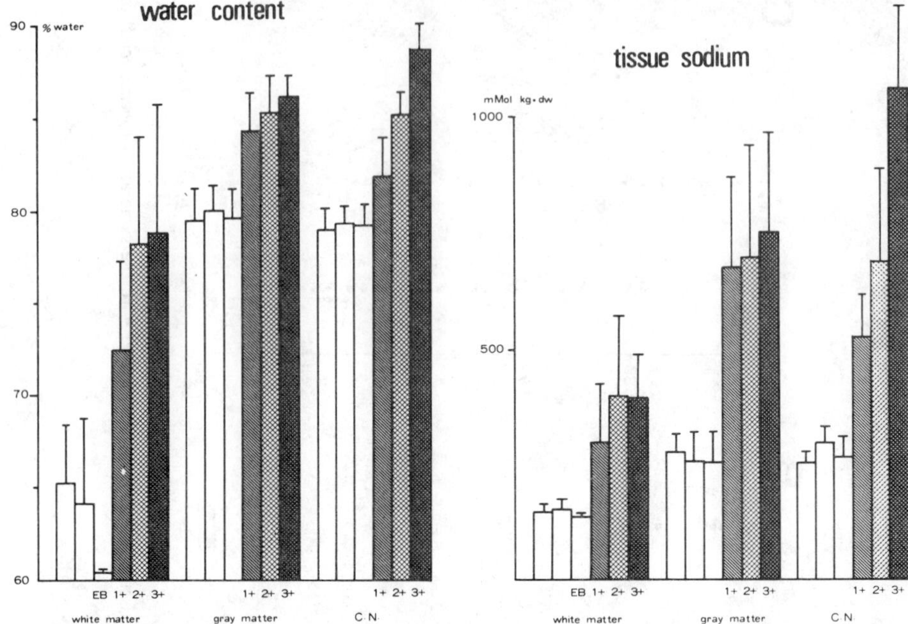

FIGURE 2. Changes of tissue sodium/potassium contents (mM/kg, dry weight) in each part of vasogenic edema in relation to the grade of EB extravasation. *White columns* indicate the opposite sides.

REFERENCES

1. ELLIOTT, K. A. C. & H. H. JASPER. 1949. Measurement of experimentally induced brain swelling and shrinkage. Am. J. Physiol. **157:** 122-129.
2. SCHUIER, F. J. & K.-A. HOSSMANN. 1980. Experimental brain infarcts in cats. II. Ischemic brain edema. Stroke **11(6):** 593-601.

In Vivo Comparative Study of Ascorbic Acid and Glucose Transport into Cerebrospinal Fluid of the Rat

J. DiMATTIO AND J. STREITMAN

New York University Medical Center
New York, New York 10016
and
New York Veterans Administration Medical Center
New York, New York 10010

The movement of molecular species from blood into cerebrospinal fluid (CSF) is a complex and unsettled subject of interest. Previous work has suggested the special handling of D-glucose and L-ascorbic acid in passage from blood into CSF.[1-3] The presence of L-ascorbic acid in brain and CSF of most species at concentrations significantly higher than plasma suggests a function, as yet unclear, for this molecule. Moreover, the molecular similarity and metabolic relationship between D-glucose and L-ascorbic acid raise further questions concerning whether transport of these molecules into CSF is carrier-mediated and if so whether it is via a single or multiple carrier scheme.

In this work, using an *in vivo* rat model, we explore the kinetics of glucose and ascorbic acid entry into CSF by determining the transport rates of labeled 3-*O*-methyl-D-glucose (mD-glu), a nonmetabolizable glucose analog that is considered in most systems to use the same carrier as D-glucose; L-glucose (L-glu)—the nonbiologic stereoisomer of D-glucose; and L-ascorbic acid (AA).

The experimental procedure includes determining plasma-CSF transport constants in male Sprague-Dawley rats (225-300 gm) where either plasma glucose or ascorbate levels have been raised. Thus, the femoral vein of an anesthetized rat was cannulated with PE 50 tubing, and mounted to a lucite board designed for the efficient removal of small samples of blood (50 μl) and the introduction of glucose and/or AA solutions. Preloading was accomplished using an LKB peristaltic pump (2-8 ml/hr) for 5-20 min prior to and during the transport experiment. Kidneys were ligated to insure steady elevated plasma concentrations.

At $t = 0$, a calibrated bolus of double-labeled test species was introduced into the circulation. Samples of blood (50-100 μl) were removed at 1,3,5,7,9, 11 and 13 min and the animal then sacrificed with pentobarbital. CSF was then quickly removed via the foramen magnum. The double-labeled samples were counted and corrected for channel spillover and quenching using the H# provided by the Beckman LS8000 scintillation counter. Blood data were fit to a double exponential curve of the form:

$$C_p = A + Be^{-b_1t} + Ce^{-b_2t} \tag{1}$$

TABLE 1. *In Vivo* Blood-CSF Glucose Transport Rates at Elevated Plasma Na-Ascorbate Concentrations

Ascorbate Concentrations (mM)		Time (min)	Transport			
			[³H]-3-0-Methyl-D-Glucose		[¹⁴C]-L-Glucose	
Plasma	CSF		C_{CSF}/C_P	K_i (min^{-1})	C_{CSF}/C_P	K_i (min^{-1})
Controls ($n = P$) (resting level)						
$0.14 \pm .03^a$	$0.39 \pm .10$	$13.8 \pm .1$	.61 (.03)^a	.050 (.003)	.015 (.003)	.00081 (.00011)
Elevated Ascorbate ($n = 5$)						
10.2	0.5	13.8	.468	.042	.0163	.00103
21.0	0.9	14.0	.395	.029	.0132	.00069
26.2	1.2	13.9	.406	.033	.0125	.00070
30.3	1.6	13.9	.424	.033	.0136	.00078
104.0	3.5	13.7	.417	.031	.0121	.00077

^{a} Values listed as means $\pm$ SD.

and a first-order CSF entry constant K_1 (min^{-1}) estimated via computer solution to a transport model of the form:

$$\frac{dC_{CSF}}{dt} = K_i Cp - K_0 C_{CSF} \qquad (2)$$

TABLE 1 illustrates *in vivo* that the D-isomer of glucose in the form of mD-glu (MW 194) enters CSF 60 times faster than the L-isomer, L-glu (MW 180), which presumably enters CSF passively. Our transport measurements at elevated AA or sodium ascorbate levels indicate that raising plasma AA decreases the rate constant

TABLE 2. *In Vivo* (³H)-L-Glucose and (¹⁴C)-L-Ascorbic Acid Transport Rates into CSF at Resting and Elevated Plasma Glucose Levels.

D-Glucose Concentrations (mM)		Time (min)	Transport			
			(¹⁴C)-L-Ascorbic Acid		(³H)-L-Glucose	
Plasma	CSF		C_{CSF}/C_P	K_i (min^{-1})	C_{CSF}/C_P	K_i (min^{-1})
Controls ($n = 10$) (resting level)						
8.1 ± 1.1^a	$4.8 \pm .9$	$13.8 \pm .3$	0.543 ($\pm$ .035)^a	.031 ($\pm$ .004)	.0153 ($\pm$ .0041)	.00078 ($\pm$ .00016)
Elevated Glucose ($n = 3$)						
40.6	6.4	13.9	0.671	.033	.0229	.00138
50.0	6.1	13.9	0.512	.026	.0367	.00219
74.4	8.1	13.8	0.361	.018	.0440	.00261

^{a} Value listed as means $\pm$ SD.

for [3H]-mD-glu only slightly. When high levels of plasma AA were achieved, CSF ascorbate levels increased only modestly, suggesting saturation of the AA transport mechanism with little effect on D-glucose transport. Similarly, elevated plasma glucose (TABLE 2) resulted in only modest increases in CSF glucose, suggesting control and carrier saturation, while rate constants for [14C]-AA entry into CSF were relatively unchanged. Thus, while both ascorbate and glucose movement into CSF are clearly carrier-mediated, the carriers do not appear to be shared or subject to competitive constraints from the other molecular species. Of possible special interest, passive L-glu movement was increased by high plasma-D-glucose, but not by ascorbate. These results suggest two distinct carriers for D-glucose and ascorbic acid.

REFERENCES

1. SPECTOR, R. & A. V. LORENZO. 1973. Ascorbic acid homeostasis in the central nervous system. Am. J. Physiol. **226:** 757-63.
2. LORENZO, A. V. 1980. Exp. Eye Res. **25** (Suppl.): 205-228.
3. HOCHWALD, G. M., M. GANDHI & S. GOLDMAN. 1983. Net transport of glucose from blood to CSF in the cat. Neuroscience **10:** 1035-40.

Proteolysis of the Calcimedin Proteins[a]

PAMELA B. MOORE

Laboratory Animal Research Center
The Rockefeller University
New York, New York 10021-6399

The calcimedin proteins comprise a set of four polypeptides of 67,000, 35,000, 33,000 and 30,000 daltons,[1,2] termed the 67K, 35K, 33K and 30K calcimedins. These proteins are obtained by calcium-dependent binding to a fluphenazine-Sepharose 4B matrix from one of several muscle tissues including chicken gizzard and bovine heart. The four proteins are separated from each other by NaCl gradient elution from DEAE-cellulose.[3] The four polypeptides elute in the order: 33K, 30K, 67K and 35K. Each protein can be rebound and re-eluted from the fluphenazine-Sepharose or phenyl-Sepharose affinity columns, also in a calcium-dependent manner, independently of the presence of the other three proteins.

Proteolytic degradation by one of several available enzymes was used to determine whether the three smaller calcimedin proteins were derived from the larger 67K calcimedin protein by degradation. Comparison of the generated fragments would also indicate relatedness between the four proteins.

Proteolysis by *S. aureus* V-8 protease was used, since this enzyme generally produces large fragments easily detected by gel electrophoretic analysis. The chymotrypsin-like serine myofibrillar binding protease was also used since this enzyme cleaves at different sites. The protein (calcimedin) was used at a concentration of 0.5 or 1.0 mg/ml. The enzyme was at a concentration of 1 μM. In some cases, the protein samples were adjusted to 1 mM $CaCl_2$ or 1 mM $MgCl_2$ prior to the addition of the enzyme. The degradation was allowed to proceed at 25° C with shaking for 15 or 30 min. The reaction was stopped by addition of sodium dodecyl sulfate to a final concentration of 1% followed by boiling for 5 min.

Proteolysis of the 30K protein by V-8 protease resulted in no discernable degradation products. This was true regardless of whether the proteolysis was performed in the presence of EGTA, 1mM $CaCl_2$ or 1mM $MgCl_2$.

The 35K calcimedin could be transiently converted into the 33K calcimedin with V-8 protease as shown in FIGURE 1. The 33K protein was further degraded to a 25-26K polypeptide upon extending the time for proteolysis. This degradation pattern occurred with or without the presence of cation.

The 67K calcimedin degradation however, was modified in the presence of calcium by the myofibrillar binding serine protease[4] and the V-8 protease (FIG. 2). In the presence of 1 mM EGTA or 1 mM $MgCl_2$ (Lanes 1 and 3) a major fragment of 57K daltons was generated. In the presence of 1 mM $CaCl_2$ plus 1 mM EGTA (10 μM free Ca^{2+}), the major fragment was 28K (see Lane 2).

[a]This work was supported by the American Heart Association.

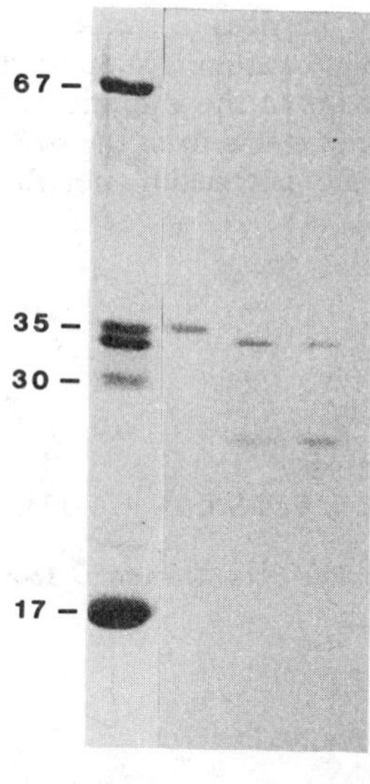

FIGURE 1. Sodium dodecyl sulfate 12% polyacrylamide gel electrophoretogram of V-8 protease digestion of the 35K calcimedin. The gel was stained with Coomassie brilliant blue R250 for visualization. *Lane* 1: Standard calcimedins: 67K, 35K, 33K and 30K. Calmodulin = 17K. *Lane* 2: Purified 35k calcimedin prior to proteolysis. *Lane* 3: 35K calcimedin digested for 15 min. *Lane* 4: 35K calcimedin digested for 30 min.

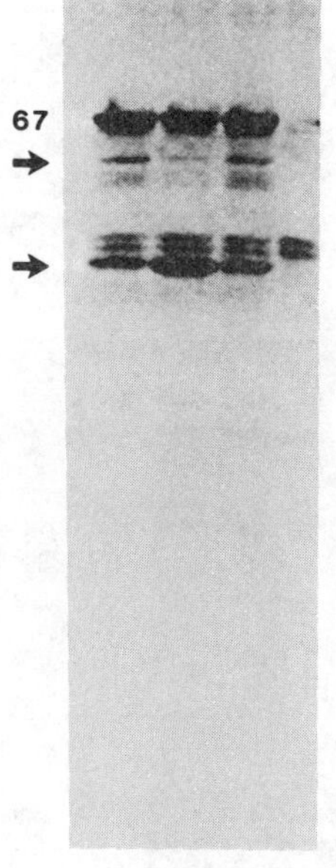

FIGURE 2. Sodium dodecyl sulfate 12% polyacrylamide gel electrophoretogram of V-8 protease digestion of the 67K calcimedin. The gel was visualized by Coomassie brilliant blue R250 staining. *Lane* 1: 67K calcimedin digested for 15 min in the presence of 1 mM EGTA. *Lane* 2: 67K calcimedin digested for 15 min with 1 mM EGTA and 1 mM $CaCl_2$ present (10 μM free Ca^{2+}). *Lane* 3: 67K calcimedin digested for 15 min with 1 mM EGTA and 1 mM $MgCl_2$ present. *Lane* 4. V-8 protease without calcimedin protein.

The calcimedin proteins bind several hydrophobic affinity matrices in a calcium-dependent manner. A conformational change in the presence of calcium is supported for the 67K calcimedin since proteolysis by two enzymes results in the alteration of the cleavage patterns with $CaCl_2$ as opposed to $MgCl_2$. The fragments from the 67K calcimedin do not correspond to those generated from the 35K calcimedin (nor the 33K calcimedin).

REFERENCES

1. MOORE, P. B. & J. R. DEDMAN 1982. J. Biol. Chem. **257:** 9663-9667.
2. MOORE, P. B., N. KRAUS-FRIEDMANN & J. R. DEDMAN. 1984. J. Cell Sci. **72:** 121-133.
3. MOORE, P. B. 1986. Biochem. J. **239:** 49-57.
4. SOHAR, I., J. W. C. BIRD & P. B. MOORE. 1986. Bioch. Biophys. Res. Commun. **134:** 1269-1274.

In Vivo Effects of Ozone on Rat Alveolar Macrophage Plasma Membrane Surface Sulfhydryl Groups

J. E. RYER, G. WITZ, B. D. GOLDSTEIN,
AND M. A. AMORUSO

*University of Medicine and Dentistry of New Jersey
Robert Wood Johnson Medical School
Piscataway, New Jersey 08854*

A number of studies have shown that mice exposed to ozone (O_3) exhibit a heightened susceptibility to pulmonary bacterial infection.[1] This may be due to an O_3-induced impairment in the function of the alveolar macrophage (AM), a resident cell in the lung responsible for pulmonary bacterial defense.[2] Alveolar macrophages kill bacteria by mechanisms which include the respiratory burst, a term describing a series of reactions leading to the production of highly reactive microbicidal intermediates. Superoxide anion radical ($O_2^{\cdot-}$), the initial species formed in the respiratory burst, is generated by a plasma-membrane-bound NADP(H)-dependent oxidase. Previous work from this laboratory has shown that alveolar macrophages obtained from mice and rats exposed to O_3 have a decrease in their ability to produce ($O_2^{\cdot-}$) upon stimulation.[3,4] Since the NADP(H)-dependent oxidase is located in the plasma membrane, it was suggested that O_3 or oxidation products formed in the lung as a result of O_3 exposure could react with AM plasma membrane sulfhydryl (SH) groups to modulate the activity of the plasma membrane oxidase. The present studies were aimed at determining whether O_3 has an effect on rat AM membrane surface SH group status and $O_2^{\cdot-}$ production at various times after *in vivo* exposure to O_3.

Sprague-Dawley female rats were exposed to either room air or 3 ppm O_3 for 3 hours. Alveolar macrophages from control rats and AM isolated from rats immediately after O_3 exposure or 2 or 24 hours after exposure were harvested by lung lavage and purified using a Ficoll-Hypaque gradient. The cells were resuspended at 1×10^7 cells/ml in balanced salt solution (BSS: 128 mM NaCl, 12 mM KCl, 1 mM $CaCl_2$, 2 mM $MgCl_2$, 2 mM glucose, 4 mM sodium phosphate buffer, pH 7).

Phorbol myristate acetate (PMA)-stimulated superoxide anion radical generation by AM in each group was assessed by measuring the superoxide dismutase (SOD)-inhibitable reduction of cytochrome *c* 30 minutes after stimulation.[3] Plasma membrane available-surface SH groups were measured using carboxypyridinedisulfide and monitoring the resultant formation of mixed disulfides through assay of thione released into the supernatant.[5] The number of SH groups per 10^6 cells was calculated using 1×10^4 M^{-1} as the extinction coefficient of the thione.

When plasma membrane SH groups from AM of O_3-exposed rats were examined immediately and 2 and 24 hours after O_3 exposure, there was a 35, 58, and 60%

TABLE 1. Change in Plasma Membrane Surface Sulfhydryl Groups in Alveolar Macrophages From Rats Exposed to Ozone[a]

Hours after Ozone Exposure	Percent Increase in Surface Sulfhydryls[b,c]
0	35
2	58
24	60

[a] Ozone exposure was 3.0 ppm for 3 hours.
[b] Control values for surface sulfhydryl groups ranged from 6.3–6.7 nmol/10^6 cells.
[c] The percent increase is the average of two experiments with four rats per group.

TABLE 2. Decrease in Superoxide Anion Radical Production in Alveolar Macrophages From Rats Exposed to Ozone[a]

Hours after Ozone Exposure	O_2^- Production (nmol/1.25×10^6 cells)	Percent Decrease in O_2^- Production[b]
Control	32.7	—
0	26.1	20
2	18.9	42
24	3.3	90

[a] Ozone exposure was 3.0 ppm for 3 hours.
[b] The percent decrease is the average of two experiments with four rats per group.

increase, respectively, in SH groups compared with controls (TABLE 1). Alveolar macrophages isolated immediately and 2 and 24 hours after O_3 inhalation exhibited a 20, 42, and 90% decrease, respectively, in O_2^- production upon stimulation (TABLE 2). These data suggest that exposure of rats to 3 ppm O_3 for 3 hours may modify AM plasma membrane SH group status, which can, in turn, alter the activity of the NADP(H)-dependent oxidase responsible for O_2^- production.

REFERENCES

1. COFFIN, D. L. 1970. AEC Symp. Ser. **18:** 257.
2. GREEN, G. M. & E. H. KASS. 1964. J. Exp. Med. **119:** 167.
3. AMORUSO, M. A., G. WITZ & B. D. GOLDSTEIN. 1981. Life Sci. **28:** 2215.
4. WITZ, G., N. J. LAWRIE, M. A. AMORUSO & B. D. GOLDSTEIN. 1985. Chem-Biol Interactions **53:** 13.
5. GRASSETTI, D. R., J. F. MURRAY & H. T. RUAN. 1985. Biochem. Pharmacol. **18:** 603.

Orotate Metabolism and Nucleotide Levels in Normal and Neoplastic Liver[a]

MICHAEL A. LEA, VIRGINIA OLIPHANT,
ALEYKUTTY LUKE, JOHN V. TESORIERO, AND
KENNETH M. KLEIN

Departments of Biochemistry, Anatomy and Pathology
University of Medicine and Dentistry of New Jersey
New Jersey Medical School
Newark, New Jersey 07103

A high-orotate diet causes an increase in the ratio of pyrimidine to purine nucleotides in rat liver[1] and can be promotional for hepatocarcinogenesis.[2] In transplanted rat hepatomas there is decreased uptake of isotope-labeled orotate.[3] Our objectives have been first, to determine whether this change in uptake can be detected in the livers of rats treated with carcinogens and, second, to investigate the extent to which dietary changes can influence nucleotide concentrations in normal liver and hepatomas. The experimental procedures for determination of nucleotides and the uptake of [^{3}H]orotate have been described previously.[4]

Male Buffalo strain rats were treated with carcinogens using either the protocol of Solt and Farber[5] or that described by Laurier *et al.*[2] No significant changes in the uptake of [^{3}H]orotate were detected in the livers of rats sacrificed in the first few months after carcinogen treatment. However, in rats sacrificed between 9 and 18 months after carcinogen treatment, the appearance of light-colored nodules or tumors was always accompanied by decreased uptake of [^{3}H]orotate into the acid-soluble tissue fraction and there was a low incorporation into RNA relative to control liver (TABLE 1). With the exception of the right lateral lobe in Rat 2, the uptake of [^{3}H]orotate was not decreased in portions of the liver that appeared grossly normal in tumor-bearing rats. There was also some decrease in the uptake of [^{3}H]uridine in primary hepatomas; this may be related to changes in circulation and/or alterations in transport activity. The data suggested that development of primary, well-differentiated hepatocellular carcinomas was accompanied by decreased uptake of orotate, which may serve as a marker for liver cancer.

We confirmed that a 1% orotate diet increases the UTP/ATP ratio in rat liver, but we detected no such change in hepatomas (FIG. 1). There were quantitative differences in the pattern of radioactive orotate metabolites in liver and hepatomas, but qualitative differences were not seen and the acid-soluble metabolites of orotate

[a] This work was supported by New Jersey Cancer Commission Grant 84-568-CCR and American Institute for Cancer Research Grant 85B75.

appeared to be similar in these tissues on control or high-orotate diet. An arginine-deficient diet was also without effect on the ratio of pyrimidine to purine nucleotides in hepatomas, but, in agreement with Hassan and Milner,[6] there was a large increase in rat liver. Much smaller changes were seen in the levels of deoxyribonucleotides. In mouse liver the levels of the nucleotides UTP, ATP, dTTP and dATP were 1371%, 29%, 229% and 54% of control values, respectively, after 7 days on an arginine-deficient diet. We have found that in rat liver and kidney the effect of a high-orotate diet on nucleotide levels cannot be related to the relative rates of orotate uptake.[4] On

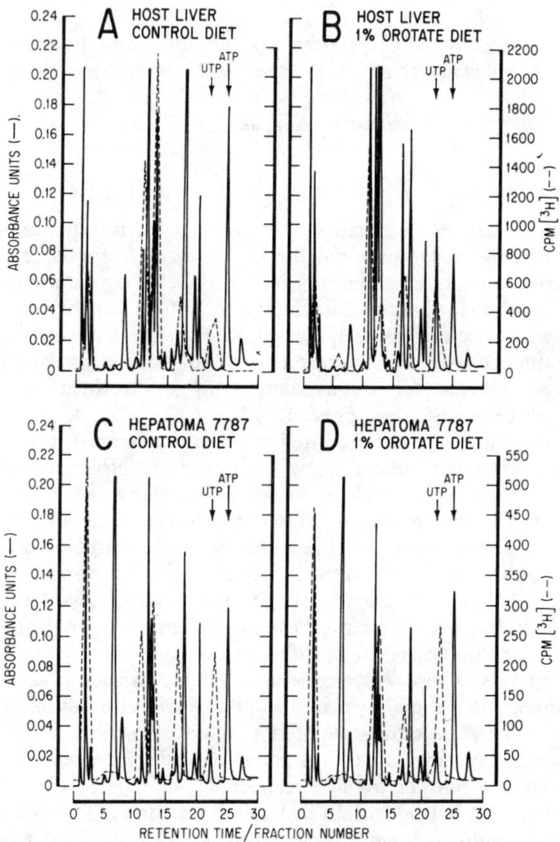

FIGURE 1. Effect of a 1% orotate diet on nucleotide concentrations and the metabolism of orotate. Absorbance profiles at 254 nm and radioactivity peaks after anion exchange chromatography of neutralized acid-soluble fractions were examined in tissue preparations from male Buffalo-strain rats bearing subcutaneous transplants of Morris hepatoma 7787. The animals were killed 10 minutes after a subcutaneous injection of [³H]orotate (50 μCi per 100 g body weight). A, B, C and D represent material from 13.7 mg host liver, 11.4 mg host liver, 12.1 mg hepatoma, and 11.6 mg hepatoma, respectively.

TABLE 1. Uptake of [³H]Orotate and Incorporation into RNA in Liver of Carcinogen-Treated Rats[a]

| | Acid-Soluble Radioactivity (dpm/100 μg tissue) | | | Incorporation into RNA (dpm/100 μg RNA) | |
| | | | | | |
Rat	Blood	Nodules or Tumors[b]	Host Liver[c]	Nodules or Tumors[b]	Host Liver[c]
1	61	197; 92	280; 311	353; 221	674; 816
2	69	93; 85	71; 225	180; 318	152; 498
3	89	97; 208	345; 267	211; 429	810; 618
4	55	52; 100	202; 242	215; 252	756; 864
5	69	70; 78	303; 325	241; 300	1166; 1369

[a] Determinations were made 15-18 months after treatment consisting of partial hepatectomy followed after 18 hours by intraperitoneal injection of dimethylhydrazine (100 mg/kg). From the first to the sixteenth week after carcinogen treatment the rats received a diet containing 1% sodium orotate. Rats were sacrificed 10 minutes after receiving a subcutaneous injection of [³H]orotate (25 μCi/100 g).

[b] The values represent tissues taken from two light-colored areas. On histologic examination these were found to be hyperplastic nodules or hepatomas.

[c] The two values represent tissue that appeared grossly normal from the right lateral and caudate lobes, respectively.

the other hand, the present data suggest that dietary changes that favor increased orotate levels have greater effects on nucleotide concentrations in normal than in neoplastic liver.

REFERENCES

1. VON EULER, L. H., R. J. RUBIN & R. E. HANDSCHUMACHER. 1963. J. Biol. Chem. **238:** 2464-2469.
2. LAURIER, C., M. TATEMATSU, P. M. RAO, S. RAJALAKSHMI & D. S. R. SARMA. 1984. Cancer Res. **44:** 2186-2190.
3. LEA, M. A., J. BULLOCK, F. L. KHALIL & H. P. MORRIS. 1974. Cancer Res. **34:** 3414-3420.
4. LEA, M. A., V. OLIPHANT & A. LUKE. 1987. Comp. Biochem. Physiol. **86B:** 581-586.
5. SOLT, D. B. & E. FARBER. 1976. Nature **263:** 702-703.
6. HASSAN, A. & J. A. MILNER. 1981. Metabolism **30:** 739-774.

Ornithine Inhibition of Arginine Uptake in Rat Brain Mitochondria

JAMES WATROUS, DAVID QUILLEN,
AND ROBERT CABRY, JR.

Biology Department
Saint Joseph's University
Philadelphia, Pennsylvania 19131

Arginine uptake by mitochondria is important since it is an allosteric effector of *N*-acetylglutamate synthetase (*N*-AGS). *N*-AGS controls the production of carbamyl phosphate, the first step in the urea cycle.[1] Freedland *et al.* have demonstrated the inhibitory effect of lysine and ornithine on arginine uptake in rat liver mitochondria, whereas Cheung and Raijman have shown that ornithine may be the limiting substrate for ornithine transcarbamylase activity.[2,3] Watrous and Keeley have shown that in rat brain mitochondria, L-arginine appears to be taken up preferentially over the D form. The addition of dinitrophenol or *m*-chlorocarbonylcyanide phenylhydrazone reduced arginine uptake while ADP or valinomycin did not.[4] This study looks at the role of ornithine and arginine as energy substrates in whole brain and establishes a pH-dependency for L-arginine transport in rat brain mitochondria.

Brains from male rats weighing 200–400 g were removed in the cold. Cerebral hemispheres were isolated for metabolic studies using constant pressure manometry and, as the source for mitochondria, using the method of Clark and Nicklas.[5] Metabolite penetration was estimated at room temperature by monitoring absorbance changes at 520 nm, using an Aminco-Chance spectrophotometer.

Using standard manometric procedures with randomized slices of cerebral hemispheres, we discovered that the addition of L-arginine (5 mM) decreased the rate of oxygen consumption by 34% ±1.3% compared with endogenous levels. FIGURE 1 shows the results of manometric experiments comparing oxygen consumption rates of L-arginine (5 mM) followed by the addition of L-ornithine (5 mM). The average decrease in respiration rate was 24.8% ±2.97%.

Arginine uptake into the mitochondrial matrix was estimated by the decrease in absorbance (mitochondrial swelling) at 520 nm. The relative rates of transport in six pH environments are presented in FIGURE 2. The pH of the incubation medium where mitochondria showed the greatest uptake rate was 7.7. At pH 7.3 and 7.6 the rates of uptake were essentially equal and showed the lowest uptake rates.

Although not all of the information concerning arginine uptake into the mitochondrial matrix is available, data from these experiments are consistent with competition between arginine and ornithine for the same transporter. Additional experiments relating the degree of inhibition and the concentration of ornithine need to be performed in order to further characterize the metabolic roles of these amino acids in brain homogenates.

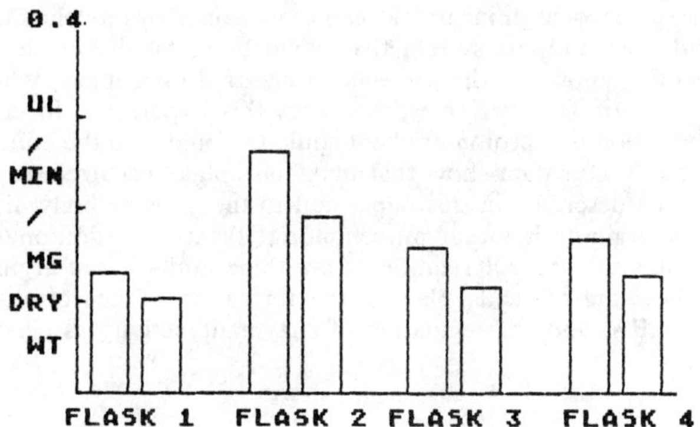

FIGURE 1. Ornithine inhibition of arginine as an energy source in rat brain mitochondria. For each pair the rate with arginine is first, followed by the ornithine rate. Constant pressure manometry flasks contained Tris PO_4 (5 mM), Tris HC1 (10 mM), EDTA (0.05 mM) and 150 mM KC1 at pH = 7. Respiration was initiated in the presence of arginine (5 mM) and followed by the addition of ornithine (5 mM) after at least 20 min. The decrease in respiration produced by ornithine was significantly different from the arginine levels at $p < 0.05$ as determined by paired t test.

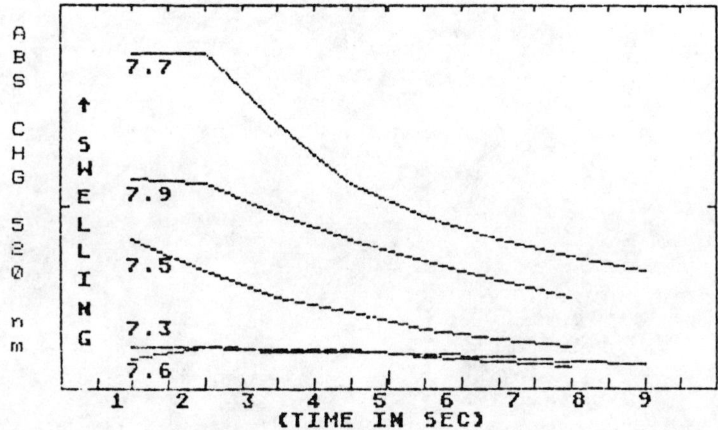

FIGURE 2. pH-dependency of arginine uptake in rat brain mitochondria. Mitochondrial swelling was monitored at 520 nm in a dual wavelength spectrophotometer. The pH of the incubation medium was maintained between 7.3 and 7.9. The rate of swelling versus time demonstrates a pH optimum at 7.7.

The pH-dependency of arginine uptake can be explained by modeling this transport system on a hydrogen antiport system that could be metabolically dependent. The reduction in arginine uptake in the presence of classical uncouplers, which has been demonstrated by Watrous and Keeley,[4] supports this hypothesis. In coupled mitochondria, the formation of a proton gradient could be coupled to the influx of arginine and the efflux of H^+. Our data show that maximal uptake occurred at pH = 7.7.

The experiments described in this paper add to the growing body of information about the role of arginine in brain mitochondrial function. Additional avenues of investigation include the characterization of ornithine inhibition of arginine metabolism, the role these amino acids play in the formation of neurotransmitters like glutamate and GABA, and the regulation of enzymatic activity associated with the urea cycle.

REFERENCES

1. MARSHALL, M., R. L. METZENBERG & P. P. COHEN. 1961. Physical and kinetic properties of carbamyl phosphate synthetase from frog liver. J. Biol. Chem. **236:** 2229-2237.
2. FREEDLAND, R. A., G. L. GROZIER, B. L. HICKS & A. J. MEIJER. 1984. Arginine uptake by isolated rat liver mitochondria. Biochem. Biophys. Acta **802:** 407-412.
3. CHEUNG, C. & L. RAIJMAN. 1981. Arginine, mitochondrial arginase and the control of carbamyl phosphate synthesis. Arch. Biochem. Biophys. **209:** 643-649.
4. WATROUS, J. & E. KEELEY. 1986. Arginine transport in rat brain mitochondria. Ann. N. Y. Acad. Sci. **494:** 291-293.
5. CLARK, J. B. & W. J. NICKLAS. 1970. The metabolism of rat brain mitochondria. J. Biol. Chem. **245:** 4724-4731.

Relationship between the Activities of β-Oxidation Enzymes and Fatty Acid Metabolism in Rat Brain[a]

SONG-YU YANG, XUE-YING HE,
AND HORST SCHULZ

Department of Chemistry
The City College of the City University of New York
New York, New York 10031

The brain relies primarily on glucose and ketone bodies as sources of energy,[1] whereas free fatty acids are the major fuel in heart and some other tissues.[2] Since all enzymes necessary for fatty acid oxidation were believed to be present in brain tissue at levels comparable to those of other tissues,[3,4] the low β-oxidation capacity of the brain was attributed to the exclusion of fatty acids from the brain by the blood-brain barrier.[3,4] This concept was supported by a number of reports[5–7] which showed that brain mitochondria could oxidize palmitate and other fatty acids at rates comparable to those obtained with mitochondria isolated from liver or heart. However, Dhopesh-warkar *et al.*[8] observed that after the intraperitoneal injection of [1-^{14}C]palmitic acid into young rats, the ^{14}C-label persisted in the carboxyl carbon of brain palmitate even after a period of 2 months. Since this observation supports the idea that the brain has a low capacity to oxidize fatty acids, we were prompted to determine the activities of β-oxidation enzymes in rat brain mitochondria and to measure the rate at which palmitoyl-L-carnitine is oxidized by rat brain mitochondria.

The activities of the β-oxidation enzymes determined in rat brain and rat heart are presented in TABLE 1. The specific activities of the brain enzymes are about 4% to 50% of those observed in heart mitochondria except for 3-ketoacyl-CoA thiolase, whose specific activity is 125 times lower in brain mitochondria than in heart mitochondria. However, the apparent K_m values and chain-length specificities of the enzymes from rat brain and heart are similar (data not shown). Although the activity of 3-ketoacyl-CoA thiolase in rat brain mitochondria is very low, the activity of acetoacetyl-CoA thiolase is high (TABLE 1). This distinctive enzyme spectrum may explain why the brain utilizes ketone bodies but not fatty acids.

The rate of long-chain fatty acid oxidation in nonsynaptic mitochondria isolated from rat brain was found to be extremely low, i.e., the addition of 30 μM palmitoyl-L-carnitine to a suspension of mitochondria stimulated the oxygen uptake by only 2.8 nanoatoms of oxygen per minute per milligram of protein at state 3 respiration. This measurement proves the rate of fatty acid β-oxidation in rat brain mitochondria to be 40 to 50 times lower than the rates observed with rat heart and liver mitochondria.

[a]This investigation was supported in part by Grants HL 18089 and HL 30847 from the National Heart, Lung, and Blood Institute, as well as by a City University of New York Faculty Research Award.

TABLE 1. Specific Activities of β-Oxidation Enzymes and Acetoacetyl-CoA Thiolase in Rat Brain Mitochondria and Rat Heart Mitochondria

| Enzyme | Substrate | Specific Activity | | Activities of Brain Enzymes Relative to Those of Heart (%) |
		Brain Mitochondria (nmol/min/mg protein)	Heart Mitochondria (nmol/min/mg protein)	
Acyl-CoA dehydrogenase	n-Butyryl-CoA[a]	1.9	50.8	4
	n-Decanoyl-CoA[b]	25.3	50.8	50
	Palmitoyl-CoA[b]	3	36.3	8
Enoyl-CoA hydratase	Crotonyl-CoA[a]	890	4590	19
	Decenoyl-CoA[a]	430	2490	17
L-3-Hydroxyacyl-CoA	Acetoacetyl-CoA[a]	140	2120	7
dehydrogenase	L-3-Hydroxydecanoyl-CoA[c]	46	500	9
3-Ketoacyl-CoA thiolase	Acetoacetyl-CoA[a]	1.3	200	0.7
	3-Ketodecanoyl-CoA[d]	3.9	500	0.8
Acetoacetyl-CoA thiolase	Acetoacetyl-CoA[e]	40	510	8

NOTE: Substrate concentrations were as follows: [a] 30 μM; [b] 50 μM; [c] 20 μM; [d] 10 μM; [e] 14 μM.

The observed rate of [1-^{14}C]palmitoyl-L-carnitine degradation catalyzed by rat brain mitochondria at 37°C was 0.45 nmol/min per mg of protein (FIG. 1). The preincubation of rat brain mitochondria with 4-bromocrotonic acid, which causes the inhibition of mitochondrial 3-ketoacyl-CoA thiolase,[9] resulted in a 90% inhibition of [1-^{14}C]-palmitoyl-L-carnitine degradation and a change in the ratio of radioactive CO_2 to acid-soluble material from 1:5 or more to 1:1 (FIG. 1). It was recently reported that the activities of carnitine palmitoyl transferase and palmitoyl-CoA synthetase are

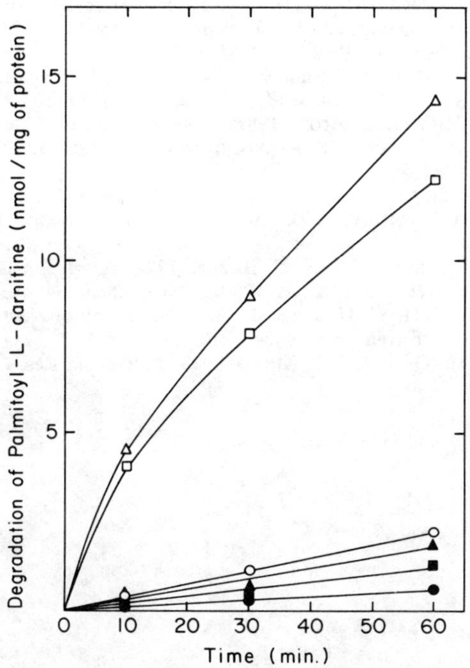

FIGURE 1. Oxidation of [1-^{14}C]palmitoyl-L-carnitine by rat brain mitochondria as a function of time. Reactions were started by the addition of rat brain mitochondria to an assay mixture which contained 60 mM Tris·HCl (pH 7.4), 120 mM KCl, 4 mM KP$_i$, 8 mM $MgCl_2$, 0.5 mM EDTA-K$_2$, fatty acid-free BSA (0.1 mg/ml), 0.05 mM malate, 40 μM[1-^{14}C]-palmitoyl-L-carnitine (160,000 cpm), and 1 mM ADP. Samples were incubated in air at 37°C. Symbols: *circle*, $^{14}CO_2$; *square*, ^{14}C-labeled acid-soluble products; *triangle*, sum of $^{14}CO_2$ and ^{14}C-labeled acid-soluble products. Data obtained with mitochondria that were preincubated for 3 min. at 37°C with 4-bromocrotonic acid are marked with *solid symbols*. Data obtained with mitochondria preincubated without the inhibitor are marked by *open symbols*.

not likely to limit the rate of fatty acid oxidation in rat brain.[10,11] Thus, the unusually low activity of 3-ketoacyl-CoA thiolase in rat brain mitochondria may be the cause for the low rates of fatty acid oxidation in rat brain mitochondria. The inability of brain to oxidize fatty acids may also explain the published observations[12,13] that the net uptake of palmitate from plasma by brains of adult dog, man and probably rat is almost zero, even though the exchange of palmitate between plasma and brain seems to be a rapid process.[13,14] We conclude that the capacity of fatty acid oxidation in rat brain is limited by the activity of the mitochondrial 3-ketoacyl-CoA thiolase as well as being constrained by the blood-brain barrier.

REFERENCES

1. PAGE, M. A., H. A. KREBS & D. H. WILLIAMSON. 1971. Biochem. J. **121:** 49-53.
2. NEWSHOLME, A. A. & C. START. 1973. Regulation in Metabolism. Wiley. New York.
3. VANNUCCI, S. & R. HAWKINS. 1983. J. Neurochem. **41:** 1718-1725.
4. HAWKINS, R. A. 1986. Fed Proc. **45:** 2055-2059.
5. VIGNAIS, P. M., C. H. GALLAGHER & I. ZABIN. 1958. J. Neurochem. **2:** 283-287.
6. BEATTIE, D. S. & R. E. BASFORD. 1965. J. Neurochem. **12:** 103-111.
7. KAWAMURA, N. & Y. KISHIMOTO. 1981. J. Neurochem. **36:** 1786-1791.
8. DHOPESHWARKAR, G. A., C. SUBRAMANIAN & J. F. MEAD. 1973. Biochim. Biophys. Acta **296:** 257-264.
9. OLOWE, Y. & H. SCHULZ. 1982. J. Biol. Chem. **257:** 5408-5413.
10. BIRD, M. I., L. A. MUNDAY, E. D. SAGGERSON & J. B. CLARK. 1985. Biochem. J. **226:** 323-330.
11. REDDY, T. S., H. SPRECHER & N. G. BAZAN. 1984. Eur. J. Biochem. **145:** 21-29.
12. PARDRIDGE, W. M. & L. J. MIETUS. 1980. J. Neurochem. **34:** 463-466.
13. HORROCKS, L. A. & H. W. HARDER. 1983. *In* Handbook of Neurochemistry. Vol. 3. A. Lajtha, Ed.: 1-16. Raven Press. New York.
14. DHOPESHWARKAR, G. A. & J. F. MEAD. 1973. Adv. Lipid Res. **11:** 109-142.

Renal Synthesis of Prostaglandin E_3 in Vivo from Dietary Eicosapentaenoate in a Female Subject

ALDO FERRETTI AND VINCENT P. FLANAGAN

Lipid Nutrition Laboratory
Beltsville Human Nutrition Research Center
Agricultural Research Service
United States Department of Agriculture
Beltsville, Maryland 20705

Ingestion of marine oil brings about several changes in plasma lipid chemistry, accompanied by reduced platelet reactivity, a lowering of systolic blood pressure, and a reduced incidence of thrombotic events. Much remains to be learned about the biochemical mechanisms underlying these effects. There is now a consensus that the hypotensive and antithrombotic effects are associated with reduced arachidonic acid metabolism through the agency of the long-chain $\omega 3$ polyunsaturated fatty acids, eicosapentaenoate (EPA, 20:5) and docosahexaenoate (DHA, 22:6),[1] two important components of marine oil.

Cyclooxygenase enzymes have a high affinity for EPA but a low capability of conversion to prostaglandin (PG) metabolites.[2] Accordingly, the issue of conversion of EPA to trienoic PGs has been a controversial one. Lands *et al.*[3] theorized that the PG metabolism of EPA requires a peroxide tone easily attainable *in vitro* but incompatible with *in vivo* conditions. Indeed, several studies demonstrated very little or no conversion of EPA to PGs *in vivo,*[4–7] whereas others showed *in vitro* cyclooxygenation of EPA in various tissues.[6,8–10] Finally, in 1983 Fischer and Weber[11] reported the first *ex vivo* conversion of dietary EPA to thromboxane A_3 (TXA_3) in human platelets. Shortly thereafter, they provided definitive evidence for the *in vivo* formation of both PGI_3 and TXA_3 in humans fed fish oil.[12]

Now we report on the presence of PGE_3 in the urine of a female subject who ingested quantities of fish oil (MaxEPA) ranging from 10 to 50 g/day (corresponding to 1.8 to 9.0 g/day of EPA and 1.3 to 6.5 g/day of DHA) during four years. Twenty-four-hr urine specimens were collected on ice in polyethylene bottles and immediately analyzed for PGE_3 by extensive modifications of previously published gas chromatographic-mass spectrometric (GC-MS) procedures. Full details will be presented elsewhere. Given the paucity of the urinary PGE_3, a full mass spectrum could not be obtained. We identified PGE_3 by comparison of chromatographic behavior of the putative natural PGE_3 with that of authentic material on OV-17 and SP-2330 gas chromatographic columns, and by selected ion monitoring-mass spectrometry. Namely, fragment ions and their relative intensities in biological samples were com-

TABLE 1. Selected Ion Monitoring Mass Spectrometry: Ion Intensity Ratios of Synthetic and Natural (Urinary) PGE_3 Converted into a PGB Derivative[a]

m/z	Synthetic PGE_3[b]	Natural PGE_3
387[c]	.014 ± .001 (4)	.013 ± .001 (2)
350	.255 ± .006 (4)	.261 ± .002 (2)
349	1.000	1.000
321	.060 ± .001 (4)	.050 ± .001 (3)
259	.104 ± .002 (3)	.092 ± .004 (3)
233	.266 ± .004 (3)	.233 ± .012 (3)
227	.453 ± .011 (4)	.442 ± .007 (3)
199	.481 ± .003 (3)	.425 ± .008 (3)
185	.840 ± .002 (3)	.820 ± .012 (3)

[a] For GC-MS analysis, natural and synthetic PGE_3 was converted to the methyl ester-trimethylsilyl ether derivative of PGB_3.[6] Intensity ratios are relative to the base peak at m/z 349.

[b] Mean ± SEM. Number of observations in parentheses.

[c] All ratios were determined with an OV-17 gas chromatographic column except those at m/z 259 and m/z 233. For these we used an SP-2330 column.

pared with those obtained from synthetic PGE_3. The results, relative to two distinct chemical derivatives, are shown in TABLE 1 and TABLE 2. The appearance of PGE_3 in the subject's urine was concomitant with a reduction of PGE_2 levels (144 versus 488 ng/24 hr) and with extensive alteration of the fatty acid profiles of plasma lipid fractions. Sixteen weeks after MaxEPA supplementation ended, no PGE_3 could be detected in the 24-hr urine specimens.

Our finding confirms that human cyclooxygenase has the ability to form prostanoid metabolites from EPA *in vivo*. Urinary primary PGs reflect renal PG synthesis,[13] and urinary PGE_3 would be expected to be an indicator of renomedullary synthesis after EPA administration. Renal primary PGs are important modulators of excretory and hemodynamic functions in the mammalian kidney. Specifically, the antihypertensive function of the inner medulla has been attributed in part to its production of large quantities of PGE_2.[14] Prostaglandin E_2 promotes Na excretion in the kidney and

TABLE 2. Selected Ion Monitoring Mass Spectrometry: Ion Intensity Ratios of Synthetic and Natural (Urinary) PGE_3 Converted into a PGE-9-Enol Derivative[a]

m/z	Synthetic PGE_3[b]	Natural PGE_3
580[c]	.021 ± .001 (2)	.020 ± .002 (2)
511	.083 ± .001 (3)	.101 ± .001 (2)
490	.066 ± .003 (3)	.063 ± .001 (2)
439	.144	.166 ± .012 (2)
421	1.000	1.000
349	.215	.246 ± .001 (2)

[a] For GC-MS analysis, natural and synthetic PGE_3 was converted to its PGE_3-9(8)-enol-tris(trimethylsilyl ether)-methyl ester derivative.[16] The full spectrum of the derivative prepared from synthetic PGE_3 will be published elsewhere. Intensity ratios are relative to the base peak at m/z 421.

[b] Mean ± SEM. Number of observations in parentheses.

[c] All ratios were determined with an OV-17 gas chromatographic column.

increases renal blood flow by vasodilation.[15] A diet-induced renal production of PGE_3, with or without concomitant reduction of PGE_2, is likely to have an effect on PG-dependent renal functions. At present little is known about the physiologic properties of PGE_3. Thus, we are not sure if a shift of the renal PGE_2/PGE_3 balance will produce physiologically favorable effects. Consequently, marine oil supplementation requires consideration, among others, of the potential renal consequences of altering that balance.

REFERENCES

1. LANDS, W. E. M. 1986. Fish and Human Health. Academic Press. Orlando, FL.: 34-46.
2. NEEDLEMAN, P., A. RAZ, M. S. MINKES, J. A. FERRENDELLI & H. SPRECHER. 1979. Proc. Natl. Acad. Sci. USA **76**: 944-948.
3. CULP, B. R., B. G. TITUS & W. E. M. LANDS. 1979. Prostaglan. Med. **3**: 269-278.
4. WHITAKER, M. O., A. WYCHE, F. FITZPATRICK, H. SPRECHER & P. NEEDLEMAN. 1979. Proc. Natl. Acad. Sci. USA **76**: 5919-5923.
5. HORNSTRA, G., E. CHRIST-HAZELHOF, E. HADDEMAN, F. TEN HOOR & D. H. NUGTEREN. 1981. Prostaglandins **21**: 727-738.
6. FERRETTI, A., N. W. SCHOENE & V. P. FLANAGAN. 1981. Lipids **16**: 800-804.
7. HANSEN, H. S. & B. JENSEN. 1983. Lipids **18**: 682-690.
8. DYERBERG, J., H. O. BANG, E. STOFFERSEN, S. MONCADA & J. R. VANE. 1978. Lancet **ii**: 117-119.
9. SMITH, D. R., B. C. WEATHERLY, J. A. SALMON, F. B. UBATUBA, R. J. GRYGLEWSKI & S. MONCADA. 1979. Prostaglandins **18**: 423-438.
10. DYERBERG, J., K. A. JORGENSEN & T. ARNFRED. 1981. Prostaglandins **22**: 857-862.
11. FISCHER, S. & P. C. WEBER. 1983. Bioch. Biophys. Res. Commun. **116**: 1091-1099.
12. FISCHER, S. & P. C. WEBER. 1984. Nature **307**: 165-168.
13. FRÖLICH, J. C., T. W. WILSON, B. J. SWEETMAN, M. SMIGEL, A. S. NIES, K. CARR, J. T. WATSON & J. A. OATES. 1975. J. Clin. Invest. **55**: 763-770.
14. DANIELS, E. G., J. W. HINMAN, B. E. LEACH & E. E. MUIRHEAD. 1967. Nature **215**: 1298-1299.
15. LEVENSON, D. J., C. E. SIMMONS & B. M. BRENNER. 1982. Am. J. Med. **72**: 354-374.
16. FISCHER, C. 1984. Biomed. Mass Spectrom. **11**: 114-117.

Endogenous Activator Regulation of the Gastric H^+,K^+-ATPase: Studies with Pure Preparation

TUSHAR K. RAY, SANDIP BANDOPADHYAY,
AND PRATAP K. DAS

Department of Surgery
The State University of New York Health Science Center at Syracuse
Syracuse, New York 13210

INTRODUCTION

In recent years, the gastric H^+,K^+-ATPase activity associated with the apical and tubulovesicular membranes of the parietal cells has been identified to be the enzymatic mechanism for the transport of gastric hydrochloric acid. Although considerable progress has been made towards the understanding of the mechanisms of H^+,K^+-ATPase reaction, our knowledge on the aspects of regulation of gastric acid secretion at the intracellular level is still meager.

Previous studies[1-4] from our laboratory implicated a cytosolic activator protein (AF) to be acting as a regulator of the gastric H^+,K^+-ATPase system. The AF has recently been purified to homogeneity from dog[5] and pig[6] fundic cells and characterized to be a dimer of two identical 40 kDa subunits in the active state. The monomer is totally inactive in H^+,K^+-ATPase activation.[5,6] Our *in vitro* studies[6] also clearly demonstrated that the AF has the characteristic qualifications of an intracellular regulator for the gastric H^+ transport process. It was suggested[6] that the AF is the terminal member of an intracellular signal-transducing cascade system where Ca^{+2} plays a critical second messenger role.

The role of Ca^{2+} was further investigated in the present study using a homogenous preparation of H^+,K^+-ATPase.[7] In addition, the question[3,4] of associability between the AF and the H^+,K^+-ATPase molecules was examined using the pure AF and the gastric microsomal H^+,K^+-ATPase preparations. The data demonstrate that while cAMP was totally ineffective, Ca^{2+} shows both activation and inhibition of the AF-dependent H^+,K^+-ATPase activity depending on the concentration. Also, the present data show a rather firm association of the AF with the H^+,K^+-ATPase molecule during activation of the enzymes.

RESULTS AND DISCUSSION

The data in FIGURE 1 demonstrate that interaction of the AF with the H$^+$,K$^+$-ATPase is fairly tight, since removal of the excess AF by centrifugation could retain stimulation of the H$^+$,K$^+$-ATPase. Such an interaction may very well be due primarily

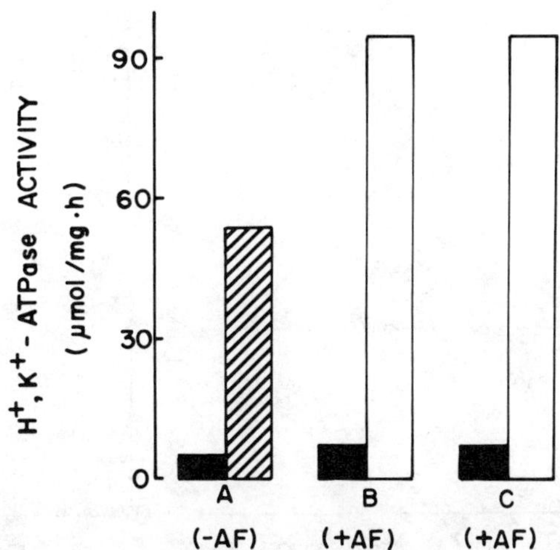

FIGURE 1. AF activation of the gastric H$^+$,K$^+$-ATPase with and without removal of the excess AF after treatment. Purified pig gastric microsomes (H$^+$,K$^+$-ATPase) were incubated without (control) and with AF (membrane to AF protein ratio was 1:1, μg/μg) for 10 min at 37°C in 50 mM Tris-HCl buffer (pH 7.4). After incubation an aliquot of the AF-treated membranes was saved and the rest was subjected to centrifugation (100,000 × *g*, for 90 min) to separate the membranes from excess AF. The control and the AF-treated membranes (before and after centrifugation) were subsequently assayed for the ATPase activity. The ATPase assay was conducted at 37°C for 15 min in 1.0 ml of the assay medium consisting of 50 mM Tris-Pipes buffer (pH 6.8), 1 mM Mg^{2+}, 10 μg membrane protein, 1 mM ATP with and without 20 mM K$^+$. After termination of the reaction by 1 ml of ice-cold 12% TCA, the Pi was assayed.[1] Appropriate blanks containing excess (10 mM) EDTA without any added Mg^{2+}, but otherwise under identical conditions, as above, were run in parallel. The Pi values obtained from the blanks were subtracted from the corresponding experimental assays. The notations are: (**A**) H$^+$,K$^+$-ATPase activity of the control membranes in the absence of the AF; (**B**) AF-pretreated membranes after removal of excess AF; and (**C**) AF-pretreated membranes without the removal of the AF by centrifugation. *Closed bars,* with Mg^{2+} as the only cation; *hatched bars,* in the presence of 20 mM K$^+$; and *open bars,* in the presence of 20 mM K$^+$ and AF.

to some ionic interaction.[6,7] These data together with our previous reports[3,4] clearly demonstrate the AF to be existing both in the soluble and membrane-bound forms.

The data in FIGURE 2 show the differential effects of the second messengers such as Ca^{2+} and cAMP on the H$^+$,K$^+$-ATPase activity. Although cAMP is known to have an important influence on gastric acid secretion,[8,9] the demonstrated (FIG. 2B) lack of effect on the H$^+$,K$^+$-ATPase suggests the cAMP effect to be distal to the H$^+$

pumping site. Calcium, on the other hand, stimulates appreciably the AF-dependent H^+,K^+-ATPase within the 1 μM range, but obliterates only the AF-dependent component rather dramatically within the 2-4 μM range (FIG. 2A). The data (FIG. 2) suggest that Ca^{2+} may be acting as a physiological switch by way of controlling the AF regulation of the H^+,K^+-transporting ATPase system. The latter conclusion is consistent with our recent reports.[10,11]

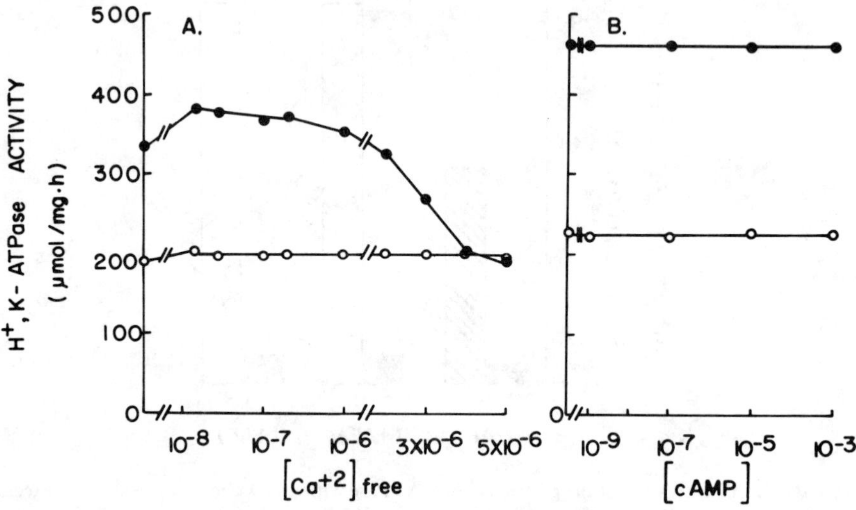

FIGURE 2. Effects of different concentrations of Ca^{2+} (A) and cyclic AMP (B) on the AF activation of the H^+,K^+-ATPase. The concentration of free Ca^{2+} was regulated by varying the amount of Ca^{2+} at a fixed (2.0 mM) concentration of EGTA. Five μg of the pure pig gastric H^+,K^+-ATPase was used in each assay and, where present, 5 μg of the pure AF was used. The details of the ATPase assay are given in FIGURE 1. The K^+-stimulated ATPase activity in the absence and presence of the pure AF was calculated by subtracting the basal rate (with Mg^{2+} as the only cation) from those containing both Mg^{2+} and K^+. The data are typical of three separate studies.

REFERENCES

1. RAY, T. K. 1978. FEBS Lett. **92:** 49-52.
2. SEN, P. C. & T. K. RAY. 1980. Arch. Biochem. Biophys. **202:** 8-17.
3. NANDI, J., M. V. WRIGHT & T. K. RAY. 1983. Biochemistry **22:** 5814-5821.
4. RAY, T. K. & J. NANDI. 1984. *In* Hydrogen Ion Transport in Epithelia. J. G. Forte & F. C. Rector, Eds.: 161-169. Wiley. New York.
5. BANDOPADHYAY, S. & T. K. RAY. 1986. Prep. Biochem. **16:** 21-32.
6. BANDOPADHYAY, S., P. K. DAS, M. V. WRIGHT, J. NANDI & T. K. RAY. 1987. J. Biol. Chem. **262:** 5665-5670.
7. NANDI, J., Z. MENG-AI & T. K. RAY. 1987. Biochemistry **26:** 4264-4272.

8. MACHEN, T. E., M. J. RUTTEN & E. B. M. EKBLAD. 1982. Am. J. Physiol. **242:** G79-G84.
9. THOMPSON, W. J., L. K. CHANG & G. C. ROSENFELD. 1980. Am. J. Physiol. **240:** G76-G84.
10. RAY, T. K., P. K. DAS & S. BANDOPADHYAY. 1987. Gastroenterology **92:** 1590.
11. BANERJEE, A., S. BANDOPADHYAY, P. K. DAS & T. K. RAY. 1987. Fed. Proc. **46**(3): 473.

Metabolic Interactions of Insulin Receptor Tyrosine Kinase[a]

PASQUALE P. VICARIO, RICHARD SAPERSTEIN,[b]
HOWARD KATZEN,[b] AND ALFRED BENNUN

Department of Biological Sciences
Rutgers University
Newark, New Jersey 07102

and

[b]*Department of Biochemical Endocrinology*
Merck & Co.
Rahway, New Jersey 07065-0900

The insulin-dependent induction of glucokinase (GK) (EC 2.7.1.2) synthesis could have feedback effects on an insulin-responsive system, e.g., insulin receptor tyrosine kinase (IRTK), an enzymatic activity associated with the B-subunit of the insulin receptor.[1] Yen and Stamm[2] reported that the activity of hepatic GK in obese insulin-resistant (ob/ob, db/db) mice was higher than in normal mice. Moreover, fasting in normal mice resulted in a decrease in GK activity, whereas fasting in their obese littermates for as long as 10 days had no effect on enzyme activity. The purpose of these studies were to investigate the interrelationship, if any, between glucokinase and IRTK.

Insulin receptor was partially purified by lectin affinity chromatography using membranes solubilized from rat liver[3] and mouse skeletal muscle.[4] IRTK activity was measured by the procedure described by Pike *et al.,*[5] using val[5]-angiotensin 11 (A11) as a substrate. Phosphorylated peptide was isolated using phosphocellulose (P81) paper.[6] Glucokinase activity was determined as described by Hentgartner and Zuber.[7] Glucokinase (*B. stearothermophilis*) was obtained commercially from Sigma Chemical Co. (St. Louis, MO).

The presence of GK in the IRTK assay resulted in a concentration-dependent inhibition of rat liver (FIG. 1A) and mouse skeletal muscle (FIG. 1B) IRTK. Half-maximal inhibition of both basal- and insulin-stimulated IRTK activity occurred at about 430 and 300 ng/ml (71 and 50 mU/ml), respectively. Glucose was required for GK-dependent inhibition of IRTK (FIG. 1C). Other substrates for GK catalytic activity were also active. Their effects, expressed as a percentage relative to glucose, are: 30% (*N*-acetyl-D-glucosamine), 21% (mannose), 16% (fructose), 14% (2-deoxyglucose), and 11% (xylose). Glucose-6-phosphate was ineffective in supporting this GK-dependent inhibition of IRTK. The product inhibition by ADP was eliminated by inclusion of an ATP-regenerating system.

[a] This work was supported by Grant PHSRR 070509-21 from the United States Public Health Service.

Hence, because GK-dependent inhibition of IRTK could be a function of either (*a*) a requirement for a substrate-dependent form of GK which could inhibitorily interact with IRTK or (*b*) a competition of GK with IRTK for available ATP, it was important to compare the effect of GK with another enzyme that also uses ATP and glucose as substrates. It was found that at a protein concentration of 10 μg/ml (1660 mU/ml) GK inhibited IRTK by 90%, whereas inhibition by hexokinase (10 μg/ml, 1400 mU/ml) was absent. In addition, phosphofructokinase (10 μg/ml, 600 mU/ml) and pyruvate kinase (10 μg/ml, 2000 mU/ml) were without effect on IRTK activity.

It was determined that inhibition of IRTK by GK was associated with changes

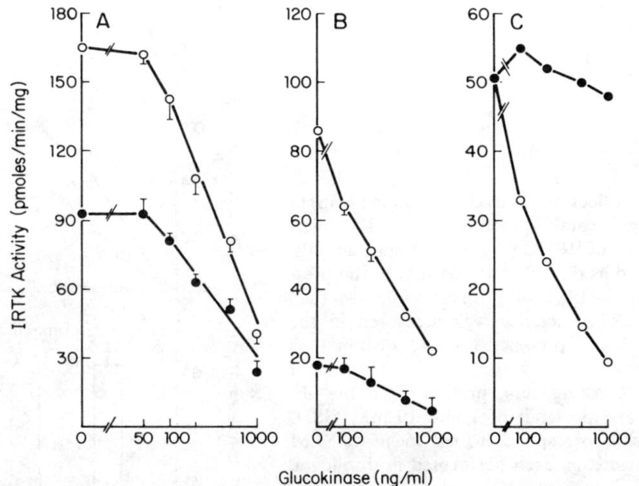

FIGURE 1. Glucokinase-dependent inhibition of insulin receptor tyrosine kinase. Portions of partially purified rat liver IRTK (**A**) (6.5 μg protein) or mouse skeletal muscle IRTK (**B**) (3.7 μg protein) were assayed for kinase activity in the absence (●) and in the presence (○) of insulin (1.0 μM) and increasing concentrations (0–1000 ng/ml, 0–166 mU/ml) of glucokinase. (**C**) Portions of rat liver IRTK, devoid of *N*-acetyl-D-glucosamine, were assayed for GK-dependent inhibition of insulin-stimulated IRTK activity in the absence (●) and presence (○) of glucose (5.5 mM). The results are expressed as a mean ± SEM of three experiments, each performed in duplicate.

in the reactivity of GK and/or IRTK sulfhydryl (SH) groups with reduced glutathione (GSH). FIGURE 2A shows that GK catalytic activity is depressed in a concentration-dependent manner by GSH. When GK was preincubated with GSH, the concentration-dependent inhibition of ITRK by GK was suppressed (FIG. 2B). Hence, it appears that SH groups on IRTK have to be oxidized for expression of activity, which could allow not only reagents like GSH and dithiothreitol (which also was effective), but also induced-fit exposed SH groups on enzymes to interact with IRTK. (The latter may be the case for GK-substrate complex.)

The interaction of GK with IRTK was found to be noncompetitive with respect to Mn^{2+}-activation, decreasing V_{max} from 200 to 111 pmol/min/mg, while having no effect on the apparent $K_{0.5}$ for Mn^{2+} (~ 0.50 mM). Similarly, GK inhibition of IRTK was noncompetitive with peptide (A11) substrate, decreasing V_{max} from 190 to 145 pmol/min/mg, while having no effect on the apparent K_m for peptide (~ 0.89 mM). Thus, GK does not appear to interact with IRTK at either the metal-dependent regulatory site on the enzyme or the peptide substrate binding site. The effect of GK on the kinetic response of IRTK to metal-ATP saturation was found to be competitive, incresing the apparent K_m for metal-ATP from 131 to 317 μM, while V_{max} was unaltered.

These data cannot discount a contribution by direct interaction between both enzymes to the mechanism of GK-dependent inhibition of IRTK, but indicates that the insulin-dependent rate of glucose uptake could lead to GK-dependent depletion

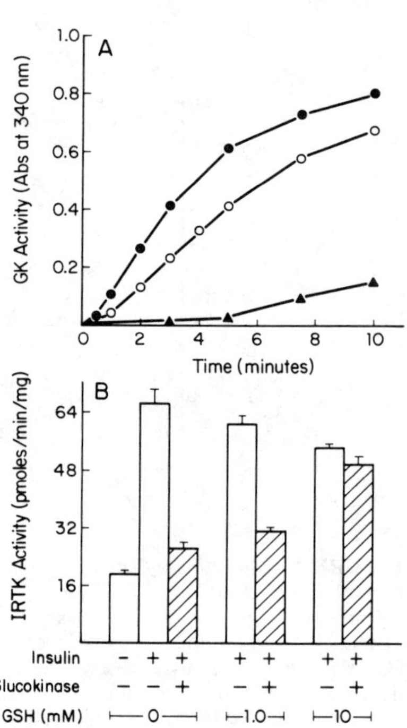

FIGURE 2. Effect of reduced glutathione (GSH) on glucokinase catalytic activity and GK-dependent inhibition of IRTK. (A) GK catalytic activity was measured as described in the text in the presence of 0 (●), 1.0 mM (○) and 10 mM (▲) GSH. (B) IRTK activity was measured in the absence (−) and presence (+) of insulin (1.0 μM) without (−) or with (+) the addition of glucokinase (500 ng/ml), pretreated in the absence (○) and presence of 1 and 10 mM GSH. IRTK activity corresponds to the mean ± SEM of two experiments, each performed in duplicate.

of ATP and thereby represent a metabolic-dependent negative feedback on IRTK. Whether these findings are associated with the obesity and insulin-resistance in animal models such as the db/db and ob/ob mouse, where persistently high levels of liver "GK" have been reported,[2] remains to be determined.

REFERENCES

1. KASUGA, M., F. A. KARLSSON & C. R. KAHN. 1982. Science 215: 185-187.
2. YEN, T. T. & N. B. STAMM. 1981. Biochem. Biophys. Acta 657: 195-202.

3. KLEIN, H. H., G. R. FREIDENBERG, R. CORDERA & J. OLEFSKY. 1985. Biochem. Biophys. Res. Commun. **127:** 254-263.
4. LEMARCHAND-BRUSTEL, Y., T. GREMEAUX, R. BALLOTTI & E. VAN OBBERGHEN. 1985. Nature **315:** 676-679.
5. PIKE, L. J., E. A. KUENZEL, J. E. CASNELLI & E. G. KREBS. 1984. J. Biol. Chem. **259:** 9913-9921.
6. ROSKOSKI, R. 1984. Methods Enzymol. **99:** 3-6.
7. HENTGARTNER, H. & H. ZUBER. 1973. FEBS Lett. **37:** 212-216.

Search for the Enaminal Derivative from the Dipalmitoyl Phosphatidyl Ethanolamine

BINAY KUMAR GHOSH,[a,b] HARRY STEINBERG,[a]
DIPAK K. DAS,[c] AND DENNIS C. SHELLY[d]

[a]Long Island Jewish Hillside Medical Center
New Hyde Park, New York 11042

[c]University of Connecticut School of Medicine
Farmington, Connecticut 06032

[d]Stevens Institute of Technology
Hoboken, New Jersey 07030

INTRODUCTION

The chemical basis of the fluorescence[1] of lipofuscin pigments in aging cells has been attributed to the N,N'-disubstituted 1-amino-3-iminopropene(I) moieties:

$$\overset{\textstyle H}{\underset{\textstyle 3 \quad 2 \quad 1}{R^4-N=CH-CH=CH-\overset{|}{N}-R^5}} \quad (I)$$

N,N'-disubstituted 1-amino-3-iminopropene

where R^4- and R^5-moieties correspond, respectively, to the remaining parts of the relevant proteins and/or amino acids.

Whether the accumulations of such pigments[2,3] in the brain cells include the fluorophore[4,5] of the propanedial (MDA) adduct from the membrane phosphatidyl ethanolamine is under consideration. The first chemical intermediate from the reaction of MDA with a primary amine, however, is expected to be an enaminal(II):

$$\underset{\text{MDA}}{OHC-CH_2-CHO} \rightarrow HO-CH=CH-CHO \overset{R-NH_2}{\longrightarrow} \underset{\text{enaminal}}{R-\overset{\textstyle H}{\overset{|}{N}}-CH=CH-CHO} \ (II)$$

$$(R = n\text{-}C_4H_9\text{-} \text{ or } C_6H_5CH_2CH_2\text{-})$$

[b]Present address: 60 East 12th Street, Apt. 7K, New York, New York 10003.

Accordingly, the nuclear magnetic resonance spectra of two enaminals(II: R- = *n*-butyl-,[6] 2-phenethyl-) were compared with those of the DL-dipalmitoyl phosphatidyl ethanolamine (DPE) and its MDA adduct (MDA-DPE).

MATERIALS AND METHODS

1,1,3,3-tetraethoxy propane and synthetic *alpha*-DPE was purchased from the Sigma Chemical Co. Cylindrical (14 mm × 4 mm) Teflon-encased magnets were used for each stirring. Acid hydrolysis[7] of 1,1,3,3-tetraethoxy propane (50 μ1) in 1 ml of 0.9% saline solution within a 20-ml glass vial provided the MDA after 20 minutes of stirring. Such fresh MDA was used in the reactions with the *n*-butyl- and 2-phenethyl amines. A $HCCl_3$ solution of DPE and neutralized MDA solution were used in the preparation of MDA-DPE. Each of the two enaminals (*n*-butyl- and 2-phenethyl-) and the MDA-DPE adduct were individually separated from the respective product mixtures after developments[8] on tlc plates. Chloroform extract of the silica gel—comprising the separated spot of the particular enaminal—isolated the desired component. The solvent for the MDA-DPE adduct extraction was a mixture of chloroform and methanol (9v:1v). Evaporation of each filtered extract under reduced pressure (38-40°C) provided the corresponding sample for use. All NMR spectra were recorded from a JEOL GX-400 (400 MHZ) spectrometer. While $DCCl_3$ was the solvent for the *n*-butyl- and the 2-phenethyl-enaminal samples, a $DCCl_3$:D_3CCOOD mixture (1v:1v) equilibrated[9] with D_2O was the solvent for the recording of the DPE spectra (FIG. 1). The same solvent was also used for the MDA-DPE.

RESULTS AND DISCUSSION

While the isolation procedure of the NMR samples from the silica gel of the respective developed tlc[8] plates added their soluble chemical binders to the otherwise pure isolates, the absorption peaks from such impurities did not interfere with those due to the three protons[10,11] of the -HC=CH-CHO moiety in any case (TABLE 1). The absorptions of the MDA-DPE adduct, however, did not show the required degrees of intensities for the -HC=CH- protons (TABLE 1). Whether this moiety had added any solvent molecule during the period between dissolution in the acidic liquid and recording of the spectra is under investigation. While the (M + NH_4^+:145) line confirms the formation of the *n*-butyl enaminal, the molecular ion line, in the mass spectra, of a new component in the MDA reaction product from 2-phenethyl amine indicated of the chemical structure:

$$C_6H_5CH_2CH_2N\diagup\diagdown\begin{matrix}CH{=}CH{-}CHO\\CH{=}CH{-}CHO\end{matrix} \qquad \text{(III)}$$

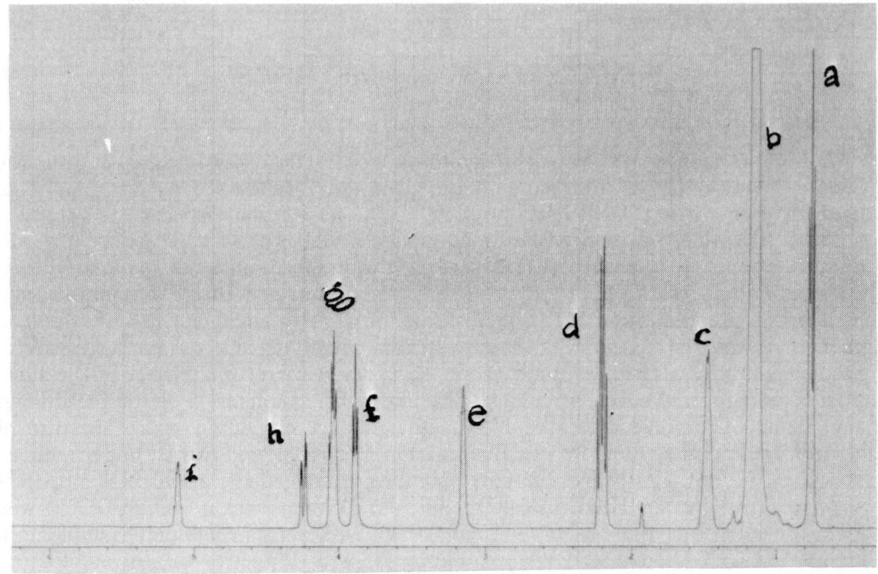

FIGURE 1. The NMR absorption peaks of the different protons in the chemical structure of the synthetic DL-*alpha*-dipalmitoyl phosphatidyl ethanolamine(DPE) are identified as follows:

$$
\begin{array}{l}
\quad\quad\quad\quad\; O \\
h \quad\quad\quad \| \\
CH_2\text{–}O\text{–}C\text{–}CH_2\text{–}CH_2\text{–}(CH_2)_{12}\text{–}CH_3 \\
| \quad\quad\quad\quad | \quad\quad | \quad\quad\quad | \quad\quad\quad | \\
| \quad\quad\quad\; O \quad d \quad\; c \quad\quad b \quad\quad a \\
|\;\; i \quad\quad\; \| \quad\; | \quad\quad | \quad\quad\quad | \quad\quad\quad | \\
CH\text{–}O\text{–}C\text{–}CH_2\text{–}CH_2\text{–}(CH_2)_{12}\text{–}CH_3 \\
| \quad\quad\quad\; O \\
| \quad\quad\quad\; \| \\
CH_2\text{–}O\text{–}P\text{–}O\text{–}CH_2\text{–}CH_2\text{–}NH_3^{(+)} \\
f \quad\quad | \quad\quad\quad g \quad\quad e \\
\quad\quad\quad O_{(-)}
\end{array}
$$

While the above identifications of a, b, c, d, e, f, g, h and i relate to the corresponding peaks above, the "g" region also incorporates[9] part of the absorptions of the "h" identification protons.

TABLE 1. Proton Nuclear Magnetic Resonance Chemical Shifts[10,11] of the
$-HC=CH-CHO$ Moieties

Enaminal	ppm to High Frequency from Tetramethyl Silane		
	$C=CH-C$	$N-CH=C$	$\overset{\displaystyle O}{\overset{\displaystyle \|}{C-C-H}}$
n-butyl-[6]	5.43	6.87	9.20
2-phenethyl-	5.47	6.72	9.21
MDA-DPE	5.49[a]	6.89[a]	8.82[b]

[a] Negligible absorption.
[b] Doubtful assignment.

REFERENCES

1. TAPPEL, A. L. 1973. Fed. Proc. **32**(8): 1873.
2. ROTHSTEIN, M. 1986. Chem. Eng. News **64**(32): 32.
3. CHIO, K. S. & A. L. TAPPEL. 1969. Biochem. **8**(7): 2826.
4. MERZLYAK, M. N. & V. B. RUMYANTSEVA. 1984. Chem. Abstr. **100**: 116792x.
5. FUJITA, T. & M. YASUDA. 1977. Chem. Abstr. **87**: 79081e.
6. KRASNAYA, ZH. A., T. S. STYTSENKO, E. P. PROKOF'EV, I. P. YAKOVLEV & V. F. KUCHEROV. 1974. Chem. Abstr. **81**: 49608r.
7. JAIN, S. K. & S. B. SHOHET. 1984. Blood. **63**(2): 364.
8. GILFILLAN, A. M., A. J. CHU, D. A. SMART & S. A. ROONEY. 1983. J. Lipid Res. **24**: 1651-1656.
9. MENA, P. L. & C. DJERASSI. 1985. Chem. Phys. Lipids **37**: 257-70.
10. KIKUGAWA, K., K. TSUKUDA & T. KURECHI. 1980. Chem. Pharm. Bull. **28**: 3325.
11. MCNAB, H. 1981. J. Chem. Soc. Perkin II: 1284.

Effect of Aminopterin on the Metabolic Fate of Hypoxanthine in Human Skin Fibroblasts

MARY JEAN C. HOLLAND

Department of Natural Sciences
Baruch College of the City University of New York
New York, New York 10010

and

Department of Psychiatry
New York University Medical Center
New York, New York 10016

INTRODUCTION

In purine metabolism, inosinate represents a branch-point from which both adenine and guanine nucleotides are derived. Human skin fibroblasts can produce inosinate by either of two pathways, *de novo* or salvage. Labeled purines produced by cells grown in the presence of [^{14}C]hypoxanthine are formed from inosinate synthesized by the salvage pathway.

In human skin fibroblasts, inosinate is rapidly converted to either adenine nucleotides or guanine nucleotides.[1-3] Of total cellular nucleotides, about 70-75% are adenine nucleotides (AMP, ADP, ATP, NAD), and the remaining 25-30% are guanine nucleotides (GMP, GDP, GTP). In the present study, the time course of labeling of adenine and guanine nucleotide pools by inosinate produced by salvage of [^{14}C]hypoxanthine was examined in human skin fibroblasts in the presence and absence of aminopterin, a potent inhibitor of purine synthesis *de novo*.[4,5]

MATERIALS AND METHODS

Normal fibroblasts were obtained from neonatal foreskins. [^{14}C]hypoxanthine was purchased from New England Nuclear Corp. Unlabeled purines were obtained from Sigma Chemical Co. All cell culture media and sera were purchased from Grand Island Biological Co.

Cultures were grown as monolayers. Extraction, separation, and identification of labeled reaction products were carried out as previously described.[3]

RESULTS AND DISCUSSION

As shown in the TABLE 1, inhibition of the *de novo* pathway by the presence of aminopterin increased the rate of nucleotide synthesis from [^{14}C]hypoxanthine, but markedly decreased the adenine:guanine labeling ratio at early time points. This change in ratio reflects an increase in guanine nucleotide synthesis from inosinate produced by salvage when the *de novo* pathway is inhibited. For example, after 1 hour the total nucleotide synthesis was increased by 43% in the presence of aminopterin, reflecting a 265% increase in guanine nucleotide synthesis, while the rate of synthesis of adenine nucleotides actually decreased slightly. By 4 hours the labeling ratios in both sets of cultures were similar.

Intracellular compartmentation is one possible explanation for these results. Inosinate, itself, may occupy more than one intracellular pool. Aspartate and glutamine, essential cosubstrates in the production of adenine and guanine nucleotides, respectively, may also be subject to intracellular compartmentation.[6] It is also possible that

TABLE 1. Labeled Ribonucleotides in Acid-Soluble Pools Extracted from Cells after Incubation in Medium Containing 100 μM [^{14}C]Hypoxanthine with (+) or without (−) 10 μM Aminopterin

	Ratios of Adenine:Guanine[a]		Total Nucleotide Synthesis[b]	
	−	+	−	+
30 min	3.68	0.98	2.79	3.88
60 min	4.52	1.16	5.11	7.32
120 min	3.85	2.97	7.84	9.61
240 min	3.28	3.78	9.48	11.29

[a] Data represent only labeled nucleotides. No correction was made for the dilution of specific activity by endogenous pools.

[b] Data are given in nanomoles of labeled hypoxanthine incorporated into nucleotides per milligram protein.

inhibition of *de novo* purine synthesis increases the availability of glutamine for the conversion of inosinate to guanine nucleotides.[7]

REFERENCES

1. BECKER, M. A. 1976. Regulation of purine nucleotide synthesis. Effects of inosine on normal and hypoxanthine-guanine phosphoribosyltransferase-deficient fibroblasts. Biochim. Biophys. Acta **435:** 132-144.
2. HOLLAND, M. J. C., A. M. DiLORENZO, J. DANCIS, M. E. BALIS, T. F. YU & R. P. COX. 1976. Hypoxanthine phosphoribosyltransferase activity in intact fibroblasts from patients with X-linked hyperuricemia. J. Clin. Invest. **57:** 1600-1605.
3. HOLLAND, M. J. C., N. C. KLEIN & R. P. COX. 1978. Experimental modulation of PRPP availability for ribonucleotide synthesis from hypoxanthine in human skin fibroblast cultures. Exp. Cell Res. **111:** 237-243.

4. BALIS, M. E. 1968. Antagonists and Nucleic Acids. North-Holland. Amsterdam.
5. BLAKLEY, R. L. 1969. The Biochemistry of Folic Acid and Related Pteridines. North-Holland. Amsterdam.
6. GARFINKEL, D. 1966. A simulation study of the metabolism and compartmentation in brain of glutamate, aspartate, the Krebs cycle, and related metabolites. J. Biol. Chem. **241:** 3918-3929.
7. CRABTREE, G. W. & J. F. HENDERSON. 1971. Rate-limiting steps in the interconversion of purine ribonucleotides in Ehrlich ascites tumor cells *in vitro.* Cancer Res. **31:** 985-991.

In Vivo Effects of Snake Venoms on Glucose Transport across the Blood-CSF Barrier of the Rat

J. DiMATTIO, S. WEINSTEIN, AND J. STREITMAN

New York University Medical Center
New York, New York 10016

Whether snake venom components can alter either carrier-mediated or passive transport properties of the blood-CSF barrier is a complex and little-explored question of toxinology.[1,2] Our previous work, in the eye, has suggested that barrier transport function could be effectively characterized by studying the rates of transport of labeled non-metabolizable test stereoisomers of D-glucose. Thus, [3H]-3-O-methyl-D-glucose (mD-glu), which is considered to use the same transport system as the biologically active D-glucose, but not to be metabolized, is used as an indication of carrier-mediated transport, and the inactive [14C]-L-glucose (L-glu) is used to measure passive movement into CSF. Rate entry constants for transport from blood into CSF were determined for both labeled test sugars in control and i.v. envenomated rats using a plasma concentration curve of the form:

$$Cp = A + Be^{-b_1 t} + Ce^{-b_2 t} \tag{1}$$

by means of methods paralleling those previously published for the eye.[3,4] To ensure that venoms were being tested at blood concentrations sufficiently high to cause pathophysiological changes in some reasonable period of time, venom dosages producing cardio-respiratory death in 1.0 hr were determined (TABLE 1). This calibrated dose was then administered via a cannulated femoral vein to anesthetized normal Sprague-Dawley rats (225-350 g) using a lucite operating board designed for this purpose and for the efficient removal of small samples of blood (50-100 μl). We reasoned that at these doses physiological changes due to venom action were well under way 20 min after i.v. venom administration and thus at this time began our procedure for measuring transport rates of radiolabeled test glucoses from blood to CSF.

TABLE 2 reports results obtained using whole lyophilized venom from three snakes: *Crotalus scutulatus* (Mojave rattlesnake); *Agkistrodon piscivorus* (Eastern cottonmouth moccasin); and *Naja nivea* (Yellow or Cape cobra). Our results in control animals clearly indicate that D-glucose enters CSF via a carrier-facilitated mechanism in that mD-glu enters CSF from blood some 38 times faster than L-glu as indicated by direct C_{CSF}/C_P measurement at the end of the 13-min transport experiment and by the calculated rate constant, K_i. Carrier specificity suggests that increases in carrier function would be reflected in mD-glu transport rates, whereas passive changes should be

TABLE 1. Dosages of Lyophilized Whole Venoms Causing Cardiac/Respiratory Death in 1.0 hr

Species	Dose (mg/kg body wt)
Mojave rattlesnake (*C. scutulatus*)	1.9
Cottonmouth moccasin (*A.p. piscivorus*)	4.0
Yellow cobra (*N. nivea*)	0.9

[a] Male Sprague-Dawley rats weighing 250-350 g.

observed with both stereoisomers, as they depend primarily on size and less on configuration.

Our results indicate that Mojave rattlesnake venom increases the passive entry of glucose into CSF. We note a 366% increase in the L-glu entry rate. This passive break in blood-CSF barrier is also reflected in a modest overall increase in mD-glu entry as well as passive movement. Thus, no alteration in D-glucose carrier function is indicated.

TABLE 2. Venom Effects on [³H]-3-O-Methyl-D-glucose and [¹⁴C]-L-glucose Transport from Blood into CSF

	n	$C_{CSF}/C_P{}^a$	K_i	% Change	p
		[³H]-3-O-Methyl-D-glucose			
Control	14	0.571 ± .030	0.0455 ± .0041		
Mojave rattlesnake (*C. scutulatus*)	7	0.608 ± .035	0.0523 ± .0053	+13	NS
Cottonmouth moccasin (*A.p. piscivorus*)	7	0.483 ± .031	0.0266 ± .0028	−42	$p < .001$
Yellow cobra (*N. nivea*)	8	0.424 ± .040	0.0268 ± .0042	−42	$p < .001$
		[¹⁴C]-L-glucose			
Control	14	0.023 ± .007	0.0012 ± .0004		
Mojave Rattlesnake (*C. scutulatus*)	7	0.107 ± .021	0.0056 ± .0020	+366	$p < .001$
Cottonmouth moccasin (*A.p. piscivorus*)	7	0.038 ± .011	0.0015 ± .0004	+20	NS
Yellow cobra (*N. nivea*)	8	0.028 ± .010	0.0018 ± .0008	+40	NS

NOTE: The statistical significance of difference from control was performed using the Student's t test and values of $p > .02$ were not considered significant.

[a] C_{CSF}/C_P is the experimentally determined ratio of concentration found in CSF and plasma at the end of the transport experiment. For each experiment the end time was between $t = 12.1$ and 13.3 min. Values are listed as mean ± SE.

Cottonmouth venom appears to decrease D-glu carrier function in that the entry rate for mD-glu is decreased significantly, whereas, passive L-glu entry is essentially unchanged. Cobra venom also appears to adversely affect carrier function and leave passive movement relatively unaffected. It is noted that decreased mD-glu entry could also be explained by a diminished blood supply to CSF sources of blood-borne substances via a loss in vascular autoregulation.

REFERENCES

1. RUSSELL, F. E. 1980. Snake Venom Poisoning. Lippencott. Philadelphia.
2. TU, A. T., ED. 1982. Rattlesnake Venoms: Their Actions and Treatment. Marcel Dekker. New York.
3. DiMATTIO, J. & J. A. ZADUNAISKY. 1984. *In vivo* entry of glucose analogs in to the ocular compartments of rat. Exp. Eye. Res. **32:** 517-532.
4. DiMATTIO, J. & J. A. ZADUNAISKY. 1982. Glucose transport across ocular barriers of the spiny dogfish, *Squalus acanthias.* J. Exp. Zool. **219:** 197-203.

Experimental Manipulation of the Blood-Retinal Barrier[a]

GARY E. KORTE,[a] MARGARET S. BURNS,[b]
AND ROY BELLHORN[c]

[a]*Department of Ophthalmology*
Montefiore Medical Center
Albert Einstein College of Medicine
Bronx, New York 10467

[b]*Department of Ophthalmology*
School of Medicine
University of California at Davis
Davis, California 95616

[c]*Department of Surgery*
School of Veterinary Medicine
University of California at Davis
Davis, California 95616

The blood-retinal barrier (BRB) is established at two sites—the retinal capillary endothelium and the retinal pigment epithelium (RPE). At both sites intercellular tight junctions prevent the penetration of blood-borne molecules into the neural retina. The absence of fenestrae and vesicular transport contributes to the BRB at the retinal endothelium and the marked structural and functional polarity of the epithelial cells contributes to the BRB at the level of the RPE. The factors controlling these properties of retinal endothelium and RPE are unknown. We have analyzed this problem by bringing retinal capillaries into apposition with RPE and examining consequent changes in the permeability related structure and function of both cell types.

When rats are exposed to fluorescent light or receive subcutaneous injections of urethane when young[1,2] the retinal photoreceptors are destroyed. When and where this occurs the retinal capillaries become embedded in the RPE; the capillaries form intraepithelial loops that are inserted between the lateral plasma membranes of adjacent RPE cells. The endothelium of the intraepithelial loop changes (FIGS. 1-3). It becomes thinner, develops fenestrae, and becomes permeable to intravenously injected horse-radish peroxidase (HRP). The endothelium on the portion of capillary remaining in the neural retina (and which is continuous with the intraepithelial segments) retains its normal thick, unfenestrated endothelium that is impermeable to HRP. This suggests that the RPE influences retinal capillary endothelium, causing it to become thin and fenestrated. This observation is interesting because the choriocapillaris, which is closely apposed to the RPE sheet and is the major nutrient source for the photoreceptors,

[a]Research for this paper was supported by Research to Prevent Blindness, Inc. and the National Eye Institute.

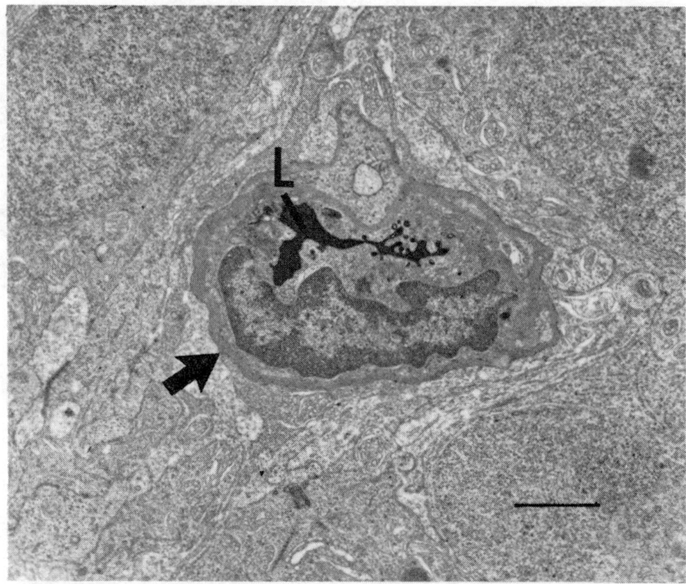

FIGURE 1. Capillary in neural retina of phototoxic rat that received HRP. Black reaction product is restricted to lumen (L) and does not reach capillary basement membrane (*arrow*). Bar denotes 5 microns.

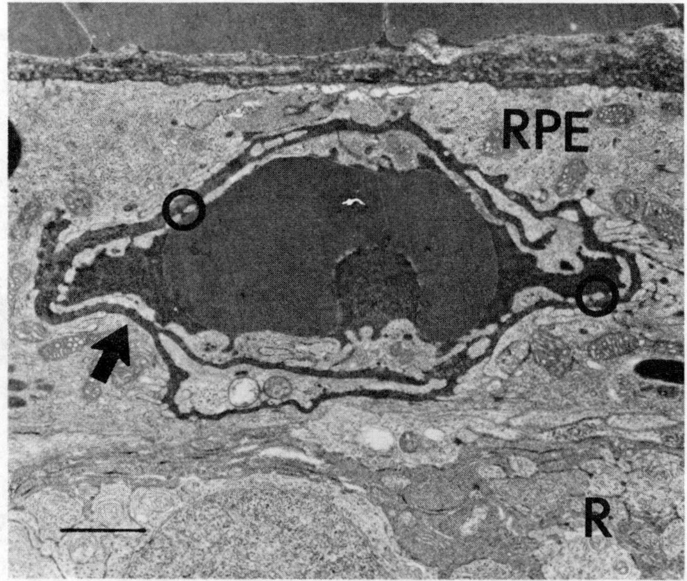

FIGURE 2. HRP penetrates intraepithelial capillaries (L, lumen) to reach basement membrane (*arrow*). Penetration occurs at fenestrae (*circle*) that are detailed in FIGURE 3. R, neural retina. Choriocapillaris and Bruch's membrane are at top of picture. Bar denotes 5 microns.

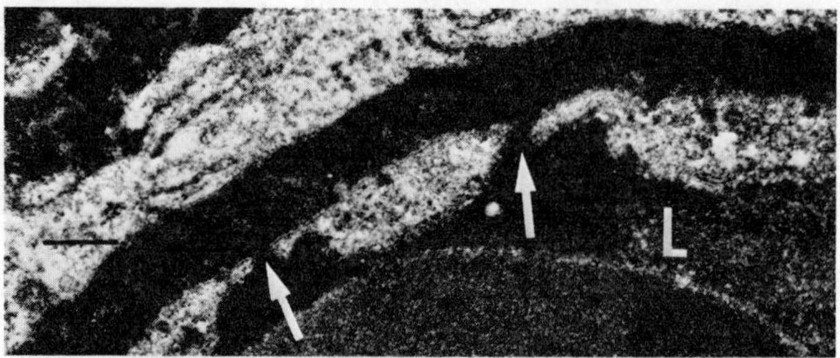

FIGURE 3. Detail of intraepithelial capillary, showing HRP reaction product in lumen (L) and basement membrane on other side of fenestrae sites (*arrows*). Fenestrae themselves are obscured by reaction product. Bar denotes 0.5 microns.

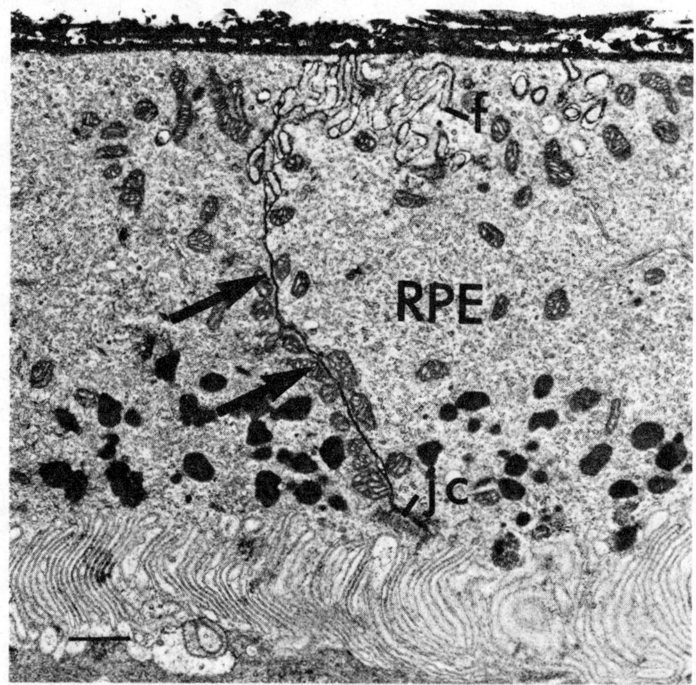

FIGURE 4. RPE where no intraepithelial capillaries occur. Lateral plasma membrane (*arrows*) is flat and undifferentiated. HRP reaction product outlines it but junctional complexes (jc) prevent penetration into the neural retina below. Note folds (f) on basal plasma membrane facing choriocapillaris, at top of picture. Bar denotes 1 micron.

normally has an endothelium that is thin, fenestrated, and permeable to HRP. Based on the observations at sites of intraepithelial capillaries, we suggest that the permeability related characteristics of choriocapillaris endothelium are due to the proximate RPE sheet. More generally, it appears that RPE, and perhaps other epithelium (such as that of the choroid plexus), can influence the structure and function of the adjacent capillaries.

The RPE facing the intraepithelial capillary segment also undergoes changes that are related to permeability across the epithelial sheet (FIGS. 4 and 5). The lateral plasma membranes of RPE cells, normally flat and undifferentiated, develop folds and tubules; form attachment sites on the capillary basement membrane; begin to endo-

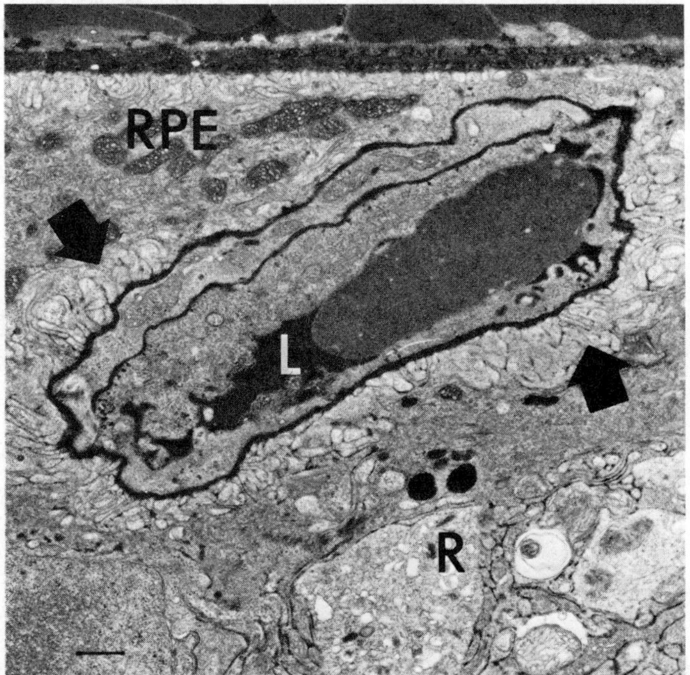

FIGURE 5. Intraepithelial capillary containing HRP reaction product in its lumen (L) and basement membrane. Note fold formation by RPE lateral plasma membrane at arrows (cf., FIG. 4). R, remnant neural retina. Choriocapillaris is at top of picture. Bar denotes 1 micron.

cytose HRP; and show evidence of basement membrane secretion. These structural and functional specializations are normally restricted to the basal plasma membrane facing the choriocapillaris. The RPE cells apposing intraepithelial capillaries have reorganized their structure and functional polarity in relation to them. This suggests that capillaries can influence RPE structure and function.

These observations show that the structural and functional specializations of retinal capillary endothelium and RPE that contribute to the BRB can be manipulated. To some extent the changes described here may be exaggerations of changes that normally modulate permeability across the BRB, especially at the RPE-choriocapillaris interface. The observations suggest that RPE cells and capillary endothelium influence each other, though the mechanism for this remains speculative (see Refs. 3 and 4 for discussion).

REFERENCES

1. BELLHORN, R. W., M. S. BELLHORN, A. H. FRIEDMAN & P. HENKIND. 1973. Urethane-induced retinopathy in pigmented rats. Invest. Ophthalmol. Vis. Sci. **12:** 65-76.
2. BELLHORN, R W., M. S. BURNS & J. V. BENJAMIN. 1980. Retinal vessel abnormalities of phototoxic retinopathy in rats. Invest. Ophthalmol. Vis. Sci. **19:** 584-595.
3. BURNS, M. S., R. W. BELLHORN, G. E. KORTE & W. J. HERIOT. 1986. Plasticity of the retinal vasculature. Prog. Retinal Res. **5:** 253-307.
4. KORTE, G. E., R. W. BELLHORN & M. S. BURNS. 1986. Remodelling of the retinal pigment epithelium in response to intraepithelial capillaries: Evidence that capillaries influence the polarity of epithelium. Cell Tissue Res. **245:** 135-142.

Differential Expression of Endothelium-specific Antigen PAL-E in Vasculature of Brain Tumors and Preexistent Brain Capillaries

R. O. SCHLINGEMANN, G. Th. A. M. BOTS,
S. G. VAN DUINEN, AND D. J. RUITER[a]

Department of Pathology
Leiden University Medical Center
2300 RC Leiden, the Netherlands

Department of Pathology
Nijmegen University
6525 GA Nijmegen, the Netherlands

In the brain, blood vessels are lined by uniquely differentiated endothelium forming a barrier of selective permeability to a wide range of solutes, the blood-brain barrier (BBB).[1] Tight junctions, a scarcity of plasmalemmal vesicles, and the absence of fenestrations in brain endothelium are cellular properties associated with this barrier function, [2] but the brain endothelium also differs from endothelium outside the brain with respect to several molecular properties related to transcellular transport like the presence of the transferrin receptor[3] and the enzyme γ-glutamyl transpeptidase.[4] In certain anatomic sites, like the choroid plexus and the pituitary gland, and some pathological conditions, including brain tumors, the BBB seems to be lacking or defective, and the endothelium of these sites resembles endothelium outside the central nervous system.

We investigated the expression of the antigen recognized by the monoclonal antibody PAL-E on endothelium in the CNS. This antibody recognizes an endothelium-specific molecule of undefined molecular weight and function, which is associated with endothelial vesicles.[5] Immunoperoxidase staining of frozen sections of normal (human, sheep, cow, rabbit) and neoplastic brain tissues (TABLE 1) with this antibody and of serial sections with the lectin Ulex Europaeus I as a vascular marker[6] revealed that PAL-E is absent from endothelium in sites with a patent BBB (TABLE 1, FIG. 1). In contrast, anatomic sites with fenestrated or permeable endothelium showed marked staining (TABLE 1, FIG. 1). A striking observation was the strong staining of endothelium in the stroma of primary and metastatic brain neoplasms (FIG. 2). This finding parallels the defective BBB observed in brain tumors using permeability tracers and

---

[a] Address for correspondence: D. J. Ruiter, M. D., Professor of Pathology, Department of Pathology, Catholic University Nijmegen, Geert Grooteplein Zuid 24, 6525 GA Nijmegen, the Netherlands.

111

ultrastructural studies on the vasculature in brain tumors showing endothelium with fenestrations containing many endothelium vesicles[7,8] like endothelium outside the CNS. The absence of the antigen in brain endothelium is another feature distinguishing this specialized endothelium from its counterpart in other tissues.

TABLE 1. Staining of Brain Endothelium by PAL-E

Tissue Tested		Number	Staining Result	Type of Endothelium
Normal Tissues				
Cerebral cortex		6	−	BBB[a]
Cerebellum		3	−	BBB
Midbrain		3	−	BBB
Medulla		3	−	BBB
Spinal cord		1	−	BBB
Eye				
Choriocapillaris		2	+ +	Fenestrated
Retina		2	−	BBB
Meninges		4	−	BBB
Dura mater		3	+ +	Not known
Arachnoidal granulations		3	−	Not known
Choroid plexus				
lateral ventricle		3	+ +	Fenestrated
fourth ventricle		1	+ +	Fenestrated
Pituitary gland		4	+ +	Fenestrated
Tumors				Variable
Primary brain tumors		28	+ +	
Oligodendroglioma	5			
Astrocytoma	8			
Meningioma	7			
Schwannoma	3			
Ependymoma	1			
Neurofibroma	2			
Craniopharyngioma	1			
Lymphoma	1			
Metastases		13	+ +	Variable
Adenocarcinoma of kidney	3			
of ovary	2			
of colon	2			
of lung	1			
of unknown sites	1			
Planocelluar carinoma of lung	2			
Melanoma	2			

[a] BBB = blood–brain barrier.

Though its function is unknown, our results may suggest a correlation of the PAL-E antigen with trans-endothelial transport. Its presence exclusively in the endothelial cell, the previously shown association with endothelial vesicles and the remarkable absence from brain endothelium in areas with BBB favor such a hypothesis. The marked staining of permeable endothelium in the brain and the absence of staining

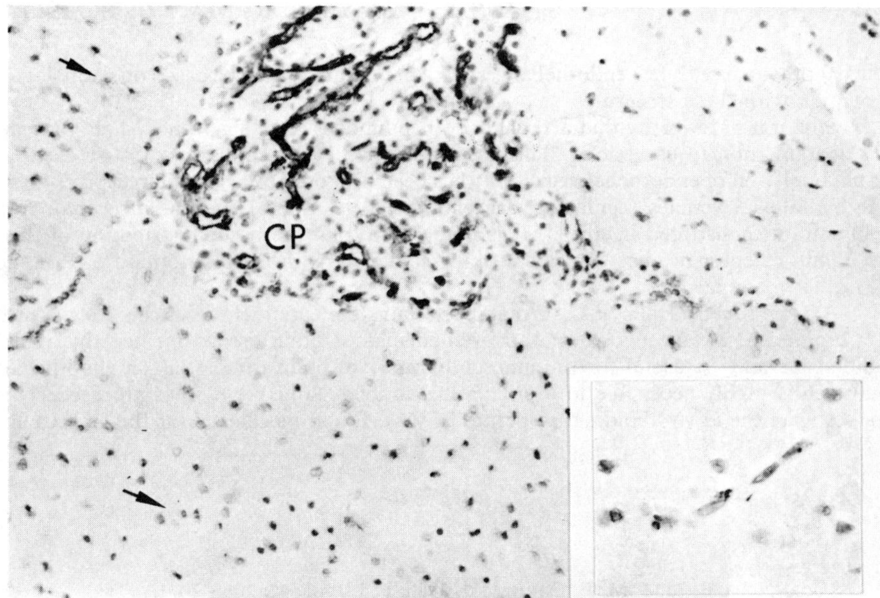

FIGURE 1. Frozen section of normal rabbit brain stained with PAL-E in an indirect immunoperoxidase staining procedure using 3-3-di-amino-benzidine as a substrate and hematoxylin counterstaining. Note marked staining of capillaries in choroid plexus (CP) and absent staining in parenchymal capillaries (*arrows* and *inset*) ($\times$160; *inset* $\times$280).

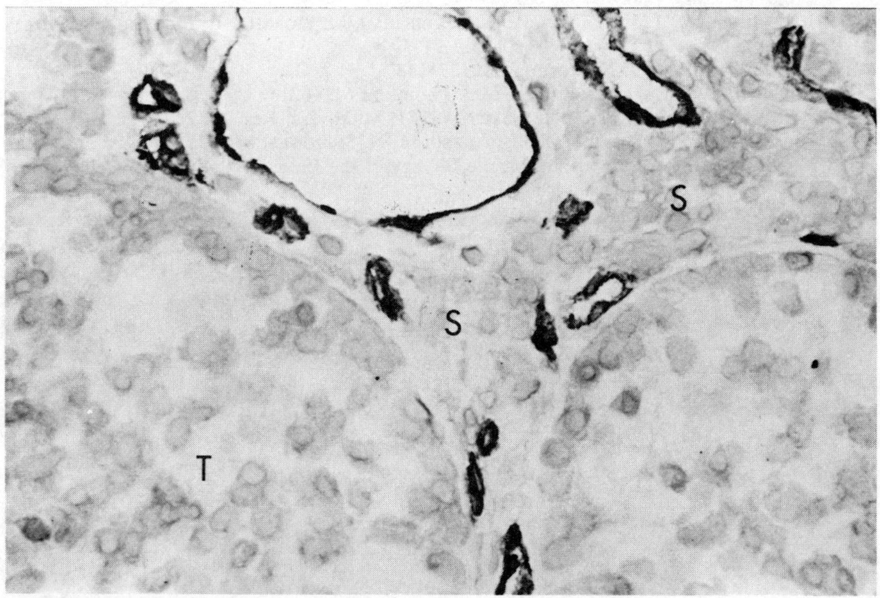

FIGURE 2. Frozen sections of metastatic adenocarcinoma stained with PAL-E. Nuclei were counterstained with neutral red. Note marked staining of vascular endothelium in the tumor stroma (S); T = tumor ($\times$400).

in endothelium with few endothelial vesicles (brain, lung)[9] may be based on a topologic relation with these structures.[9]

Still, it is noteworthy that arterial endothelium does not stain with PAL-E, though it contains numerous vesicles. This suggests that this determinant is restricted to a subpopulation of endothelial vesicles. It has been proposed that specific receptors exist in endothelial vesicles regulating transendothelial transport,[9] such as a receptor for albumin demonstrated in mice.[10] It is noteworthy that the tissue distribution of this albumin receptor in mice shows a striking similarity with the distribution of PAL-E in other mammals.

As mentioned earlier, PAL-E markedly stains the endothelium of the vasculature in brain tumors; adjacent normal or even edematous brain tissues are negative. This difference may be useful for immunoscintigraphy of brain tumors, as an endothelial antigen is highly accessible to a circulating antibody. Further studies are needed to investigate the *in vivo* binding properties of PAL-E and its efficacy for this approach.

REFERENCES

1. CORNFORD, E. M. & M. E. CORNFORD. 1986. Fed. Proc. **45:** 2065-2071.
2. THORGEIRSSON, G. 1983. *In* Biochemical Interactions at the Endothelium. A. Cryer, Ed.: 5-39. Elsevier. Amsterdam.
3. JEFFERIES, W. A., M. R. BRANDON, S. V. HUNT, A. F. WILLIAMS, K. C. GATTER & D. Y. MASON. 1984. Nature **312:** 162-163.
4. DeBAULT, L. E. & P. A. CANCILLA. 1980. Science **207:** 653-655.
5. SCHLINGEMANN, R. O., G. M. DINGJAN, J. BLOK, J. J. EMEIS, S. O. WARNAAR & D. J. RUITER. 1985. Lab. Invest. **51:** 71-76.
6. HOLTHOFER, H., I. VIRTANEN, A. L., KARINIEMI, M. HORMIA, E. LINDER & A. MIETTINEN. 1982. Lab. Invest. **47:** 60-65.
7. LONG, D. M. 1970. J. Neurosurg. **32:** 127-145.
8. HIRANO, A. & T. MATSUI. 1975. Hum. Pathol. **6:** 611-621.
9. WAGNER, R. C. & J. R. CASLEY-SMITH. 1981. Microvasc. Res. **21:** 267-298.
10. GHITESCU, L., A. FIXMAN, M. SIMIONESCU & N. SIMIONESCU. 1986. Proceedings of the 4th International Symposium on the Biology of the Vascular Endothelial Cell, Noordwijkerhout, the Netherlands.: 131.

Does Nicotine Alter the Metabolic and Histochemical Profile of Developing Rat EDL Muscle?[a]

KENNETH J. ROSE, EDITH A. GOLDIE,
AND FLEUR L. STRAND[b]

Biology Department
New York University
New York, New York 10003

Nicotine has the dual property of being both an acetylcholine agonist and a lipid-soluble molecule that readily penetrates cell membranes. As an acetylcholine agonist, nicotine mimics the action of this neurotransmitter at the neuromuscular junction.[1] As a lipophilic molecule, nicotine is able to pass through both the placenta[2] and the blood-brain barrier.[3]

The action of nicotine on the maturational process can be viewed by using the developing extensor digitorum longus (EDL) muscle of the rat as a model. This muscle, which in the adult possesses fast-twitch characteristics,[4] passes through several developmental stages. Early stages are characterized by an immature active state and a rudimentary metabolic enzyme system.[5] The rate of maturation of this muscle can be analyzed by using an indicator for cellular respiration.

Previous electrophysiological data from our laboratory have shown that nicotine, administered during gestational or early postnatal life, accelerates the maturational rate of EDL muscle development.[6] This study was designed to confirm our previous data by biochemical and histochemical means.

Sprague-Dawley rats were treated either prenatally and/or postnatally with nicotine tartrate salt (Kodak; 0.25 mg/kg twice a day, i.p. to pregnant dams; 0.05 mg/kg per day, s.c. to pups) or saline solution (0.9% NaCl; equal volume), and measurements of EDL respiratory or histochemical characteristics were made 2 or 3 weeks after birth. For biochemical data, EDL muscles were removed from anesthetized (sodium pentabarbital; 40 mg/kg, i.p.) 2-week old pups and incubated in solutions of 1% 2,3,5-triphenyl tetrazolium chloride (TCl) in 0.05 M phosphate buffer at 36°C for 1 hour. The muscles were then bathed in acetone for 15 min, after which time the acetone-formazan red dye (the product of TCl upon reduction) solution was analyzed in a Spectronic 20 spectrophotometer (435-nm setting) and the absorbance/EDL dry weight values were recorded. For the histochemical analysis, EDL muscles were removed from anaesthetized pups, frozen in 95% ethanol cooled with dry ice,

[a]Research for this paper was supported by The Council for Tobacco Research.

[b]Address for correspondence: Professor Fleur L. Strand, Department of Biology, 1009 Main Building, New York University, Washington Square, New York, New York 10003.

and placed in a cryostat ($-20°C$). Thick sections (10 μm) were cut serially and picked up on slides coated with egg-white albumen. These sections were then incubated in a solution of NADH, nitroblue tetrazolium, and 0.1 M phosphate buffer (pH 7.3) for 30 min at room temperature, rinsed in distilled water, dehydrated with graded ethanol series, cleared in xylene, and mounted.

The results show that prenatal administration of nicotine increases the metabolic properties of the developing EDL muscle as viewed at 2 weeks after birth versus saline-treated controls. Absorbance/EDL dry weight values of all nicotine treatment groups were higher than those of control animals, significantly so in prenatally treated pups (FIG. 1). Histochemical studies using the indicator for NADH diaphorase showed no differences in fiber-type distribution at 2 or 3 weeks of age. However, the muscles of prenatally nicotine-treated pups appeared to show deeper staining versus that of controls.

FIGURE 1. Cellular respiratory rate as indicated by absorbance/EDL dry weight measurements using TCl. Absorbance/EDL dry weight values for group treated prenatally with nicotine (nicpre) are significantly larger ($p <$ 0.02) than those of saline-treated control animals (salcon). nicpost = postnatal nicotine treatment; nicprepost = pre- and postnatal nicotine treatment. $n = 15$ for each group; * $p < 0.02$ versus control using Student's t test. Values are means $\pm$ SEM.

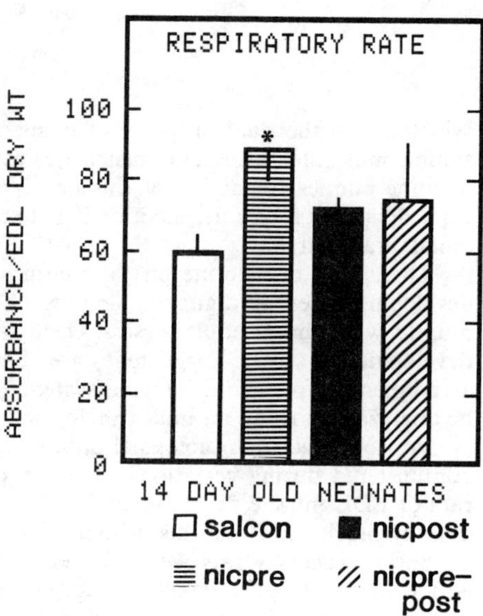

The results from the biochemical study confirm our previous electrophysiological data. We suggest, then, that nicotine administered prenatally may have a profound effect on later muscle metabolism, and possibly accelerates muscle maturation.

REFERENCES

1. McGeer, P. L., J. C. Eccles & E. G. McGeer. 1978. Molecular Neurobiology of the Mammalian Brain. Plenum Press. New York.
2. Fabro, S. & S. M. Sieber. 1969. Caffeine and nicotine penetrate the pre-implanation blastocyst. Nature **223:** 410-411.

3. SERSHEN, H. & A. LAJTHA. 1979. Cerebral uptake of nicotine and of amino acids. J. Neurosci. Res. **4:** 85-91.
4. CLOSE, R. 1964. Dynamic properties of fast and slow skeletal muscles of the rat during development. J. Physiol. **173:** 74-95.
5. ZUBRZYCKA-GAARN, E. & M. G. SARZALA. 1980. Sarcoplasmic reticulum and sarcolemma during development. *In* Plasticity of Muscle. D. Pette, Ed.: 35-52. De Gruyter. Berlin.
6. ROSE, K. J. & F. L. STRAND. 1985. Comparison of prenatal and postnatal nicotine administration on neuromuscular maturation. Soc. Neurosci. Abstr. **11**(2): 1214.

Macroscopic Investigations of the Human Stratum Corneum Evaluating a Method for Heat Separation of the Epidermis from the Dermis

CAROL SIGNOR, MARGURETE WOODWORTH,
LINDA DeNOBLE, AND
TAMIE KURIHARA-BERGSTROM

Ciba-Geigy Corporation
Basic Pharmaceutics Research
Ardsley, New York 10502

Heat has been used for many years as a method to separate epidermis from dermis.[1] Heat-separated epidermal tissue has been used for investigating various substances, such as prostaglandins,[2] cyclic AMP,[3] and phospholipids.[4] This technique is also routinely employed in the investigations of the physicochemical properties of pharmaceutically effective compounds' transport across human epidermis. Heating the skin at 60°C for 2 min results in a clear separation of epidermis from dermis.[5] Studies with hairless mouse skin on the effect of scalding at 60°C on permeation have been reported.[6] However, to the present, no studies concerning the influence of heat separation of the human epidermis on percutaneous absorption have been reported. In this study we used *in vitro* skin transport studies, histologic observations, and differential scanning calorimetry (DSC) to examine the influence of heat separation on epidermal properties.

METHODS

For skin transport studies full-thickness human cadaver skin (back area) was placed in saline solution at 60°C for from 0.5 min to 15 min. The epidermis was carefully removed and mounted on diffusion cells. The donor side of the cell was charged with salicylic acid solutions at pH 3.5 or pH 7.4 and the receiver side filled with saline solution. The quantity of salicylic acid transported across the epidermis and collected in the receiver compartment was assayed by HPLC. The slope of the steady-state region of the amount-versus-time plot was used to calculate the flux value. Permeability coefficients were calculated. Thermal analysis was performed on hydrated stratum corneum using the Perkin-Elmer DSC-4 differential scanning calorimeter from

118

30°C to 120°C at a scan rate of 5°C/min. The lipid transitions are identified by their thermal reversibility. The lipid-protein and protein transitions are identified by their thermal irreversibility.

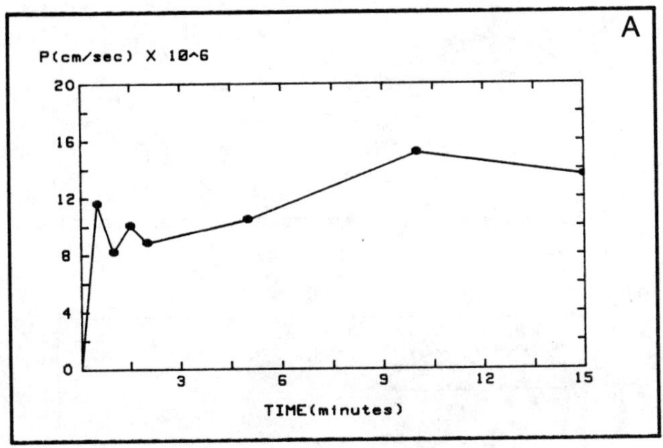

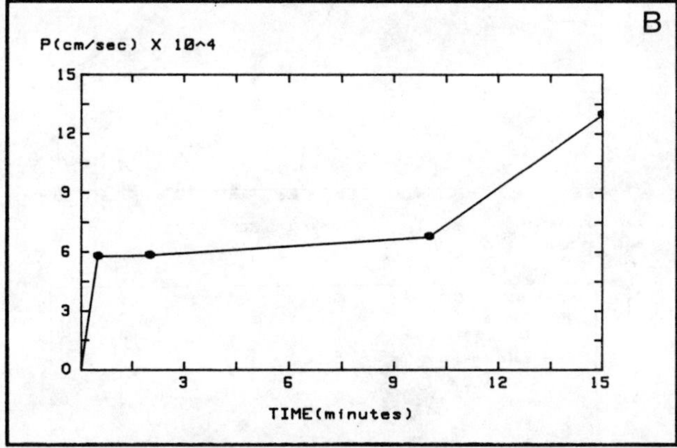

FIGURE 1. Average permeability coefficient of salicylic acid across human epidermis as a function of time on heat separation of epidermis at 60°C. (A) pH 3.5 and (B) pH 7.4 ($n = 2$).

RESULTS AND DISCUSSION

When salicylic acid was used in skin permeation studies, an increase in transport was observed after heat treatment at 60°C for 10 min (FIG. 1). The average permeability coefficient from a pH 3.5 aqueous solution of salicylic acid at 2-min heat separation is 8.81×10^{-6} cm/sec, while at 10-min heat separation the permeability coefficient

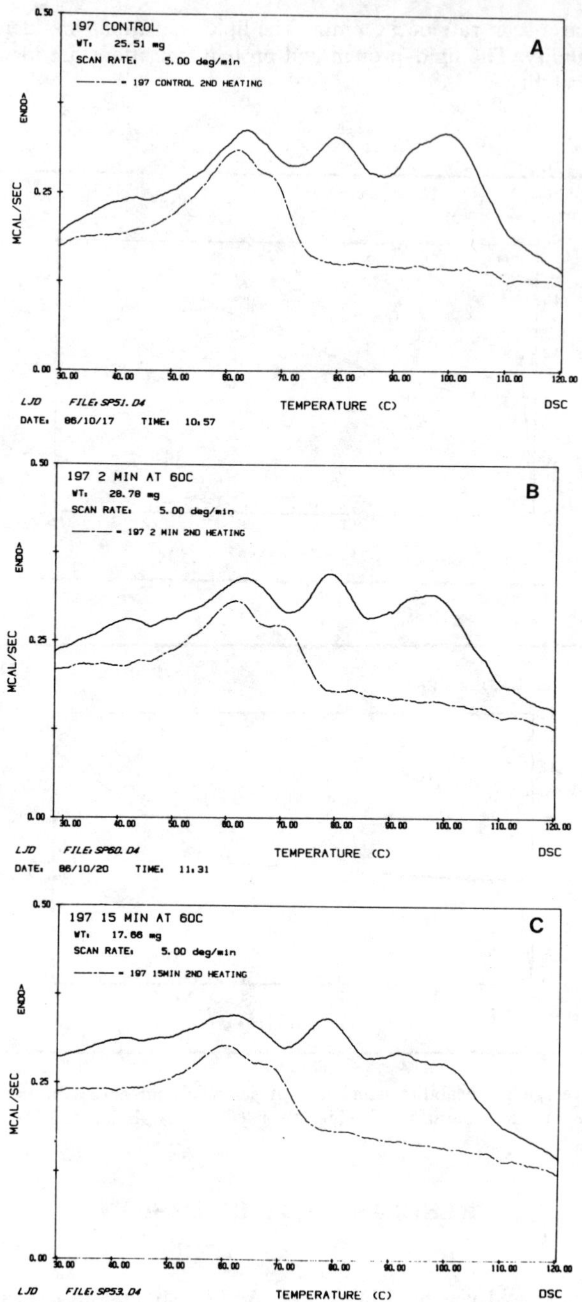

FIGURE 2. DSC scans of 60°C heat-separated epidermis. (**A**) no heat control; (**B**) 2-minute heat separation; and (**C**) 15-minute heat separation.

is 1.51×10^{-5} cm/sec. A similar increase in permeability was seen at 15-min heat separation from pH 7.4 aqueous solutions, 5.8×10^{-4} cm/sec at 2 min and 1.30×10^{-3} cm/sec at 15 min. Up to and including 5 min no increase in permeability was observed. Since the stratum corneum is the main barrier to cutaneous transport, frozen sections of heat-separated epidermis were examined microscopically to correlate stratum corneum damage with duration of scalding at 60°C. The stratum corneum appeared normal and intact up to and including 15-min heat treatment. Thermal analysis shows no structural changes in the lipid and lipid-protein complex of the stratum corneum after heat separation at 60°C up to 15 min (FIG. 2). The broadening of the protein transition peak after 15 min heat at 60°C may indicate a structural change in the protein component of the stratum corneum; however, more molecular level studies are required to investigate further these structural changes. There were no significant changes in partition coefficients with increased duration of scalding at 60°C. These data indicate that the properties of stratum corneum are not affected by scalding at 60°C up to 5 min.

Up to and including 2-min heat separation the epidermis was easily removed from the dermis, but with increasing time it became progressively more difficult to separate.

It has been concluded from this study that heat separation of the epidermis for up to 2 min at 60°C is a reliable method for percutaneous transport studies, since this procedure does not appear to alter stratum corneum properties. Also concluded from this study is that stratum corneum properties have been altered by heat separation at 60°C for 10 min, since an increase in permeation was observed.

REFERENCES

1. BAUMBERGER, J. P., V. SUNTZEFF & E. V. COWDRY. 1942. J. Natl. Cancer Inst. **2:** 413–416.
2. KASSIS, V. & J. SONDERGAARD. 1982. Arch. Dermatol. Res. **273:** 301–306.
3. WADSKOV, S. & J. SONDERGAARD. 1978. Acta Derm. Venerol. (Stockholm) **58:** 191–195.
4. TSAMBAOS, D. & G. MAHRLE. 1979. Arch. Dermatol. Res. **266:** 177–180.
5. KLIGMAN, A. M. & E. CHRISTOPHERS. 1963. Arch. Dermatol. **88:** 702–705.
6. BEHL, C. R., G. L. FLYNN, T. KURIHARA, W. SMITH, O. GATMAITAN, W. HIGUCHI, N. F. H. HO & C. L. PIERSON. 1980. J. Invest. Dermatol. **75:** 340–345.

Hyperglycemia Blocks the Polarization Response of Confluent Endothelial Cells to Wounding

R. N. MASCARDO[a] AND J. TESTA

Research Service
Veterans Adminstration Medical Center
Newington, Connecticut 06111

Glucose toxicity has been implicated in the pathogenesis of chronic structural and functional abnormalities found in diabetes mellitus. For instance, hyperglycemia induces neural tube fusion abnormalities in human[1] and mouse[2] embryos, presumably by affecting cellular locomotion adversely. Furthermore, hyperglycemia induces delayed replication, disruption of the cell cycle, and accelerated death of human umbilical vein endothelial cells *in vitro*.[3] Since the endothelial cell lining of arteries is subjected continually to the traumatic, shearing action of the circulating blood, and to elevated, ambient glucose concentrations in the diabetic state, it was important to determine if hyperglycemia could affect the locomotor response of endothelial cells to injury. To address this question, we used an endothelial cell monolayer model of wounding[4] to determine the effect of high glucose levels associated with uncontrolled diabetes (2-8 mg/ml or 11-44 mM) on the polarization[5] and locomotor[6] responses of endothelial cells at the border of an experimental wound produced by scraping off an area of the monolayer.

METHODS

Confluent pulmonary artery endothelial cell (CCL-209) monolayers on coverslips or in wells were incubated in Dulbecco's modified Eagle medium with 20% fetal bovine serum (DMEM + FBS), and different glucose concentrations (1-20 mg/ml) for 2-168 hr. Cell monolayers were wounded by scraping a 25-50-cell-wide path off the surface of the coverslip or well bottom. Polarization response (PR) of the border cells was quantitated by observing the perinuclear location of the centrosome, a microtubule organizing center (MTOC), 15 min after wounding by immunofluorescence microscopy. We have shown previously that the MTOCs of 70-82% of border cells exhibit a positive orientation or PR (*i.e.,* the MTOC is located between the nucleus of the cell and the wound path) in the presence of serum or specific growth

[a]Address for correspondence: Renato N. Mascardo, M.D., 43-203, Research Building, Veterans Administration Medical Center, 555 Willard Avenue, Newington, CT 06111.

factors.[4,5] In contrast, serum-deprived cells exhibit random MTOC orientation, with approximately 50% of cells showing a positive MTOC orientation and the other 50% having a negative response (i.e., the MTOC is oriented away from the wound). Results were expressed as the net positive PR (% PR of experimental cells minus % PR of serum-deprived control cells).

Cell migration response to wounding was quantitated by time lapse videorecording of wound closure. Results were expressed as the % change in wound width 3 hr after wounding. We have observed previously that significant wound narrowing during this period occurs only in the presence of serum or insulin.[6]

TABLE 1. Effect of Glucose, Mannitol and Glyburide on the Polarization Response of Endothelial Cell Monolayer to Wounding

Condition	Mean Change in Polarization Response (PR, %) $\pm$ SE (% PR of Experimental Cells $-$ % PR of Serum-deprived Control Cells)
1. DMEM($-$)FBS	0
2. DMEM($+$)FBS, G(1)	30 $\pm$ 1.5
3. G(2)	22 $\pm$ 2.0[a]
4. G(4)	9 $\pm$ 1.0[b]
5. G(8)	6 $\pm$ 2.0[b]
6. G(20)	1 $\pm$ 0.8[b]
7. M(2)	23 $\pm$ 2.3[a]
8. M(4)	16 $\pm$ 2.0[b]
9. M(8)	4 $\pm$ 0.4[b]
10. M(20)	1 $\pm$ 0.5[b]
11. G(1), GLB	27 $\pm$ 2.0
12. G(2), GLB	28 $\pm$ 2.0
13. DMEM($+$)FBS, G(4), GLB	26 $\pm$ 1.5

NOTE:. Confluent CCL-209 cells on coverslips were incubated in Dulbecco's modified Eagle medium with 20% fetal bovine serum (DMEM + FBS), and a humidified 10% CO_2 atmosphere at 37°. Different concentrations of glucose (G) or mannitol (M) were added 48 hr before the monolayers were wounded. Glyburide (GLB, 10^{-4}M), was added 16 hr before wounding. The polarization response of 150-200 border cells was counted 15 min after wounding by fixation of the cells with methanol, and immunofluorescence localization of the centrosome in relation to the nucleus and the wound path. Results were expressed as the mean % change in positive polarization response (PR) (% PR of experimental cells $-$ % PR of serum-deprived control cells) $\pm$ SE of triplicate experiments. Numbers in parentheses represent concentrations in mg/ml.

[a] Different from DMEM + FBS, G(1), $p < 0.05$.
[b] Different from DMEM + FBS, G(1), $p < 0.01$.

RESULTS

CCL-209 cells exposed to a physiological glucose concentration (1 mg/ml) exhibited a brisk net positive PR (28-32 %). Increasing the glucose level (2 mg/ml) blunted the PR significantly. The inhibitory action of hyperglycemia was concentration-dependent, and total inhibition of the PR was observed at a glucose level of 20 mg/ml (TABLE 1). A time course study revealed that the inhibitory action of 2 mg/

ml of glucose was observed initially after 24 hr of exposure. At higher levels (4-8 mg/ml), the inhibitory effect of glucose was observed within 2 hr of incubation. Mannitol (2-20 mg/ml) had a similar inhibitory effect on the PR of border cells. However, replacement of the culture medium containing mannitol (2-4 mg/ml) with medium containing physiological glucose resulted in a complete recovery of the PR within 15 hr. In contrast, replacement of the hyperglycemic medium with medium containing normal glucose failed to induce a similar recovery of PR. Hyperglycemia also blunted wound closure in a concentration-dependent manner (TABLE 2). Glyburide, an oral hypoglycemic sulfonylurea, induced a reversal of the inhibitory effect of hyperglycemia on the PR (TABLE 1) and wound narrowing (TABLE 2).

TABLE 2. Effect of Glucose, Mannitol, and Glyburide on Cell Migration Response of Endothelial Cells to Wounding

Condition	Mean Change in Wound Width (WW, % $\pm$ SE) at 3 hr [$100 - (WW$ at $Time_n/WW$ at $Time_0 \times 100)$]
1. DMEM($-$)FBS	2 ± 0.8
2. DMEM($+$)FBS, G (1)	36 ± 2.0
3. G (2)	18 ± 2.2
4. G (4)	10 ± 1.5
5. M (2)	26 ± 1.5
6. M (4)	18 ± 1.8
7. G (1), GLB	38 ± 2.0
8. G (2), GLB	36 ± 1.0^a
9. DMEM($+$)FBS, G (4), GLB	25 ± 2.0^b

NOTE: Confluent endothelial cells were incubated in DMEM + FBS, and different concentrations of glucose (G) or mannitol (M) for 60 hr. Glyburide (GLB, 10^{-4}M) was added 16 hr before the cells were wounded. The width of a specific segment of the wound was monitored and recorded immediately after wounding and 3 hr later with an Olympus IMT inverted microscope equipped with phase-contrast optics, videocamera, TV monitor, and time-lapse recorder. Results were expressed as the mean change in wound width after 3 hr $\pm$ SE of triplicate experiments. Numbers in parentheses represent concentrations in mg/ml.

^{a} Different from condition 3, $p < 0.01$.
^{b} Different from condition 4, $p < 0.01$.

CONCLUSIONS

On the basis of these data, we conclude that hyperglycemia inhibits cellular polarization and migration responses to wounding in endothelial cells. Even a relatively slight glucose elevation (2 mg/ml) was sufficient to retard the locomotor response of cells to wounding. Furthermore, the sulfonylurea agent, glyburide, reversed this toxic effect of hyperglycemia.

REFERENCES

1. MILLS, J. L., L. BAKER & A. S. GOLDMAN. 1979. Diabetes **28:** 292–293.
2. GOLDMAN, A. S., L. BAKER, R. PIDDINGTON, *et al.* 1985. Proc. Natl. Acad. Sci. USA **82:** 8227–8231.
3. LORENZI, M., E. CAGLIERO & S. TOLEDO. 1985. Diabetes **34:** 621–627.
4. MASCARDO, R. N., & P. SHERLINE. 1984. Ann. N.Y. Acad. Sci. **435:** 451–453.
5. MASCARDO, R. N. & P. SHERLINE. 1984. Diabetes **33:** 1099–1105.
6. MASCARDO, R. N. & P. SHERLINE. 1986. Diabetes **35:** 204A.

Neural Effects of ACTH Peptide Treatment in the Developing Rat Neuromuscular Junction[a]

RUTH E. FRISCHER AND FLEUR L. STRAND

Biology Department
New York University
Washington Square
New York, New York 10003

The organization of the developing neuromuscular junction (NMJ) can be influenced by many factors. Two of these factors include the fragments ACTH 4-10 and ACTH 4-9 (ORG 2766). ACTH 4-10 is 1000 times less potent than its analogue ORG 2766,[1] but both ACTH fragments are capable of influencing the developmental rate of neuronal regeneration[2,3] and neuromuscular maturation.[4]

Using scanning electron microscopy (SEM), we have previously shown that the NMJs of the extensor digitorum longus muscle of 15-day-old rats exhibited increased nerve arborization with postnatal administration of ACTH 4-10 (10 μg/kg/day, s.c) and a diminution of nerve terminal branching with ORG 2766 treatment at the same dosage.[5] We believe this to be a dosage effect with ORG 2766 exerting an inhibitory influence at this high concentration.

The present study was undertaken using light microscopy and morphometric analysis to ascertain whether there was quantitative evidence for our qualitative SEM observations. Furthermore, the study was designed to test whether the two peptide fragments evoked similar morphological changes at comparable dosages.

Sprague-Dawley rat pups were treated with either ACTH 4-10 (10 μg/kg/day, s.c.), ORG 2766 (0.01 μg/kg/day, s.c.), or an equivalent volume of saline solution from birth to 14 days of age.

Light microscopy, using a combined silver-cholinesterase stain,[6] facilitated the measurements of parameters that included muscle fiber diameter, total length of nerve terminal branching, and cholinesterase-positive endplate area and perimeter. Experimental preparations were observed with an Olympus BH 100$\times$ oil immersion objective and all measurements were performed with a Bioquant image analysis system. Student's *t* test was used for statistical evaluation. Data are expressed as mean and standard error of the mean (TABLE 1).

At two weeks of age, Org 2766 treatment resulted in a highly significant ($p <$ 0.001) increase in nerve terminal branching compared to saline-treated controls. In addition, a similarly significant increase ($p < 0.01$) in nerve terminal branching was demonstrated in ACTH 4-10-treated neonates vs. saline-treated controls. There were no significant differences in the measurements for muscle fiber diameter, endplate

[a]Research for this paper was supported by the Council for Tobacco Research.

TABLE 1. Measurements from Combined Silver and Cholinesterase Stains of Endplates from EDL Muscles of 14-Day-Old Neonatal Rats ($n = 40$) Treated with Saline Solution, ACTH 4-10,[a] or ACTH 4-9[b]

Parameters	Saline Solution	ACTH 4-10	ACTH 4-9 (ORG 2766)
Muscle fiber diameter (μm)	38.2 (0.9)[c]	38.2 (0.9)	38.5 (0.9)
Endplate area (μm^2)	1556.8 (82.9)	1631.9 (63.6)	1792.9 (104.8)
Endplate perimeter (μm)	160.5 (4.0)	159.0 (2.9)	166.9 (4.8)
Nerve terminal branching (μm)	146.4 (6.7)	172.4 (8.0)[d]	192.7 (7.7)[e]

[a] 10μg/kg/day, s.c.
[b] ORG 2766, 0.01 μg/kg/day s.c.
[c] Mean ± SEM
[d] $p < 0.01$.
[e] $p < 0.001$.

area, and perimeter between the ACTH 4-10, ORG 2766, and the saline-treated controls.

Our results suggest a neuronal susceptibility to these peptide fragments during development. It is possible, then, that ACTH peptides act as triggers for axonal sprouting by influencing neuronal responsiveness to a variety of growth factors.

REFERENCES

1. GREVEN, H. M. & D. DE WIED. 1973. The influence of ACTH on performance. Structure activity studies. Prog. Brain Res. **39:** 429-442.
2. STRAND, F. L. & T. T. KUNG. 1980. ACTH accelerates recovery of neuromuscular function following crushing of peripheral nerve fibres. Peptides **1:** 135-138.
3. SAINT-COME, C. & F. L. STRAND. 1985. ACTH 4-10 improves motor unit performance during peripheral nerve regeneration. Peptides **6** (Suppl. 1): 77-83.
4. FRISCHER, R. E., N. M. EL-KAWA & F. L. STRAND. 1985. ACTH peptides as organizers of neuronal patterns in development: Maturation of the rat neuromuscular junction as seen by scanning electron microscopy. Peptides **6** (Suppl. 2): 13-19.
5. ACKER, G. R., R. E. FRISCHER & F. L. STRAND. 1984. ACTH peptide modulation of the developing mammalian neuromuscular system seen through three different perspectives. Ann. N.Y. Acad. Sci. **435:** 370-375.
6. PECOT-DECHAVASSINE, M., A. WERNIG & H. STOVER. 1979. A combined silver and cholinesterase method for studying exact relations between the pre- and postsynaptic elements at the frog neuromuscular junction. Stain Technol. **54:** 25-27.

Muscle Insertions and Basicranial Morphology in Rodents

JOHN H. WAHLERT

Department of Natural Sciences
Baruch College
City University of New York
New York, New York 10010

SHARON L. SAWITZKE

Graduate Program in Anthropology
Graduate School and University Center
City University of New York
New York, New York 10036

Knowledge of the relationship between soft tissues and the marks they leave on bones is essential for paleontologists and mammalogists who attempt to interpret the significance of osteological details and differences. Our dissections of recent rodents for the purpose of gaining such knowledge revealed differences in the insertions of the longus capitis and rectus capitis anterior muscles on the basicranium; morphology of the bone in this region corresponded only in part with the muscle attachments.

The paired longus capitis muscle arises on the second or third through sixth or seventh cervical vertebrae and inserts on the basicranium between the auditory bullae; a low, median keel separates the pair of insertions; the muscle flexes the atlanto-occipital joint and draws the neck downward.[1] The paired rectus capitis anterior muscle arises on the ventral surface of the atlas and inserts on the basioccipital posterior and posterolateral to the insertion of the longus capitis; it, too, flexes the atlanto-occipital joint.[1] The insertions in rodents are clearly illustrated for the wood rat *Neotoma*[2] (FIG. 1A) only. They are described in general terms for the cricetine[3] and dipodoid[4] rodents and roughly illustrated for the rat.[5]

In the dormouse *Eliomys* (FIG. 1D), family Gliridae, which was the chief subject of our dissections, the longus capitis inserts in a narrow zone on lateral flanges of the basioccipital that turn ventrally where the bone abuts the auditory bullae; the posterior end of this zone can be identified on skulls. The insertion continues anteriorly onto the basisphenoid, where it makes no mark. The rectus capitis anterior muscle was not found. The difference between the insertion in *Eliomys* and *Neotoma*[2] spurred us to dissect the region in other rodents.

The insertions of the two muscles in an Old World mouse, *Mus musculus* (FIG. 1C), are again different from those in the New World genus, *Neotoma*, although both belong to the family Muridae. In *Mus* the longus capitis insertion extends anterior to marked depressions in the basioccipital onto the basisphenoid. The insertion of the rectus capitis anterior, which is posterolateral to it, is unmarked on the basioccipital; lateroventral extension of the basioccipital is very slight in this region.

Tamias striatus (FIG. 1B), the eastern chipmunk, represents the family Sciuridae. The insertion of the longus capitis extends anteriorly to the suture between the basioccipital and basisphenoid, beyond the anterior ends of paired depressions on the

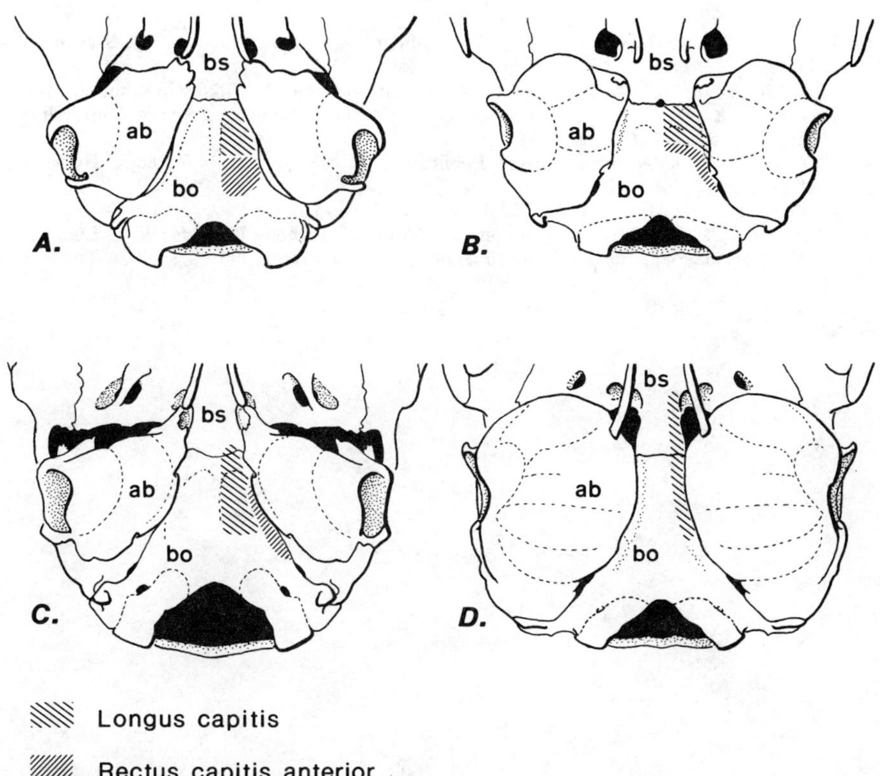

FIGURE 1. Insertions of longus capitis and rectus capitis anterior muscles on basicrania of rodents: **A:** *Neotoma cinerea* (after Howell,[2] FIG. 23B); **B:** *Tamias striatus,* American Museum of Natural History (AMNH) 12566; **C:** *Mus musculus,* AMNH 237511; **D:** *Eliomys quercinus,* AMNH 42505. Only left half of muscle insertions shown; crania not drawn to scale. Abbreviations: ab, auditory bulla; bo, basioccipital; bs, basisphenoid.

basioccipital. Strong lateral flanges of the basioccipital turn ventrally against the bullae and widen the area of insertion of the muscle. The insertion of the rectus capitis anterior is posterior and posterolateral to that of the longus capitis.

These observations demonstrate that there is great variation in insertions of the longus capitis and rectus capitis anterior muscles on the rodent basicranium. Further study of more specimens and taxa may yield important phylogenetic information. The dormice, which are of uncertain relationships among rodents,[6] appear to be simply different again at this stage of the investigation. Since only the lateral flanges of the basioccipital are related to muscle insertions, fossil specimens will yield limited information about basicranial musculature.

REFERENCES

1. EVANS, H. E. & G. C. CHRISTENSEN. 1977. Miller's Anatomy of the Dog. W. B. Saunders. Philadelphia, PA.
2. HOWELL, A. B. 1926. Anatomy of the Wood Rat. Williams & Wilkins. Baltimore, MD.
3. RINKER, G. C. 1954. Miscellaneous Publications of the Museum of Zoology, University of Michigan. **83:** 1-124, FIGS. 1-18.
4. KLINGENER, D. 1964. Miscellaneous Publications of the Museum of Zoology, University of Michigan. **124:** 1-100.
5. GREENE, E. C. 1935. Trans. Am. Phil. Soc., n.s. **28:** 1-370.
6. VIANEY-LIAUD, M. 1985. *In* Evolutionary Relationships among Rodents. W. P. Luckett & J.-L. Hartenberger, Eds.: 277-309. Plenum Press. New York.

A Method for Studying Small Intestinal Transit in the Rat

S. LeROY, S. C. SUTTON, AND P. T. BEALL

Basic Pharmaceutics Research
Ciba-Geigy Corporation
Ardsley, New York 10502

Transit time studies of particles through the small intestine of the rat have often been complicated by the variables of fasting, gastric emptying, and trauma to the gastrointestinal (GI) tract, thus causing an increase in transit time. A method utilizing a chronic indwelling cannula, plug, and introducer allows the study of *natural* transit time in the rat small intestine.

The cannula is a modification of the Babkin's Method.[1] A 2-cm piece of PE-390 tubing was heat-molded over an open flame into the shape shown (FIG. 1). The introducer was made by cutting off the end of a positive displacement pipet. The plug was made by heat sealing one end of a pipet tip and heat molding a knob on the other end.

Surgery was performed on Sprague-Dawley rats weighing 350–400 g. The rats were anesthetized with a ketamine/acepromazine mixture and the cannula was implanted into the duodenum 1 cm away from the stomach. The cannula exited through the abdominal wall, and was plugged. The skin was closed over the cannula with autoclips. The animal was permitted to recover for 1 week under reverse lighting conditions (6 PM, lights on; 6 AM, lights off).

Two days prior to the transit experiment, the rats were given a liquid diet. The experiment began at 11:00 AM to ensure maximum intestinal activity.[2] A premeasured sample of the test particles was loaded into the introducer with an aliquot of $NA^{51}CrO_4$-labeled 0.9% saline solution to serve as an internal control. The rat was lightly anesthetized with ether, an incision was made in the abdominal skin, and the cannula was located. After the plug was removed, the tip of the introducer was inserted into the cannula, and the test material was injected. The introducer was cut to serve as a plug and the incision was closed. The animal was awake by the end of the procedure. During the test period the animal was in darkness and had access to liquid diet. After the specified time, the animal was anesthetized with ether and the intestines were exposed, clamped, and removed.

The large particles (diameter: 500 μ–1 mm) were readily visible when the intestine was cut open longitudinally. Smaller particles (diameter: 50 μ) required magnification or radiolabeling for quantification of results. When radiolabeled particles were used, the intestines were quickly frozen with an ethanol/dry-ice mixture and cut into 2-cm segments for quantification in the gamma counter.

The ^{51}Cr-labeled saline fluid front traveled through 59%, 73%, and 81% of the small intestine at 30 min, 1-hr, and 2-hr intervals, respectively. Moreover, there appeared to be a trend toward some separation of the fluid phase and the particles (FIG. 2).

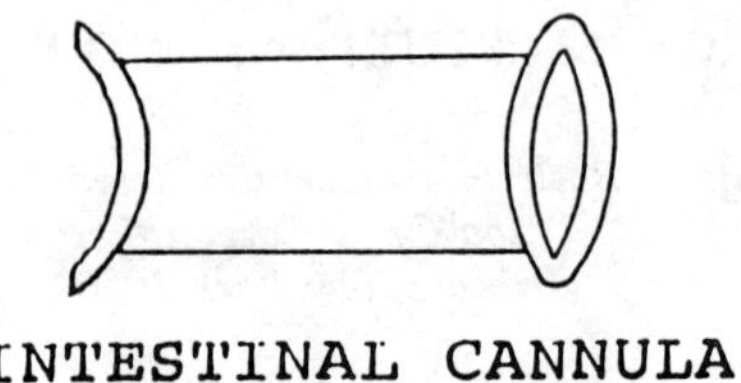

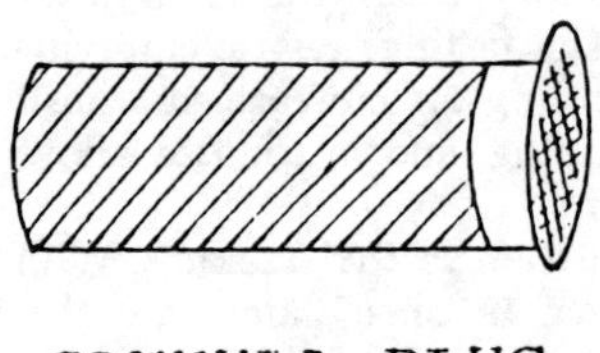

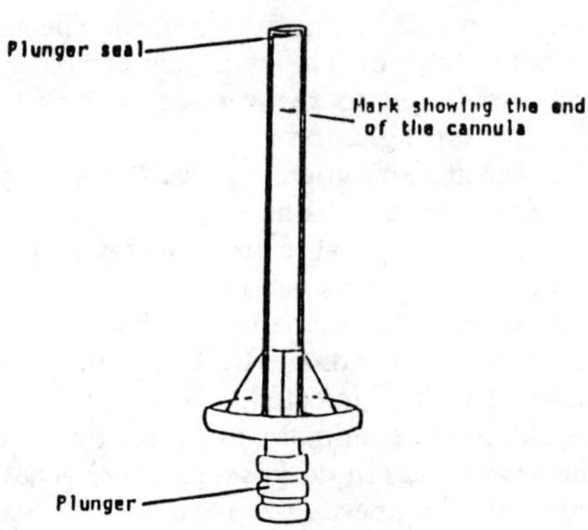

FIGURE 1. Construction of the cannula, plug and introducer.

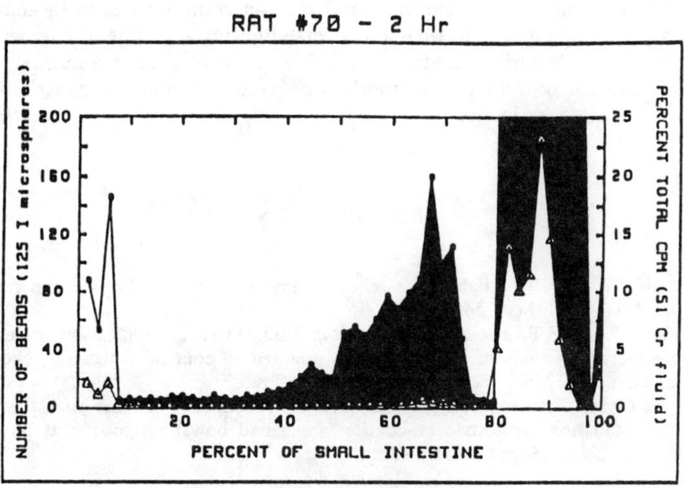

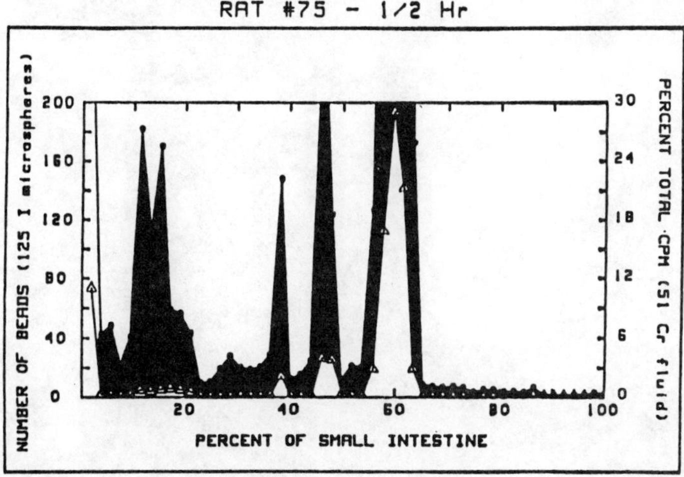

FIGURE 2. Two examples of the fluid phase and microsphere separation. △ transit time of fluid; ● number of 50-μ microspheres per 2-cm segment.

This method yielded reproducible results in studying the transit time of fluid and particles through the small intestine. Because of the anatomic position of the cannula, the test material can be loaded directly into the duodenum, thereby avoiding the variable of gastric emptying. The introducer provided for complete delivery of the test material without causing the intestinal trauma commonly seen in acute surgical techniques.[3] Since the method utilized a reverse lighting schedule, the experiment was done during the rat's active (feeding) period, and avoided the artifactual increase in transit time caused by fasting. This method provides a significant contribution to the research of transit time in the rat.

REFERENCES

1. LOPUKHIN, Y. M. 1976. Babkin's method of intestinal fistula. *In* Experimental Surgery: 85-86. MIR Publishers. Moscow.
2. POTTER, V. R., E. F. BARIL, M. WATANABE & E. D. WHITTLE. 1968. Systematic oscillations in metabolic functions in liver from rats adapted to controlled feeding schedules. Fed. Proc. **27**: 1238-1245.
3. OHRN, P. G., H. JOHANSSON & F. NILSSON. 1976. Influence of ether anaesthesia and orogastric intubation on gastric evacuation and small bowel propulsion after laparotomy. Acta Chir. Scan. [Suppl.] **461**: 5-16.

Effect of Lipids on the Permselective Properties of Gastric Mucus[a]

A. PIASEK, H. TSUKADA, A. SLOMIANY,
AND B. L. SLOMIANY[b]

Gastroenterology Research Laboratory
Department of Medicine
New York Medical College
Valhalla, New York 10595

INTRODUCTION

The adherent layer of mucus which covers the epithelial surfaces of alimentary tract is a complex mixture of proteins, glycoproteins and lipids in the form of a gel imbibed with water and electrolytes.[1,2] In the stomach, this layer constitutes part of a barrier that protects the underlying mucosa from the damaging effects of intraluminal acid and pepsin. Although the ability of gastric mucus to retard the diffusion of hydrogen ions has been recently demonstrated,[3,4] the permeability properties of this protective layer are not well defined. Here, we report on the permeability of pig gastric mucus to molecules of different sizes and on the effect of lipids on this property.

MATERIALS AND METHODS

Gastric mucus was obtained by instillation of the freshly dissected pig stomach with 2 M NaCl in 10 mM sodium phosphate buffer, pH 7.0. The recovered instillate was filtered through millipore HA (0.45 μm) filter, dialyzed against distilled water, and lyophilized. The dried mucus was dissolved in 6 M urea and subjected to delipidation with butanol-isopropyl ether (2:3, v/v). The delipidated mucus was then dialyzed against distilled water and lyophilized. For permeability measurements, the

[a]This work was supported by U.S.P.H.S. Grants AM 21684-10 and AA 05858-05 from the National Institutes of Health.

[b]Author to whom correspondence should be addressed.

intact and delipidated mucus preparations were dissolved at 30 mg/ml in 0.15 M NaCl. The permeability measurements were performed in a specially designed two-compartment chamber connected through the cylindrical sample port.[3] The sample port was filled either with saline solution (control) or studied preparation of mucus, while one of the compartments contained 0.15 M NaCl, pH 6.0, and the other was filled with the solution containing the test substance. The following test solutions were used: 49 mM HCl, 26 mM sucrose in 0.15 M NaCl, 16 μM pepsin in 0.15 M NaCl, and 5 μM horseradish peroxidase in 0.15 M NaCl. The amount of substance diffusing through the saline solution or mucus barrier was measured in the NaCl compartment at 15-min intervals for up to 2 hr. HCl was measured using a pH electrode, sucrose by phenol-sulfuric acid method, pepsin by proteolytic activity assay, and peroxidase enzymatically.[4,5]

RESULTS AND DISCUSSION

The chemical characteristics of pig gastric mucus used in the permeability studies are given in TABLE 1. The mucus contained 64.2% protein, 15.1% carbohydrate, 17.6% lipids, and 0.3% covalently bound fatty acids. Among the major lipid classes, the neutral lipids accounted for 71% of the total lipids, glycolipids 19.5%, and phospholipids 9.5%. The carbohydrates consisted mainly of galactose, N-acetylglucosamine, N-acetylgalactosamine, and fucose, while the covalently bound fatty acids were represented by hexadecanoic and octadecanoic acids. Comparisons of the diffusion values of hydrogen ion, sucrose, pepsin, and peroxidase through the intact and delipidated mucus with that through the unstirred layer of 0.15 M NaCl (control) indicated that permeability of the NaCl layer to sucrose is about seven times lower than that to HCl. A 65-fold reduction in permeability was observed with pepsin and 94-fold reduction with peroxidase. Substitution of the NaCl layer with intact gastric mucus had a profound detrimental effect on the diffusion rate of the investigated substances. The diffusion rate of HCl decreased by 52%, and a 65% reduction in permeability was obtained for sucrose, 91% for pepsin, and 93% for peroxidase. Extraction of lipids caused a 25% increase in permeability of the delipidated mucus to HCl, but had only negligible effect on the permeability to sucrose, pepsin and peroxidase (FIG. 1).

The results indicate that the extent of resistance offered by mucus gel to different molecules depends upon the mucus composition and size of the penetrating molecules. While mucus gel *in vivo* can be expected, at best, only to slow down the diffusion of hydrogen ions, its capacity to impede penetration by larger molecules is considerably

TABLE 1. Chemical Composition of Pig Gastric Mucus

Component	Relative Weight (%)
Protein	64.2 ± 7.5
Carbohydrate	15.1 ± 1.9
Lipids	17.6 ± 2.2
Covalently bound fatty acids	0.3 ± 0.1

NOTE: Each value represents the means ± SD of triplicate analyses.

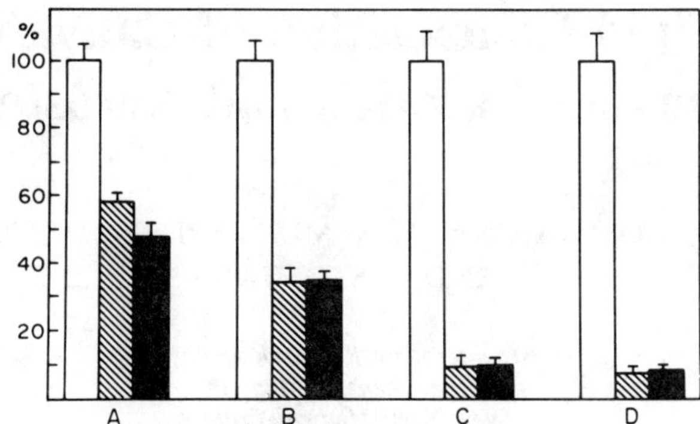

FIGURE 1. Comparison of the permeability of HCL, sucrose, pepsin and horseradish peroxidase through the unstirred layer of 0.15M NaCl (☐) with that through the intact (■) and delipidated (▨) gastric mucus. **A,** HCl; **B,** sucrose; **C,** pepsin; **D,** peroxidase. Values represent the means ± SD of six experiments performed in duplicate.

greater. Although earlier studies suggested that gastric mucus is already not permeable to molecules of the size of myoglobin (MW 17,000),[6] the data presented in this report clearly show that the mucus layer is even permeable to molecules as large as pepsin (MW 38,000) and horseradish peroxidase (MW 44,000). This process, although slow under normal conditions, could play a significant role in stomach disease, where the composition of the protective mucus layer is altered.

REFERENCES

1. GLASS, G. B. J. & B. L. SLOMIANY. 1977. *In* Mucus in Health and Disease. M. Elstein & D. V. Parke, Eds.: 311–347. Plenum Press. New York.
2. SLOMIANY, A., S. YANO, B. L. SLOMIANY & G. B. J. GLASS. 1978. J. Biol. Chem. **253:** 3785.
3. SAROSIEK, J., A. SLOMIANY, A. TAKAGI & B. L. SLOMIANY. 1984. Biochem. Biophys. Res. Commun. **118:** 523.
4. SLOMIANY, B. L., W. LASZEWICZ, V. L. N. MURTY, M. KOSMALA & A. SLOMIANY. 1985. Comp. Biochem. Physiol. **82C:** 311.
5. STEINMAN, R. M. C. & Z. A. COHEN. 1972. J. Cell Biol. **55:** 186.
6. ALLEN, A. 1981. *In* Basic Mechanisms of Gastrointestinal Mucosal Cell Injury and Protection. J. W. Harmon, Ed.: 351–367. Williams and Wilkins. Baltimore, MD.

Lipid Composition of Saliva in Recurrent Parotitis and Xerostomia[a]

B. L. SLOMIANY, V. L. N. MURTY, I. D. MANDEL,[b]
AND A. SLOMIANY

Gastroenterology Research Laboratory
Department of Medicine
New York Medical College
Munger Pavilion
Valhalla, New York 10595

[b]*School of Dental and Oral Surgery*
Columbia University
New York, New York 10032

INTRODUCTION

Saliva, which is the product of major and minor salivary glands located in the oral cavity, plays an important role in the defense of teeth against abrasion and caries, and protects the oral mucosa from mechanical, chemical, and microbial insults.[1,2] Under normal physiological conditions, the protective qualities of saliva are maintained by the flow rate and the concentration of protective components. This delicate balance is, however, disturbed in diseases affecting the function of salivary glands. We report here on the content and composition of lipids in parotid saliva of a patient with unilateral recurrent parotitis and in submandibular saliva of a patient with xerostomia.

MATERIALS AND METHODS

Samples of parotid saliva (5 ml per gland) were obtained from normal and affected parotid salivary glands with the aid of a Teflon cup device.[3] Submandibular saliva, collected with a basic plastic collector,[3] was obtained from a patient with xerostomia associated with "dry mouth" condition, while saliva from healthy individuals served as control. During collections, salivary flow rate was stimulated with 2% citric acid applied three times per minute to the lateral border of the tongue. The individual samples were immediately treated with buffered EDTA, filtered through a Millipore (HA 0.45 μm) filter, dialyzed exhaustively against distilled water, and lyophilized.[3]

[a]This work was supported by U.S.P.H.S. Grant DE 05666-08 from the National Institute of Dental Research of the National Institutes of Health.

The dry samples were extracted with chloroform-methanol (2:1, v/v) and filtered through grade-F sintered glass funnels to retain insoluble protein residues. Fractionation of the extracted lipids into neutral lipids, glycolipids and phospholipids was accomplished by column chromatography on silicic acid.[4] Neutral lipids were then separated into individual components by thin-layer chromatography, identified by co-chromatography with the standard compounds and quantitated.[3,4] The glycolipids were separated on SEAE-Sephadex into neutral and acidic fractions, differentiated to glycosphingolipids and glyceroglucolipids and quantitated by gas-liquid chromatography.[3] The phospholipids were subjected to two-dimensional thin-layer chromatography and the individual compounds were quantitated by measuring their phosphorus content.[4]

TABLE 1. Content of Protein and Lipids in the Parotid and Submandibular Saliva of Patients with Unilateral Recurrent Parotitis and Xerostomia

Constituent	(mg/100 mg) Parotid Saliva		(mg/100 mg) Submandibular Saliva	
	Parotitis	Normal	Xerostomia	Normal
Protein	116.10	144.90	214.60	132.41
Total lipids	8.41	7.93	14.21	9.43
Free fatty acids	1.62	2.78	1.68	2.34
Mono- and diglycerides	0.12	0.10	0.15	0.19
Triglycerides	1.38	1.15	1.74	1.34
Cholesterol	0.47	1.10	0.69	0.51
Cholesteryl esters	0.10	0.66	1.29	1.26
Glycosphingolipids	0.50	0.10	0.92	0.11
Glyceroglucolipids	3.81	1.61	5.30	2.23
Phospholipids	0.12	0.10	0.31	0.20

NOTE: Each value represents means of triplicate analyses. The values for normal individuals were obtained from 10 subjects.

RESULTS AND DISCUSSION

The content of protein and lipids in the parotid saliva of a patient with unilateral parotitis and in the submandibular saliva of a patient with xerostomia is given in TABLE 1. Parotid saliva from the affected gland, while exhibiting the total lipid content similar to that of saliva from gland with normal function, contained 1.7 times less neutral lipids, 2.6 times more glycolipids and 1.2 times more phospholipids. Neutral lipids from each type of sample were of similar composition. However, parotid saliva of the affected gland exhibited 42% lower content of free fatty acids, 57% lower content of cholesterol and 85% lower content of cholesteryl esters (TABLE 1). Less apparent differences were found in the composition of phospholipids (TABLE 2).

The xerostomic saliva, in comparison to that of normal submandibular saliva, exhibited 2.7 times greater content of glycolipids and 1.5 times greater content of phospholipids, but the values for neutral lipids were quite similar to those in normal saliva (TABLE 1). The glycolipids of xerostomic saliva were enriched in glucosyl- and lactosylceramide but its glyceroglucolipids composition did not differ from that of normal saliva. The phospholipids of xerostomic saliva like those of normal saliva were rich in phosphatidycholine, phosphatidylethanolamine, sphingomyelin, and lysophosphatidylcholine (TABLE 2). The normal saliva had higher values for phosphatidylethanolamine, sphingomyelin, and lysophosphatidylcholine, while the levels of diphosphatidylglycerol, phosphatidic acid, and phosphatidylinositol were higher in xerostomic saliva.

The results presented above together with the earlier-obtained data on salivary lipids in cystic fibrosis and Sjögren's syndrome,[5,6] may well be of direct relevance to

TABLE 2. Phospholipid Composition of the Parotid and Submandibular Saliva of Patients with Unilateral Recurrent Parotitis and Xerostomia

| | % Total Lipid Phosphorus | | | |
| | Parotid Saliva | | Submandibular Saliva | |
Constituent	Parotitis	Normal	Xerostomia	Normal
Phosphatidylethanolamine	17.7	15.4	16.2	21.6
Phosphatidylcholine	28.3	27.6	20.6	22.8
Phosphatidylserine	7.0	8.5	7.5	7.1
Phosphatidylinositol	2.1	2.0	8.4	2.5
Sphingomyelin	12.4	14.0	9.9	14.2
Lysophosphatidylcholine	15.3	13.7	9.8	16.1
Phosphatidic acid	3.6	2.2	6.5	2.3
Phosphatidylglycerol	5.3	3.8	6.9	1.8
Diphosphatidylglycerol	4.5	3.4	8.0	4.1
Unidentified	3.8	9.4	6.8	7.5

NOTE: Each value represents means of triplicate analyses. The values for normal individuals were obtained from 10 subjects.

the understanding of the factors involved in the etiology of oral disease and could expand the usefulness of saliva for diagnostic purposes.

REFERENCES

1. MANDEL, I. D. & S. WOTMAN. 1976. Oral Sci. Rev. **8:** 25.
2. SLOMIANY, B. L., V. L. N. MURTY & A. SLOMIANY. 1986. Progr. Lipid Res. **24:** 311.
3. SLOMIANY, B. L., V. L. N. MURTY, M. AONO, A. SLOMIANY & I. D. MANDEL. 1982. Arch. Oral Biol. **27:** 803.
4. SLOMIANY, B. L., V. L. N. MURTY, E. ZDEBSKA, A. SLOMIANY, K. GWOZDZINSKI & I. D. MANDEL. 1986. Arch. Oral Biol. **31:** 187.
5. SLOMIANY, B. L., M. AONO, V. L. N. MURTY, A. SLOMIANY, M. J. LEVINE & L. A. TABAK. 1982. J. Dent. Res. **61:** 1163.
6. SLOMIANY, B. L., M. KOSMALA, C. NADZIEJKO, V. L. N. MURTY, K. GWOZDZINSKI, A. SLOMIANY & I. D. MANDEL. 1986. Arch. Oral Biol. **31:** 699.

Lemur Systematics and Hemoglobin Phylogeny[a]

D. H. COPPENHAVER,[b] P. J. HOPKINS,[c]
M. M. EHRHARDT,[d]
AND L. K. DUFFY[d]

[b]Department of Microbiology
University of Texas Medical Branch
Galveston, Texas 77550

[c]Department of Human Genetics
University of Michigan Medical School
Ann Arbor, Michigan 48109

[d]Microsequencing Laboratory
Center for Neurologic Research
Brigham & Women's Hospital
Boston, Massachusetts 02115

There can be only one actual phylogenetic history for any group of organisms. The use of biochemical data is one of several possible means to try to estimate the genetic relationships between organisms. Neontological data, including macromolecular sequences, become particularly relevant in groups with poor or poorly understood fossil records. Hence, we have chosen to study globin evolution in prosimian primates. The dynamics of the evolution of this family of molecules is particularly well suited to contribute to an understanding of prosimian phylogeny.[1,2]

Results of analyses of β-globin sequences are beginning to produce information bearing on a number of important questions in the phylogeny of prosimian primates. We have determined the amino acid sequences of β globins from *Lemur catta,*[2] *Lemur variegatus,*[3] and *Hapalemur griseus* (in preparation). These proteins have been analyzed in comparison to the other known prosimian β-globin sequences: *Lemur fulvus,*[4] *Nycticebus coucang,*[5] and *Loris tardigradus.*[6] The latter two species belong to the superfamily Lorisoidea, while the remaining species belong to the superfamily Lemuroidea. The phenetic distance between the β-globin sequences for each species is shown in TABLE 1. A number of generalizations are supportable based upon these data. (A) All of the lemuroid sequences are clearly and equally separated from the lorisoid sequences. This is also supported by prosimian α-globin data.[7] (B) As a group the lemuroid sequences are much more divergent from each other than are the lorisoid sequences. (C) Within the Lemuroidea, the *L. catta* and the *L. variegatus* β globins are closely related. (D) *H. griseus* β globin appears approximately equidistant from the three *Lemur* sequences. (E) *L. fulvus,* β globin, in contrast, shows fewer changes from that of *H. griseus* than it does from the *catta/variegatus* cluster.

[a]Duke University Primate Center Publication No. 404.

TABLE 1. Minimum Mutation Matrix for Prosimian β-Globin Chains

	L. catta	L. variegatus	L. fulvus	H. griseus	N. coucang	L. tardigradus
Lemur catta		8	22	14	34	34
Lemur variegatus	6		19	13	32	32
Lemur fulvus	18	15		8	27	27
Hapalemur griseus	11	9	8		21	21
Nycticebus coucang	26	24	23	25		4
Loris tardigradus	25	24	23	25	4	

The number of amino acid substitutions fixed between each pair of β globins is given below the diagonal. The minimum number of nucleotide substitutions needed to account for the amino acid substitutions is given above the diagonal.

The interpretation of simple substitution matrices derived from sequence data is complicated by the appreciable but variable amount of homoplasy which is hidden in the data. Thus, we analyzed the available sequence data cladistically using a maximum parsimony tree-building procedure.[8] The minimum replacement length solution is shown in FIGURE 1. These data appear robust in that all possible starting branch orders for the lemuroid species produced the same final tree. This reconstruction supports the first three conclusions outlined above, and clarifies the seemingly contradictory statements (D) and (E). *H. griseus* and *L. fulvus* β globins are depicted as sharing a common ancestor after the divergence of the *catta/variegatus* clade. These

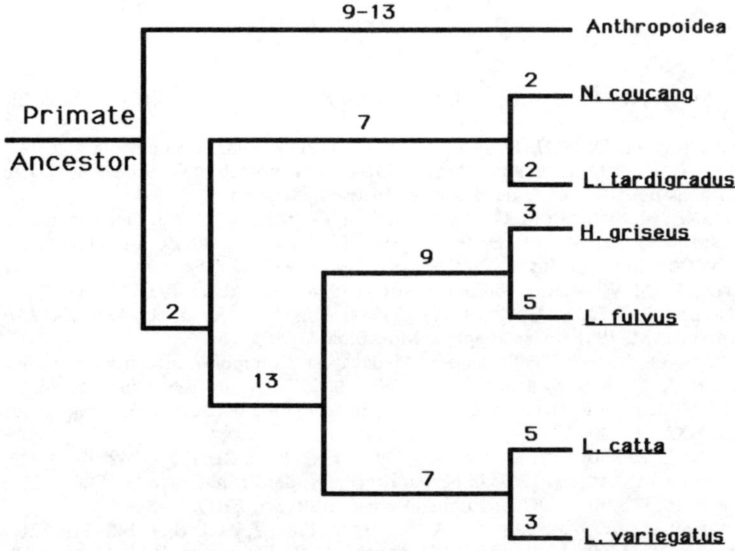

FIGURE 1. Dendrogram of relationships of prosimian β globins. The A solution maximum parsimony reconstruction[8] is shown. The dendrogram was rooted using the rabbit β-globin sequence as the outgroup (not shown). The number of nucleotide substitutions fixed along each branch of the dendrogram is shown. The substitutions shown for the branch leading to higher primates represents the range of substitutions seen using three anthropoid β-globin sequences: a hominoid (*Homo sapiens*), a cercopithecoid (*Macaca fuscata*), and a ceboid (*Cebus apella*). Owing to the markedly uneven rate of fixation of substitutions in this globin lineage, the branch points should not be used to estimate divergence times for the species shown.

data lend support to the suggestion that reanalysis of the phylogeny of the genus *Lemur* is needed,[3] and weigh strongly against the reclassification of *variegatus* in the monospecific genus "*Varecia,*" while *L. catta* and *L. fulvus* remain congeneric.[9]

Our findings are also relevant to the evolution of the globins. If interpreted strictly according to molecular clock theory,[10] the evolutionary divergence detected between the four lemuroid species would give unacceptably ancient divergence times for these four species.[2,3,9] Indeed, examination of the branch lengths in FIGURE 1 shows that lemuroid β-globin evolution proceeded at 2-3 times the rate detected in either the lorisoid or anthropoid clades, conforming our previous observations of rapid evolution of lemuroid β globins.[2,3,11-13] These data thus provide further compelling evidence that molecular clock theory is inadequate for describing globin evolution.[3,8] It is unknown at present if gene conversions within the β-globin complex contributed to the rapid evolution of the lemuroid hemoglobins. The observation that lemur β-globin sequences show some similarities to higher primate δ globins[11-13] along with the finding that the δ locus in *Lemur fulvus* is a pseudogene[14,15] may suggest undetected gene conversions in this evolutionary lineage, producing lemuroid β globins which are partially paralogous to the β globins of other primates.

REFERENCES

1. HOLMQUIST, R., T. H. JUKES, M. GOODMAN & G. W. MOORE. 1976. J. Mol. Biol. **105:** 39-74.
2. COPPENHAVER, D. H., J. D. DIXON & L. K. DUFFY. 1983. Hemoglobin **7:** 1-14.
3. DUFFY, L. K. & D. H. COPPENHAVER. 1984. *In* Hemoglobin. G. Schnek & C. Paul, Eds. Editions de l'Universite de Bruxelles. Brussels, Belgium.
4. MAITA, T., M. SETOGUCHI, G. MATSUDA & M. GOODMAN. 1979. J. Biochem. **85:** 755-764.
5. MATSUDA, G., T. MAITA, B. WATANABE, H. OTA, A. ARAYA, M. GOODMAN & W. PRYCHODKO. 1973. Int. J. Pept. Protein Res. **5:** 419-421.
6. MAITA, T., M. GOODMAN & G. MATSUDA. 1978. J. Biochem. **84:** 377-383.
7. DUFFY, L. K. & D. H. COPPENHAVER. 1984. Ann. N.Y. Acad. Sci. **435:** 254-257.
8. GOODMAN, M. 1981. Prog. Biophys. Mol. Biol. **37:** 105-164.
9. TATTERSALL, I. 1982. The Primates of Madagascar. Columbia University Press. New York.
10. WILSON, A. C., S. S. CARLSON & T. J. WHITE. 1977. Annu. Rev. Biochem. **46:** 573-639.
11. HILL, R. L., J. BUETTNER-JANUSCH & V. BUETTNER-JANUSCH. 1963. Proc. Natl. Acad. Sci. USA **80:** 885-893.
12. HILL, R. L. & J. BUETTNER-JANUSCH. 1964. Fed. Proc. **21:** 1236-1242.
13. BUETTNER-JANUSCH, J. 1967. *In* Neueu Ergebnisse der Primatologie. D. Starck, R. Schneider & H. J. Kuhn, Eds. G. Fischer Verlag. Stuttgart, FRG.
14. BARRIE, P. A., A. J. JEFFREYS & A. F. SCOTT. 1981. J. Mol. Biol. **149:** 319-336.
15. JEFFREYS, A. J., P. A. BARRIE, S. HARRIS, D. H. FAWCETT, Z. J. NUGENT & A. C. BOYD. 1982. J. Mol. Biol. **156:** 487-503.

Vegetation Replacement Models for Southern Nassau County, New York

JESS PAUL HANKS

Department of Biology
The City College of the City University of New York
New York, New York 10031

Ground-water levels in southern Nassau County have declined by 3.6 m due to the installation of sanitary sewers and ground-water pumping.[1] As sewer building continues to completion in the area (about 1990) ground-water levels will decline further by 0.2—5.8 m. If ground-water pumping patterns are altered, according to the Master Water Plan, ground-water levels may rebound in southwestern areas by 4 m and decline further in southeastern areas by 4 m. These changes in ground-water levels should influence the plant habitats and communities. If ground-water levels change sufficiently at a site the existing vegetation will be replaced by other vegetation that the new ground-water regime will support. Based on field-sampling in the area and past vegetation studies, descriptive habitat and community replacement models have been developed to portray the kinds and magnitude of vegetation change that could occur in Nassau County given changing ground-water levels.[2] The replacement models are implemented here for two areas in southwestern Nassau County to illustrate the potential vegetation changes that could occur.

In general, to assess the vegetation replacement sequence after perturbation, first the existing vegetation at a site is determined. Next, the predicted water level change is determined from U. S. Geological Survey simulations. The habitat that the new ground-water levels will support is determined. Over successional development time, the vegetation will be replaced by those types of communities, within the habitat, that the new ground-water levels will support. Three variables in the vegetation replacement process are presented in FIGURE 1: the communities (excluding streams and managed parks), the ground-water gradient supporting each community, and the development time necessary to produce each community via secondary succession. The stream communities are omitted in the figure because their replacement process has been described.[3] Managed parks are upland sites controlled by human development and are not likely to increase or decrease substantially given changing ground-water levels. The models predict that southern Nassau County will experience vegetation replacement toward drier types of vegetation after the installation of sanitary sewers (TABLE 1). If possible changes in ground-water pumping patterns are implemented, southwestern areas may then experience water-table rebounds and replacement sequences toward more mesic types of vegetation, while southeastern areas will become even drier. The result will be a decline in community and species diversity in the area and an increase in communities with broad ranges of tolerance to environmental perturbation.

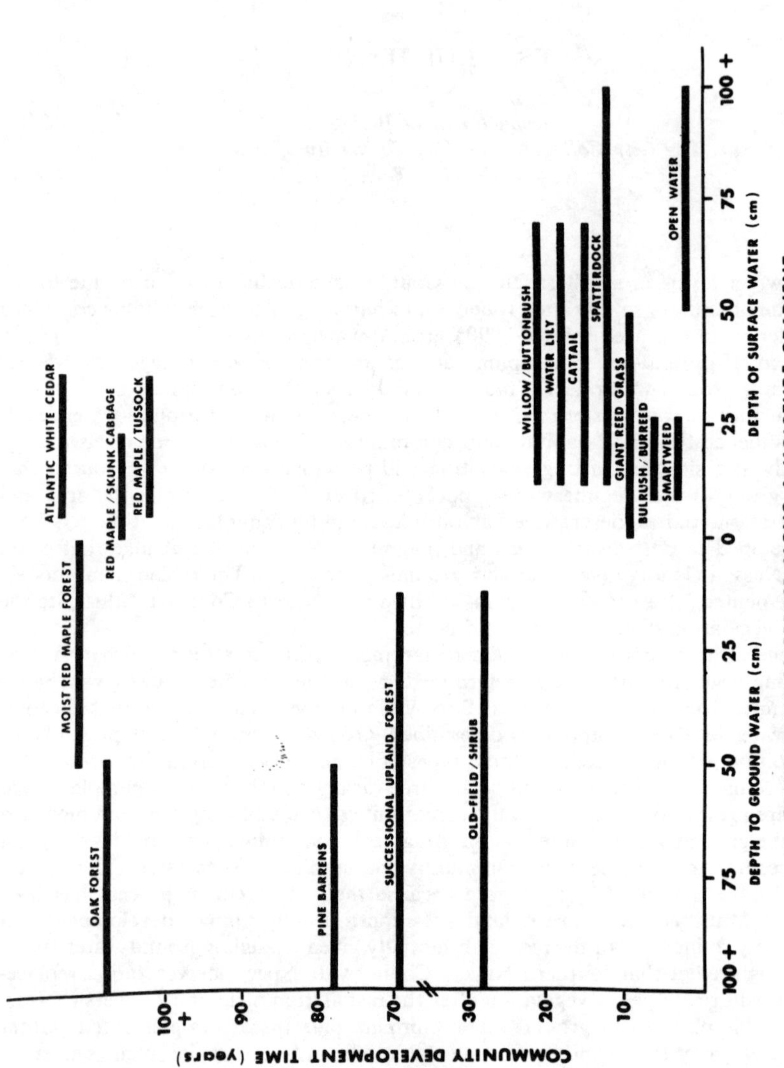

FIGURE 1. The position of the plant communities in the area along a water table gradient and the development time required to produce each community through the process of secondary succession. Stream communities and managed parks are excluded.

TABLE 1. The Occurrences of the Plant Communities in Valley Stream and Mill Creek under Baseline, No Action, and Master Water Plan (MWP) Ground-water Regimes[a]

Habitat and Community	Valley Stream			Mill Creek		
	Baseline	No Action	MWP	Baseline	No Action	MWP
Wetlands						
1. Willow/Buttonbush	—	—	—	10	3	12
2. Spatterdock	1	—	2	5	3	8
3. Smartweed	—	—	—	1	—	—
4. Waterlily	—	—	—	2	—	2
5. Bulrush/Burreed	1	—	—	2	—	2
6. Giant reed grass	—	—	—	2	6	4
7. Cattail	—	—	—	2	—	2
Swamps						
8. Atlantic white cedar	—	—	—	—	—	—
9. Red maple/Tussock	—	—	—	1	—	1
10. Red maple/Skunk cabbage	—	—	2	3	—	4
Uplands						
11. Moist red maple forest	2	2	—	1	5	—
12. Old field/Shrub	1	—	—	9	—	—
13. Succession. Upland forest	1	—	—	3	—	—
14. Pine barrens	—	—	—	—	—	—
15. Managed park	4	4	4	2	2	2
16. Oak forest	2	4	4	4	25	9
Streams						
17. Starwort/Jewelweed	1	—	6	4	—	4
18. Pondweed	—	—	—	—	—	—
19. Mugwort/Jewelweed	5	10	—	1	10	—
Pond						
20. Open water	5	—	5	4	2	6

[a] Values represent the number of occurrences for each community in each stream system under these three regimes. Baseline conditions show the existing communities in 1980. No action vegetation assumes that ground-water levels will decline and no mitigation activities are implemented. MWP values assume well fields are abandoned in southwestern areas and moved to eastern sections of the county.

REFERENCES

1. GARBER, M. S. & D. J. SULAM. 1976. Factors affecting declining water table levels in a sewered area of Nassau Co., N.Y. J. Res. U. S. Geol. Survey **4:** 255-265.
2. HANKS, J. P. 1985. Plant communities and ground-water levels in southern Nassau County, Long Island, New York. Bull. Torrey Bot. Club. **112:** 79-86.
3. HANKS, J. P. 1987. Present and future vegetation in Nassau Co., N.Y. Ann. N.Y. Acad. Sci. **494:** 427-429.

Morphologic Interactions between Cells of the Anteriad Migrating Fold during *Rana pipiens* Gastrulation

J. LeBLANC,[a] M. YODER,[b] AND I. BRICK[b]

[a]The College of Staten Island
Biology Department
City University of New York
Staten Island, New York 10301

[b]Department of Biology
New York University
New York, New York 10003

During amphibian gastrulation, morphogenetic movements result in the primitive body plan. Epithelial movement, during this embryonic period, is responsible for much of the reorganization that establishes the definitive gastrula anatomy. If cells are to form an epithelium and then exhibit epithelial translocations, sufficient intercellular cohesion, adhesions along their adjacent edges, would be required to maintain integrity.

In *Xenopus laevis,* experimental evidence has indicated that involution of the deep region of the marginal zone is responsible for the formation of the mesodermal mantle and involution of the superficial layer of the marginal zone for the formation of the endodermal roof of the archenteron.[1,2] Nakatsuji[3] states that experimental observations suggest that mesodermal cells are actively moving during amphibian gastrulation and that these cells are loosely grouped with large intercellular spaces and are not a cohesive epithelial sheet. The above observations are based on developmental studies in a limited number of amphibian species and, therefore, should not necessarily be accepted as universal.

In this study, cellular interactions at the dorsal and lateral lip areas of the blastopore (FIG. 1a) in *Rana pipiens* embryos at various stages during gastrulation have been investigated by transmission electron microscopy.

The following types of adjoining cell surfaces were examined: (1) between lip cells lining the blastoporal groove; (2) between lip cells at the surface of the groove and underlying lip cells; (3) between cells of the yolk plug lining the blastoporal groove; (4) between yolk plug cells at the surface of the groove and underlying yolk plug cells; and (5) between cells of the lip and yolk plug. Areas of unspecialized apposition and intercellular junctions that have the appearance of desmosomes, maculae adherens, with characteristic intercellular densities and intracellular dense plaques with, in some instances, tonofilament-appearing arrays were observed between all the adjoining cell surfaces listed above.

FIGURE 1b and c, exhibits desmosome-appearing junctions between yolk plug cells at the surface of the blastoporal groove and underlying yolk plug cells, and FIGURE 2 shows those between lip cells within the surface layer lining the blastoporal groove. Desmosomes are theorized to be adhesive structures that, owing to their connections

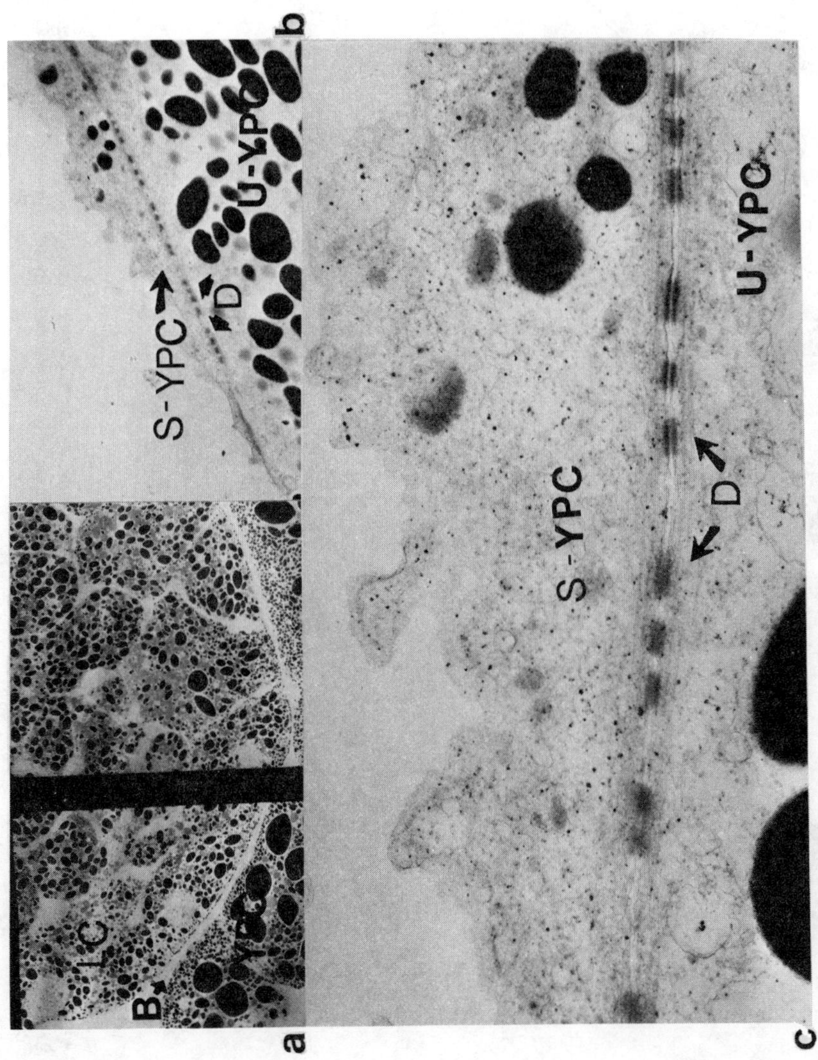

FIGURE 1. Stage 11½. (a) Sagittal section through dorsolateral lip cells (LC) of blastopore (B) and yolk plug cells (YPC); ×703. (b,c) Apparent desmosome junctions (D) between yolk plug cells at the surface of the blastoporal groove (S-YPC) and underlying yolk plug cells (U-YPC); (b) ×7980; (c) ×43,100. Entire figure reduced to 58% of original size.

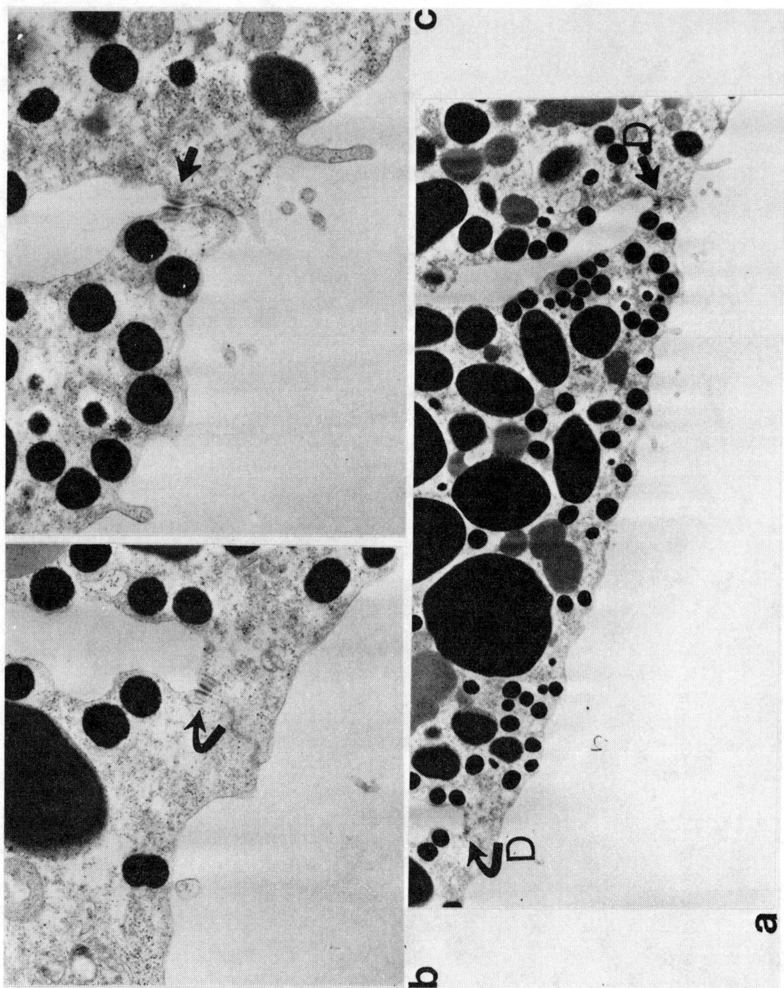

FIGURE 2. Stage 11½. (a) Apparent desmosome junctions (D) between lip cells lining the blastoporal groove; ×9,576. (b) Enlargement of desmosome junction indicated by curved arrow in (a); ×32,376. (c) Enlargement of desmosome junction indicated by straight arrow in (a); ×32,376. Entire figure reduced to 58% of original size.

to tonofilaments, may distribute force throughout an epithelium. The cellular arrangements and intercellular junctions suggest that the layer of lip cells (presumptive endodermal roof of archenteron) and the layer of yolk plug cells that line the blastoporal groove are both an epithelium. Desmosome-appearing junctions have, as yet, not been observed between the more loosely arrayed internal lip cells that lie between the surface of the blastoporal groove and the surface of the embryo (mostly presumptive mesoderm) or between the more loosely arrayed internal yolk plug cells. The areas of unspecialized apposition and desmosome-appearing junctions between the surfaces of cells of the lip and yolk plug, which are moving past each other at the blastopore, may be required for an orderly and oriented inward movement.[4]

REFERENCES

1. KELLER, R. E. 1975. Vital dye mapping of the gastrula and nerula in *Xenopus laevis*. I. Prospective areas and morphogenetic movements of the superficial layer. Dev. Biol. **42:** 222–241.
2. KELLER, R. E. 1976. Vital dye mapping of the gastrula and nerula in *Xenopus laevis*. II. Prospective areas and morphogenetic movements of the deep layer. Dev. Biol. **51:** 118–137.
3. NAKATSUJI, N. 1984. Cell locomotion and contact guidance in amphibian gastrulation. Am. Zool. **24:** 615–627.
4. BRICK, I., B. E. SCHAEFFER, H. E. SCHAEFFER & J. F. GENNARO, JR. 1974. Electrokinetic properties and morphologic characteristics of amphibian gastrula cells. Ann. New York Acad. Sci. **238:** 390–407.

Resource Partitioning in a Stream Fish Community[a]

ANTONIOS PAPPANTONIOU, JOSEPH W. RACHLIN,
AND BARBARA E. WARKENTINE

Department of Biological Sciences
Herbert H. Lehman College
City University of New York
Bronx, New York 10468

A first-order approximation for understanding how members of a fish community allocate available food resources can be achieved through an assessment of the degree of diet similarity found within the community. One method of studying this is to sample a discrete community over a collecting season and to evaluate the degree of diet overlap among the members. The purpose of this study was to examine the diet of concurrently collected fish from a single pool over a collecting season, in order to assess the degree of dietary overlap and similarity, and to evaluate this as a measure of the potential for competition among closely related species.

To accomplish this, a section of the Waccabuc River in Westchester County, New York was chosen for the study. The study area is characterized by well-defined pools and riffles. One such well-defined pool was chosen as the study site and sampled monthly from March–November. In addition to fish collections, reference collections of potential invertebrate food resources were made using Surber nets[1] and Dendy plates.[2] These plates were placed in the pool and harvested monthly. Fish specimens were collected using a backpack electroshocker. Both fish and invertebrates were preserved and returned to the laboratory for analysis. The stomach contents of each fish was identified to the lowest possible taxon and enumerated. Dietary overlap among the members of the fish community sampled from the pool study site was then evaluated using the Morisita overlap index as modified by Horn,[3] and the Shoener similarity index.[4]

TABLE 1 presents the diet, and food item proportions, for each of the 12 species of fish comprising the discrete freshwater pool community of this study. From these data, pairwise comparisons of diet overlap and similarity were calculated and these data are presented in TABLE 2. The study pool contained three congeneric centrarchids, *Lepomis auritus, L. macrochirus,* and *L. gibbosus.* The data in TABLE 1 indicate interesting differences in their diets. All pairwise comparisons of these three species (TABLE 2) indicates no significant overlap and minimal diet similarity during the study period. With the suggested value of 0.6 as the cut-off for significant overlap,[5] it can be seen from the data that the redbreast sunfish and the fallfish have significant diet overlap; the same is true for the bluegill and the large-mouth bass; and also for

[a]The research upon which this paper is based was supported in part by PSC-CUNY Research Grants 6-64141, 6-65144, and 6-65152.

TABLE 1. Stomach Contents of 12 Fish Species Representing a Discrete Freshwater Pool Community

Fish Species (No.)	Stomach Contents	Proportion
Lepomis auritus (9) (Redbreast sunfish)	Insecta	0.02
	Diptera	0.22
	Chironomidae	0.05
	Simuliidae	0.09
	Heptageniidae	0.42
	Apidae	0.02
	Perlidae	0.03
	Taeniopterygidae	0.05
	Hydropsychidae	0.12
L. macrochirus (4) (Bluegill sunfish)	Ephemeroptera	0.06
	Baetidae	0.61
	Heptageniidae	0.11
	Corixidae	0.22
L. gibbosus (5) (Pumpkinseed)	Insecta	0.12
	Chironomidae	0.12
	Simuliidae	0.62
	Heptageniidae	0.12
Alosa pseudoharengus (2) (Alewife)	Chironomidae	0.92
	Hydropsychidae	0.08
Micropterus dolomieui (1) (Smallmouth bass)	Lepomis sp.	1.00
M. salmoides (7) (Largemouth bass)	Insecta	0.05
	Coleoptera	0.08
	Diptera	0.21
	Chironomidae	0.05
	Baetidae	0.34
	Heptageniidae	0.05
	Corixidae	0.03
	Veliidae	0.10
	Taeniopterygidae	0.03
	Trichoptera	0.05
Catostomus commersoni (4) (White sucker)	Coleoptera	0.02
	Chironomidae	0.49
	Simuliidae	0.08
	Heptageniidae	0.24
	Taeniopterygidae	0.04
	Trichoptera	0.02
	Hydropsychidae	0.12
Semotilus corporalis (7) (Fallfish)	Oligochaeta	0.06
	Simuliidae	0.17
	Heptageniidae	0.50
	Gerridae	0.06
	Plecoptera	0.17
	Hydropsychidae	0.06

TABLE 1.—Continued

Fish Species (No.)	Stomach Contents	Proportion
Rhinichthys atratulus (6)	Chironomidae	0.71
(Blacknose dace)	Heptageniidae	0.21
	Hydropsychidae	0.07
Esox americanus (18)	Cambaridae	0.19
(Redfin pickerel)	Diptera	0.12
	Baetidae	0.06
	Aeshnidae	0.06
	Trichoptera	0.06
	Asellidae	0.06
	Etheostoma olmstedi	0.25
	Micropterus salmoides	0.06
	Notropis cornutus	0.12
Etheostoma olmstedi (15)	Chironomidae	0.83
(Tessellated darter)	Simuliidae	0.06
	Baetidae	0.02
	Heptageniidae	0.07
	Plecoptera	0.01
	Hydropsychidae	0.01
Exoglossum maxillingua (22)	Dryopidae	0.02
(Cutlips minnow)	Elmidae	0.01
	Chironomidae	0.11
	Simuliidae	0.39
	Baetidae	0.02
	Heptageniidae	0.09
	Sialidae	0.08
	Odonata	0.01
	Taeniopterygidae	0.06
	Trichoptera	0.03
	Hydropsychidae	0.11
	Polycentropodidae	0.01
	Ancylidae	0.04

the pumpkinseed and the cutlips minnow. The data of TABLE 2 further show that among the other members of the pool community, significant overlap existed between the alewife and both the white sucker and the blacknose dace; between white sucker, blacknose dace, tessellated darter, and alewife.

Examination of the data presented in TABLES 1 and 2 indicates that there is sufficient difference in the diets of the three congeneric centrarchids so that there is no significant dietary overlap and only minimal diet similarity. This is entirely consistent with the findings of Keast,[6] Werner and Hall,[7] and Mittelbach.[8] In his study Keast[9] found that the bluegill, *Lepomis macrochirus* was a generalist in its dietary habits. By contrast we found that the bluegill was not a generalist, but focused on mayflies, which made up 72% of its diet. The generalist in this study was the redbreast sunfish. The pumpkinseed, also had a rather narrow diet of mostly blackflies (74%). It would seem from these data that these closely related species have sufficient differences in their diets so as to preclude the possibility of any meaningful competition, even with scarcity of resources.

TABLE 2. Pairwise Comparisons of Diet Overlap and Similarity among 12 Fish Species Representing a Discrete Freshwater Pool Community

Fish Species	Morisita Overlap Index	Schoener Similarity Index
Micropterus dolomieui vs. all others	0	0
Lepomis auritus vs. *L. macrochirus*	0.13	12.64
L. auritus vs. *L. gibbosus*	0.24	27.88
L. auritus vs. *Micropterus salmoides*	0.33	35.10
L. auritus vs. *Alosa pseudoharengus*	0.09	12.30
L. auritus vs. *Catostomus commersoni*	0.51	51.67
L. auritus vs. *Semotilus corporalis*	0.82	58.81
L. auritus vs. *Esox americanus*	0.14	12.49
L. auritus vs. *Rhinichthys atratulus*	0.32	33.18
L. auritus vs. *Etheostoma olmstedi*	0.15	23.37
L. auritus vs. *Exoglossum maxillingua*	0.42	38.69
Lepomis macrochirus vs. *L. gibbosus*	0.03	11.11
L. macrochirus vs. *M. salmoides*	0.70	42.11
L. macrochirus vs. *A. pseudoharengus*	0	0
L. macrochirus vs. *C. commersoni*	0.07	11.12
L. macrochirus vs. *S. corporalis*	0.15	11.10
L. macrochirus vs. *E. americanus*	0.13	6.25
L. macrochirus vs. *R. atratulus*	0.05	11.11
L. macrochirus vs. *E. olmstedi*	0.04	9.24
L. macrochirus vs. *E. maxillingua*	0.08	22.34
Lepomis gibbosus vs. *M. salmoides*	0.06	15.79
L. gibbosus vs. *A. pseudoharengus*	0.18	12.50
L. gibbosus vs. *C. commersoni*	0.37	32.84
L. gibbosus vs. *S. corporalis*	0.44	29.16
L. gibbosus vs. *E. americanus*	0	0
L. gibbosus vs. *R. atratulus*	0.23	25.00
L. gibbosus vs. *E. olmstedi*	0.26	25.10
L. gibbosus vs. *E. maxillingua*	0.85	59.56
Micropterus salmoides vs. *A. pseudoharengus*	0.09	5.27
M. salmoides vs. *C. commersoni*	0.16	17.09
M. salmoides vs. *S. corporalis*	0.10	5.26
M. salmoides vs. *E. americanus*	0.30	24.02
M. salmoides vs. *R. atratulus*	0.13	10.53
M. salmoides vs. *E. olmstedi*	0.12	13.48
M. salmoides vs. *E. maxillingua*	0.11	20.34
Alosa pseudoharengus vs. *C. commersoni*	0.78	56.72
A. pseudoharengus vs. *S. corporalis*	0.01	5.55
A. pseudoharengus vs. *E. americanus*	0	0
A. pseudoharengus vs. *R. atratulus*	0.94	78.57
A. pseudoharengus vs. *E. olmstedi*	0.99	84.04
A. pseudoharengus vs. *E. maxillingua*	0.21	18.92
Catostomus commersoni vs. *S. corporalis*	0.43	36.92
C. commersoni vs. *E. americanus*	0	1.96

TABLE 2.—Continued

Fish Species	Morisita Overlap Index	Schoener Similarity Index
C. commersoni vs. *R. atratulus*	0.93	77.60
C. commersoni vs. *E. olmstedi*	0.84	62.47
C. commersoni vs. *E. maxillingua*	0.47	45.19
Semotilus corporalis vs. *Esox americanus*	0	0
S. corporalis vs. *R. atratulus*	0.25	26.98
S. corporalis vs. *E. olmstedi*	0.09	14.28
S. corporalis vs. *E. maxillingua*	0.45	31.20
Esox americanus vs. *R. atratulus*	0	0
E. americanus vs. *E. olmstedi*	0	4.40
E. americanus vs. *E. maxillingua*	0.02	6.74
Rhinichthys atratulus vs. *E. olmstedi*	0.96	79.00
R. atratulus vs. *E. maxillingua*	0.28	27.37
Etheostoma olmstedi vs. *E. maxillingua*	0.28	28.06

Significant diet overlap and similarity was found to occur between the redbreast sunfish, *L. auritus,* and the fallfish, *Semotilus corporalis;* between the bluegill, *L. macrochirus,* and the largemouth bass, *Micropterus salmoides,* and between the pumpkinseed, *L. gibbosus* and the cutlips minnow, *Exoglossum maxillingua.* Significant dietary overlap and similarity was also found to exist among the bottom feeding white sucker, *Catostomus commersoni,* blacknose dace, *Rhinichthys atratulus,* and the tessellated darter, *Etheostoma olmstedi.* Since these fish, showing significant dietary overlap, tend to occupy similar positions in the water column, there is indication that if food resources become diminished they may become competitors. However, TABLE 1 indicates, there is sufficient variety in their respective diets to allow for adequate prey switching in the face of diminished food abundance, and therefore, maintenance of the pool community structure through competition avoidance.

REFERENCES

1. SURBER, E. W. 1936. Trans. Am. Fish. Soc. **66:** 193-202.
2. HESTER, F. E. & J. S. DENDY. 1962. Trans. Am. Fish. Soc. **91:** 420-421.
3. HORN, H. S. 1966. Am. Nat. **100:** 419-424.
4. SCHOENER, T. W. 1970. Ecology **51:** 408-418.
5. ZARET, T. M. & A. S. RAND. 1971. Ecology **52:** 336-342.
6. KEAST, A. 1978. Environ. Biol. Fish. **3:** 7-31.
7. WERNER, E. E. & D. J. HALL. 1979. Ecology **60:** 254-256.
8. MITTELBACH, G. G. 1984. Ecology **65:** 499-513.
9. KEAST, A. 1985. Can. J. Fish. Aquat. Sci. **42:** 1114-1126.

Feeding Preference of Sympatric
Hake from the Inner
New York Bight[a]

JOSEPH W. RACHLIN AND
BARBARA E. WARKENTINE

Department of Biological Sciences
Lehman College
City University of New York
Bronx, New York 10468

A total of 156 *Urophycis regia,* 133 *U. chuss,* and 145 *U. tenuis* originally collected from the Brigantine Shoals area off Little Egg Harbor, New Jersey, and part of the permanent fish collection of the Department of Ichthyology of the American Museum of Natural History, New York, were examined for their dietary diversity, evenness, overlap, and preference. The size ranges of these fish were, respectively, 5.3-20.6 cm, 6.0-32.4 cm, and 4.7-13.4 cm standard length. Length-frequency histograms and scale analyses indicated that all fish were juveniles. Diet diversity was computed by the Shannon Information Index,[9] evenness by the Heip Evenness Function,[1,2] overlap by both the Schoener Similarity Index[8] and the Morisita Overlap Index,[4,5] and finally preference by the Relativized Electivity (E^*) of Vanderploeg and Scavia:[10,11] $E^* = (W - (1/n))/(W + (1/n))$, where W is the Manly Preference Index[3] and $n =$ the number of kinds of food items. Stomachs were removed from all fish and the stomach contents were identified to the lowest practicable taxon and enumerated. An estimate of the environmental food resource base available to these fish was achieved by pooling the stomach contents of all 434 animals.

Dietary data from the pooled stomach contents indicated that the resource base consisted of mysids, *Neomysis americana,* the decapod crustacean, *Crangon septemspinosa,* the amphipods, *Aeginella longicornis, Calliopus laeviusculus, Gammarus annulatus, G. marinus, Microdeutopus gryllotalpa,* and *Pontogenia inermis,* the isopod, *Edotea montosa,* the copepod, *Acartia* sp., unidentified nematodes and the fish, *Etrumeus teres* (round herring). While these were the dominant food items, the fish also fed on the mysid, *Mysidopsis bigelowi,* and larve of the decapods, *Libinea emarginata, Cancer borealis, C. irroratus,* the copepod *Temora* sp., and the polychaete, *Nereis virens.*

TABLE 1 shows both the diet of the three hake species and the similarity and overlap in their feeding patterns. From these data it can be seen that *U. regia* and *U. tenuis* have a 48.4% similarity and a 0.64 overlap in diet. *U. regia* and *U. chuss* have a 70.8% similarity and a 0.95 overlap, while *U. tenuis* and *U. chuss* have a 57%

[a] The research upon which this paper is based was supported in part by PSC-CUNY Research Grants 6-65144 and 6-66311.

TABLE 1. Diet, Similarity and Overlap Among the Three Species of Sympatric Hake

A: Urophycis Diet

	U. tenuis	*U. chuss*	*U. regia*
Mysids	42%	66%	62%
Decapods	54%	12%	8%
Amphipods	2%	2%	8%
Isopods	0	4%	0
Copepods	1%	4%	20%
Nematodes	1%	9%	1%
Fish	0	3%	1%

B: Similarity and Overlap

	Schoener Similarity	Morisita Overlap
U. regia vs. *U. tenuis*	48.385	0.642
U. regia vs. *U. chuss*	70.835	0.948
U. tenuis vs. *U. chuss*	57.050	0.738

similarity and an overlap value of 0.74. Using the percent similarity data and the proposal of Zaret and Rand[12] that an overlap of 0.60 or greater is taken to indicate biologically meaningful overlap, we conclude that all three species of hake have biologically meaningful overlap in their diet and therefore potentially compete for the same resource base. This would lead to interspecific competition, especially when the resource base is reduced. The question then becomes, have these species developed any mitigating feeding behavior to reduce potential competition? Feeding preference might provide such mitigation.

To assess feeding preference we followed the procedure of pooling the stomach contents of all the hake as an estimate of the resource base available to and actually being fed on by these animals (Rachlin and Warkentine[6,7]). The proportional representation of the food items in the stomachs of each of the hake species was then evaluated and compared with the proportional representation of these same food items in the resource base. TABLE 2 presents these data. An examination of TABLES 1 and 2 shows the presence of nonpreferred food items in relatively high numbers in the

TABLE 2A. Environmental Resource Base Estimated From Pooled Stomach Content Data of the Sympatric Hake

Environment Resource Base	Percentage
Mysids	57%
Decapods	25%
Amphipods	4%
Isopods	1%
Copepods	8%
Nematodes	4%
Fish	1%

TABLE 2B. Feeding Preference of the Three Species of Sympatric Hake

| Diet | Urophycis regia | | Urophycis chuss | | Urophycis tenuis | |
	E*	Evaluation	E*	Evaluation	E*	Evaluation
Mysids	+0.0564	P	−0.1181	NP	+0.1515	P
Decapods	−0.5017	NP	−0.5043	NP	+0.6007	P
Amphipods	+0.3435	P	−0.4941	NP	−0.0441	NP
Isopods	−1.0000	NP	+0.3403	P	−1.0000	NP
Copepods	+0.4212	P	−0.5093	NP	−0.6397	NP
Nematodes	−0.5636	NP	+0.2488	P	−0.3339	NP
Fish	−0.1316	NP	+0.2076	P	−1.0000	NP

NOTE: P = Mathematical preference as computed by the relativized E* index; NP = Nonmathematical preference.

diets of these animals, and this is an indication of opportunistic feeding during periods of high abundance of the resource base. More importantly, TABLE 2 shows, that except for the preference of mysids by both *U. regia* and *U. tenuis,* there is complete nonoverlap in the preferred food items among the three hake species. This would allow for the separation of feeding and the avoidance of intense interspecific competition during periods of reduction of the resource base, and indicates resource partitioning by these fish.

REFERENCES

1. HEIP, C. 1974. A new index measuring evenness. J. Mar. Biol. Ass. U.K. **54:** 555-557.
2. HEIP, C. & P. ENGLES. 1974. Comparing species diversity and evenness indices. J. Mar. Biol. Assoc. U.K. **54:** 559-563.
3. MANLY, B. F. J. 1974. A model for certain types of selection experiments. Biometrics **30:** 281-294.
4. MORISITA, M. 1959a. Measuring of the dispersion of individuals and analysis of the distributional patterns. Memoirs of the Faculty of Science, Kyushu University, Series E (Biology), **2:** 215-235.
5. MORISITA, M. 1959b. Measuring of interspecific association and similarity between communities. Memoirs of the Faculty of Science, Kyushu University. Series E (Biology) **3:** 65-80.
6. RACHLIN, J. W. & B. E. WARKENTINE. 1986. Dietary preference of the spotted hake, *Urophycis regia,* from the inner New York Bight. Ann. N.Y. Acad. Sci. **494:** 434-437.
7. RACHLIN, J. W. & B. E. WARKENTINE. 1987. The use of museum ichthyological holdings for initial diet studies. Copeia **1:** 214-216.
8. SCHOENER, T. W. 1970. Nonsynchronous spatial overlap of lizards in patchy habitats. Ecology **51:** 408-418.
9. SHANNON, C. E. & W. WEAVER. 1962. The Mathematical Theory of Communication. The University of Illinois Press. Urbana, Ill.
10. VANDERPLOEG, H. A. & D. SCAVIA. 1979. Two electivity indices for feeding with special reference to zooplankton grazing. J. Fish. Res. Board Can. **36:** 362-365.
11. VANDERPLOEG, H. A. & D. SCAVIA. 1979. Calculation and use of selectivity coefficients of feeding: Zooplankton grazing. Ecol. Modelling **7:** 135-149.
12. ZARET, T. M. & A. S. RAND. 1971. Competition in tropical stream fishes: Support for the competitive exclusion principle. Ecology **52:** 336-342.

G-Banding of the Chromosomes of *Apogon maculatus* and *A. pseudomaculatus* (Perciformes: Apogonidae)[a]

KENNETH A. RIVLIN, JOSEPH W. RACHLIN, AND
BARBARA E. WARKENTINE

Department of Biological Sciences
Lehman College
City University of New York
Bronx, New York 10468

In a continuation of studies on the cytogenetics of members of the Apogonidae[2,3] two further methods of G-banding were applied to chromosome preparations of gill epithelial tissue of *Apogon maculatus* and *A. pseudomaculatus*. A modification of the trypsin G-banding procedure of Arrighi and Hsu[1] was applied to the chromosomes of *A. pseudomaculatus* collected from the Florida Keys. The method outlined in Rivlin *et al.*[2] involves colchicine treatment (0.05 mg colchicine per gram of body weight), hypotonic pretreatment (0.4% KCl) of the gill arches, and absolute methanol-glacial acetic acid (3:1) fixation, dabbing the gill filaments onto water-coated precleaned glass slides followed by rapid air-drying. The freshly prepared slides were "aged" for G-banding by heating on a slide warmer at 90°C for 15 minutes. After cooling to room temperature the slides were placed in a 0.5% Trypsin-Dulbecco phosphate-buffered saline solution (pH 7) at 4°C for 5-30 seconds followed by two rinses of 0.9% saline solution and were then stained by flooding with 2% Giemsa-Sorensen's 1/15 M phosphate buffer (pH 6.8) for 4 min. The time of the trypsin treatment is critical and variable, and must be determined by trial and error for each batch of slides. Too short a time gives uniform staining and too long a time gives the chromosomes a fuzzy appearance.

FIGURE 1 shows the G-banded karyotype of *A. pseudomaculatus* produced by this procedure, as well as a diagrammatic representation of the banding pattern. *A. pseudomaculatus* has a chromosome number of 36 with 30 metacentric-submetacentric and 6 subtelocentric-telocentric chromosomes. The five pairs of large chromosomes showed multiple G-bands along their length allowing matching of the homologues.

Gill epithelial tissue of *Apogon maculatus* was treated with a replication banding technique involving the incorporation of bromodesoxyuridine into the replicating DNA during S phase. This is followed by treating metaphase chromosomes with 33258 Hoechst solution and then staining with Giemsa. The staining pattern produced represents regions of early and late chromosome replication which correlates with G-

[a]This research was supported in part by PCS-CUNY Research Award 6-66311.

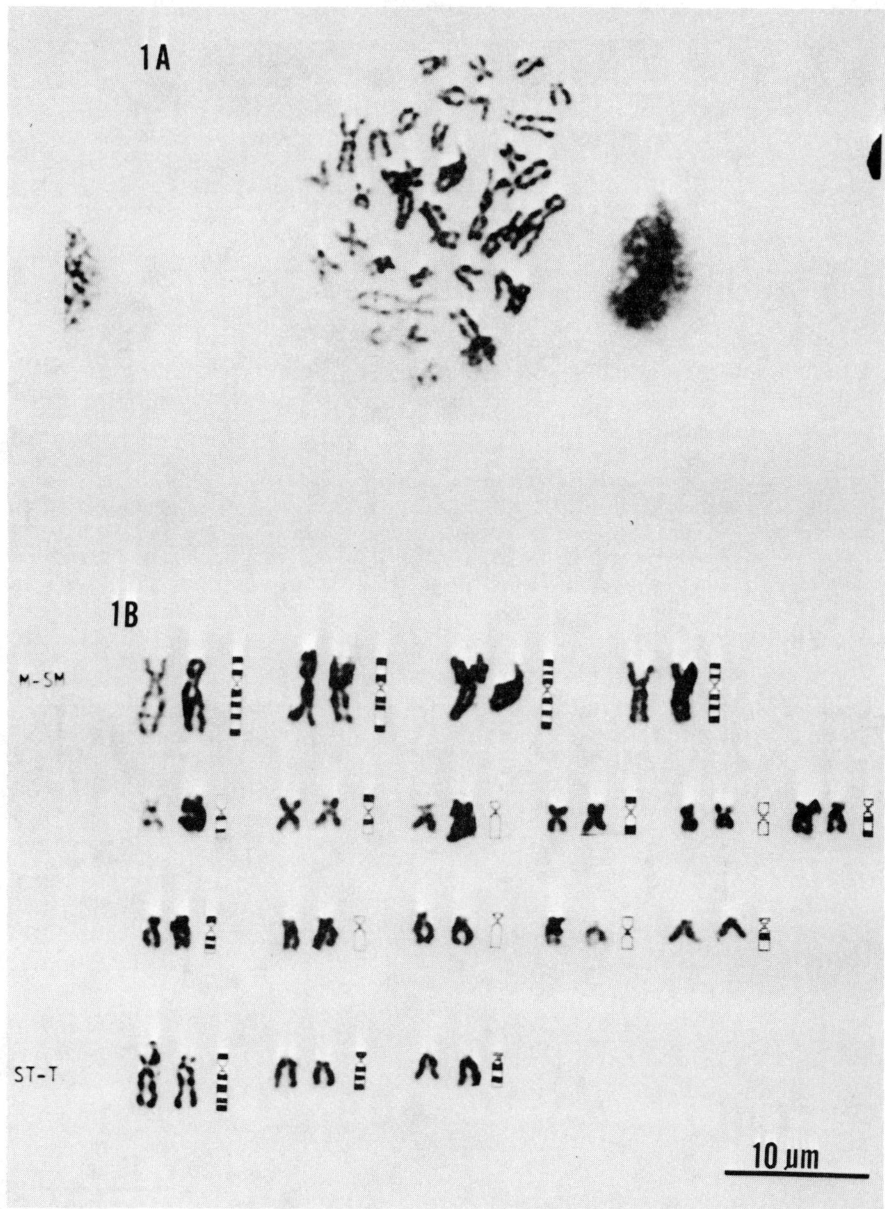

FIGURE 1. (A) G-banded karyotype of *Apogon pseudomaculatus*. (B) Paired arrangement of homologous chromosomes of *A. pseudomaculatus*, with diagrammatic representation of banding pattern for clarity. *Bar:* 10 μm.

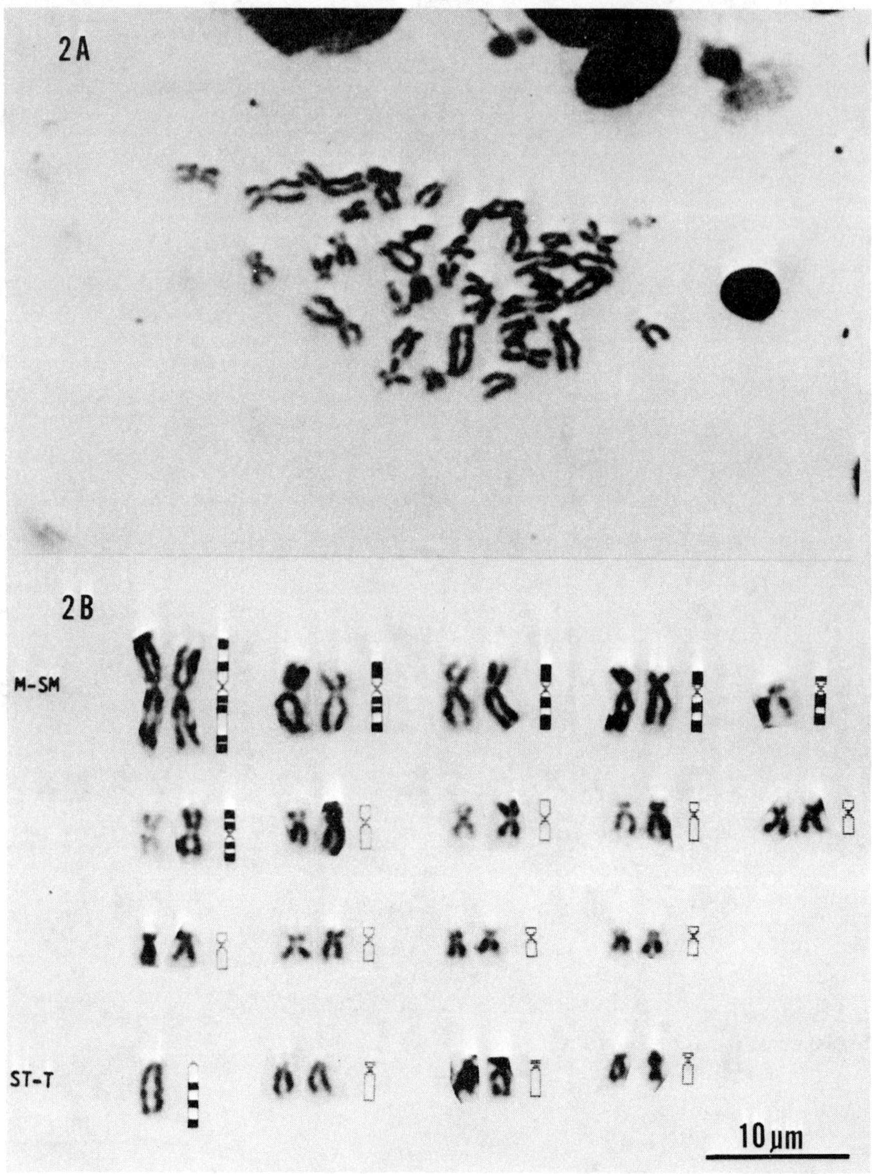

FIGURE 2. (A) Replication banded karyotype of *Apogon maculatus.* (B) Paired arrangement of homologous chromosomes of *A. maculatus,* with diagrammatic representation of banding pattern for clarity. *Bar:* 10 μm.

bands. The procedure used is a modification of the amphibian chromosome technique of Schempp and Schmid.[4] The fish, collected in Puerto Rico, received subcutaneous injections of 0.25 ml of a 0.025% fluorodesoxyuridine (FdU) and 0.5% bromodesoxyuridine (BrdU) [20 mg BrdU and 1 mg FdU in 4 ml of 0.9% NaCl] solution, and 0.25 ml of a 0.05% colchicine solution. After injection the fish were placed in a well-aerated holding tank for 17 hours. The procedures for cell harvest and slide preparations are described in Rivlin *et al.*[2] The freshly prepared slides were "aged" by heating on a slide warmer at 90°C for 15 minutes. Slides were stained with 0.5 μg/ml 33258 Hoechst solution for 10 minutes and then rinsed in a buffered solution (0.05M NaCl, 0.001M KCl, and 0.003M KH_2PO_4 adjusted to pH 5.5 and 0.1M NaOH). The slides were then placed horizontally, flooded with the buffered solution, and exposed to ultraviolet light (254 nm) for 30 minutes at a distance of 10 cm. This was followed by incubation in 2 × SSC (0.3M NaCl and 0.03M trisodium citrate) at 60°C for 90 min. The slides were stained by flooding with 5% Giemsa in 1/15M pH 6.8 Sorensen phosphate buffer and air-dried.

FIGURE 2 shows the replication-banded karyotype of *A. maculatus* and a diagrammatic representation of the banding pattern. *A. maculatus* has a chromosome number of 34, with 27 metacentric-submetacentric chromosomes and 7 subtelocentric-telocentric chromosomes. Two chromosomes are unmatched and may represent sex chromosomes. The five pair of large chromosomes show multiple bands along their length allowing matching of homologues. The BrdU-substituted chromosomal segments, when stained with 33258 Hoechst and Giemsa, stain pale, whereas the thymidine-incorporated regions stain dark, and because of the longitudinal variation in the replication time of chromosomal segments, show a differential staining pattern.

ACKNOWLEDGMENTS

We wish to thank Art Heine and Ed Murphy of Marine Tropical Imports for their help in obtaining specimens of the fish species used in this study. We also wish to thank Mary Lou Gaeta for her invaluable assistance.

REFERENCES

1. ARRIGHI, F. E. & T. C. HSU. 1974. Staining constitutive heterochromatin and Giemsa crossbands of mammalian chromosomes. *In* Human Chromosome Methodology. J. J. Yunis, Ed.: 59-72. Academic Press. New York.
2. RIVLIN, K. A., J. W. RACHLIN & G. DALE. 1985. A simple method for the preparation of fish chromosomes applicable to field work, teaching, and banding. J. Fish Biol. **26:** 267-272.
3. RIVLIN, K. A., G. DALE & J. W. RACHLIN. 1986. Karyotypic analysis of three species of cardinal fish (Apogonidae) and its implications for the taxonomic status of the genera *Apogon* and *Phaeoptyx*. Ann. NY Acad. Sci. **463:** 211-213.
4. SCHEMPP, W. & M. SCHMID. 1981. Chromosome banding in amphibia. VI: BrdU-replication patterns in *Anura* and demonstration of XX/XY sex chromosomes in *Rana esculenta*. Chromosoma **83:** 697-710.

Analysis of the Dietary Preference of the Sand Flounder, *Scophthalmus aquosus*, from the New Jersey Coast[a]

BARBARA E. WARKENTINE AND
JOSEPH W. RACHLIN

Department of Biological Sciences
Lehman College
City University of New York
Bronx, New York 10468

The sand flounder, *Scophthalmus aquosus,* is a consistently present and ubiquitous member of the marine benthic fish community along the New Jersey coast of the New York Bight. *Scophthalmus aquosus* is presently considered a trash fish and as a result, except for one paper by Moore,[1] little detailed information has been focused on the ecology of this species. However, there are a number of studies that provide very descriptive lists of the food items that constitute the diet of the sand flounder.[1-7] Although these studies provide useful data bases, they contain no information on the spatial dietary preference of this species. The purpose of this study is to determine whether the dietary preference of *S. aquosus* varies over a portion of its distributional range along the New Jersey coast.

Scophthalmus aquosus were collected by otter trawl during the month of June from six locations along the New Jersey coast: (1) Absecon Inlet; (2) Little Egg Inlet; (3) Beach Haven Crest; (4) Seaside Heights; (5) Sea Girt; and (6) Elberon Ground. Upon capture, stomachs were excised from all fish collected in the nets and their contents identified to species when possible and enumerated. To assess the resource base, at each collecting site the stomach contents from all fish collected at that site were pooled. This method of assessing the resource base has been shown to yield a less biased estimate of the food available to the piscine community than is determined by conventional sampling gear (plankton nets and bottom grabs).[8,9] Feeding preferences for *S. aquosus* collected from each station were determined using two mathematical models—the Manly Preference Index[10] and the Relativized Electivity Index (E_*) of Vanderploeg and Scavia,[11] which evaluate the proportional representation of food items in the diet of the fish against the proportional representation of the same food items in the resource base.

[a]This research was supported in part by the Women's Research and Development Fund of C.U.N.Y. and by PSC-C.U.N.Y. Grants No. 6-65144 and 6-66311.

TABLE 1 presents the percent occurrence of each food item in the guts of *S. aquosus* from each station. One can see from these data that the most dominant food item consumed at all stations was the shrimp *Neomysis americana*. These data also show that sand flounders do not exhibit uniform diets over this sampling range. Fish collected from station 5 (Sea Girt) had the greatest variety of food items in their guts, whereas those from station 6 (Elberon Ground) had the lowest variety. Using the models of Manly and Vanderploeg and Scavia, dietary preference was determined for sand flounders at each station and the results are presented in TABLE 2. It can be seen that *Neomysis americana* is the preferred food item for *S. aquosus* collected from all stations except Beach Haven Crest. In addition to this, sand flounders collected from

TABLE 1. Percent Occurrence of Each Food Item in the Diet of *Scophthalmus aquosus* from Six Locations within the New York Bight

Food Item	Station Numbers[a]					
	1	2	3	4	5	6
Neomysis americana	88.9	98.5	84.0	97.7	98.2	88.9
Mysidopsis bigelowi	—	—	—	0.8	0.1	—
Crangon septemspinosa	—	0.2	—	—	0.2	—
Gammarus annulatus	8.8	0.9	6.7	0.7	—	—
Gammarus marinus	1.7	—	7.8	—	0.1	—
Cancer megalops larvae	0.2	0.1	—	—	0.4	—
Shrimp zoea	—	—	—	—	0.3	—
Nematoda	0.2	0.3	1.5	0.3	0.3	—
Hydroids	0.2	—	—	0.5	0.3	—
Ammodytes hexapterus	—	—	—	0.1	0.1	11.1

[a] 1 = Absecon Inlet; 2 = Little Egg Inlet; 3 = Beach Haven Crest; 4 = Seaside Heights; 5 = Sea Girt; and 6 = Elberon Ground.

Absecon Inlet showed preference for the amphipod, *Gammarus annulatus*, nematodes and campanularian hydroids. Those collected from Little Egg Inlet preferred *G. annulatus*. *Scophthalmus aquosus* from Beach Haven Crest showed preference for *G. annulatus* and *G. marinus*. Sand flounders from Seaside Heights exhibited preference for the mysid, *Mysidopsis bigelowi*, the amphipod, *G. annulatus*, and campanularian and halecium hydroids. Sand flounders from Sea Girt preferred *M. bigelowi*, shrimp zoea, and campanularian hydroids. These results clearly indicate that the sand flounder, *S. aquosus*, exhibits spatial dietary preference within this section of the New York Bight. Also, the fact that a prey item makes up a large percentage of the diet (*Neomysis americana* was 84.0% of the diet at station 3) and is not preferred may indicate that the fish are feeding opportunistically.

TABLE 2. Dietary Preference of *Scophthalmus aquosus* Collected from Six Locations within the New York Bight

Food Item	Manly Preference	E*	
Absecon Inlet (39°22.55′N:74°19.79′W)			
Neomysis americana	0.1813	0.0420	P
Gammarus annulatus	0.1855	0.0535	P
Gammarus marinus	0.1645	−0.0064	NP
Cancer megalops larvae	0.0162	−0.8232	NP
Nematoda	0.2262	0.1516	P
Campanularian hydroids	0.2262	0.1516	P
Little Egg Inlet (39°28.27′N:74°15.44′W)			
Neomysis americana	0.3953	0.3281	P
Crangon septemspinosa	0.0990	−0.3380	NP
Gammarus annulatus	0.3958	0.3286	P
Cancer megalops larvae	0.0660	−0.5039	NP
Nematoda	0.0440	−0.6395	NP
Beach Haven Crest (39°37.10′N:74°10.40′W)			
Neomysis americana	0.0877	−0.4807	NP
Gammarus annulatus	0.3519	0.1693	P
Gammarus marinus	0.4105	0.2430	P
Nematoda	0.1499	−0.2504	NP
Seaside Heights (39°55.70′N:74°03.66′W)			
Neomysis americana	0.1543	0.0386	P
Mysidopsis bigelowi	0.1735	0.0968	P
Gammarus annulatus	0.1735	0.0968	P
Nematoda	0.1084	−0.1370	NP
Ammodytes hexapterus	0.0434	−0.5342	NP
Campanularian hydroids	0.1735	0.0968	P
Halecium hydroids	0.1735	0.0968	P
Sea Girt (40°08.00′N:74°00.90′W)			
Neomysis americana	0.1951	0.3223	P
Mysidopsis bigelowi	0.1591	0.2281	P
Crangon septemspinosa	0.0660	−0.2004	NP
Gammarus marinus	0.0020	−0.9600	NP
Shrimp zoea	0.2938	0.4921	P
Cancer megalops larvae	0.0054	−0.8969	NP
Nematoda	0.0592	−0.2562	NP
Unidentifiable fish	0.0227	−0.6296	NP
Campanularian hydroids	0.1164	0.0759	P
Garland hydroids	0.0796	−0.1138	NP
Elberon Ground (40°18.58′N:73°53.10′W)			
Neomysis americana	0.8980	0.2847	P
Ammodytes hexapterus	0.1020	−0.6610	NP

NOTE: P = preferred; NP = nonpreferred.

REFERENCES

1. MOORE, E. 1947. Studies on the marine resources of southern New England. VI. The sand flounder, *Lophopsetta aquosa* (Mitchell): A general study of the species with special emphasis on age determination by means of scales and otoliths. Bull. Bingham Oceanogr. Collect. Yale Univ. **11**(2): 1-79.

2. DE SYLVA, D. P., F. A. KALBER, JR. & C. N. SHUSTER, JR. 1962. Fishes and ecological conditions in the shore zone of the Delaware River estuary, with notes on other species collected in deeper water. Univ. Del. Lab. Inf. Ser. Publ. **5**: 1-164.

3. RICHARDS, S. W. 1963. The demersal fish population of Long Island Sound. Bull. Bingham Oceanogr. Collect., Yale Univ. **18**(2): 1-101.

4. STICKNEY, R. R., G. L. TAYLOR & R. W. HEARD, III. 1974. Food habits of Georgia estuarine fishes. 1. Four species of flounder (Pleuronectiforms: Bothidae). Fish. Bull., U.S. **72**: 515-525.

5. HICKEY, C. R. 1975. Fish behavior as revealed through stomach contents analysis. N. Y. Fish Game J. **22**: 148-155.

6. LANGTON, R. W. & R. E. BOWMAN. 1981. Food of eight Northwest Atlantic pleuronectiform fishes. NOAA-NMFS-SSRF-**749**: 1-16.

7. MORRIS, T. L., JR. 1981. Mouth structure relative to food habits for seven northwest Atlantic pleuronectiform fish species. Conseil International pour l'Exploration de la Mer. France.: 1-10.

8. RACHLIN, J. W. & B. E. WARKENTINE. 1986. Dietary preference of spotted hake, *Urophycis regia,* from the inner New York Bight. Ann. N. Y. Acad. Sci. **494**: 434-437.

9. RACHLIN, J. W. & B. E. WARKENTINE. 1987. The use of museum ichthyological holdings for initial diet studies. Copeia **1**: 214-216.

10. MANLY, B. F. J. 1974. A model for certain types of selection experiments. Biometrics **30**: 281-294.

11. VANDERPLOEG, H. A. & D. SCAVIA. 1979. Two electivity indices for feeding with special reference to zooplankton grazing. J. Fish. Res. Board Can. **36**: 362-365.

Response of Sodium Flux in Goldfish to Acute Changes in Ambient Temperature

BRENDA P. MOFFITT AND HOLLY ROBERTS

Department of Biological Sciences
Lehman College
City University of New York
Bronx, New York 10468

LARRY I. CRAWSHAW

Department of Biology
Portland State University
Portland, Oregon 97207

Teleost fish maintain a constant internal osmotic pressure. When living in a hypo-osmotic environment, osmotic gain of water and diffusional loss of sodium are compensated for by urinary loss of water and active uptake of sodium by the gills and digestive tract.[1]

Metabolic rate and heart rate of carp have been shown to have Q10 values of about 2.5 in response to acute thermal changes,[2,3] and diffusion, a physical process, might be expected to have a Q10 of 2.0. Thus, as net sodium flux is the algebraic sum of diffusional efflux and energy-dependent influx, it might be expected to alter in response to temperature change. This has been shown in fish acclimated to the temperature at which the measurements were being made.[4,5] Acute temperature changes affected both sodium and water fluxes in the eel, *Anguilla anguilla.*[6] The present investigation was undertaken to see whether acute temperature change affected sodium flux in the goldfish. The uptake of sodium-22 by thermally acclimated fish was studied before and immediately after an acute change in ambient temperature.

PROCEDURE

Goldfish (*Carassius auratus* L), average weights 12.4 g and 55.4 g, were acclimated to 22.0 ± 1.0°C for a minimum of two weeks. Food was withheld for 2 days prior to transferring weighed fish to individual stainless steel chambers contained within a single closed waterjacket through which water at 22.0 ± 1.0°C was continuously recirculated. Translucent covers were placed on the chambers, so that the animals were not disturbed by external events, but did experience the natural photoperiod. During a 24-hour stabilization period, there was a gradual replacement of the water in each bath by distilled water. Immediately prior to the experiment, this flow was stopped, and the volume of the water in each bath was adjusted to give a fish-to-water volume ratio of about 1:10. (The actual water volume was measured at the end of the experiment.) "Cold" NaCl was added to each bath to give a nominal concentration of 120 mEq/L, and was followed by Na-22 (0.25 μCi/100 ml). Aliquots were removed immediately, and at 15-min intervals thereafter. After 60 min, the temperature of the water in the waterjacket was changed rapidly by 7.0°C or −7.0°C, this change being complete in 3 to 5 min. The new temperature was maintained for 75 min and then it was restored to 22.0°C for a further 60 min. During confinement, the fish rested quietly in the chambers and seemed unaffected by the thermal changes. The radioactivity of samples from each aliquot was measured using a well-type counter, and the rate of Na-22 disappearance from the water in the baths was calculated.

TABLE 1. Average Rates of Na-22 Disappearance from the Baths

Temperature Change (°C)	Time Period (minutes)	Rate of Na-22 Loss (counts · min^{-1} · g^{-1})
+7	0–60	1756
	75–135	1565
	145–190	992
−7	0–60	92
	75–135	303
	145–190	677

RESULTS AND DISCUSSION

In FIGURE 1, the disappearance of sodium-22 from the baths indicates sodium influx. Because two distinct weight classes of fish were used, their data are plotted separately, but it can be seen that the rate of sodium-22 uptake per gram was the same for both groups (solid and open circles), so a single regression line was calculated. Small changes in slope were obtained when the ambient temperature was changed (TABLE 1), but the data can be expressed equally well by single regression lines covering the whole 3 hours of the experiment. Neither raising nor lowering the ambient temperature acutely appeared to alter the rate of sodium-22 loss from the baths.

Although the goldfish is a highly eurythermal animal, a change in body temperature of 7.0°C must be considered significant, and other enzymatically catalyzed processes have demonstrable sensitivity to thermal displacements as small as 2.0°C in the carp,[2,3]

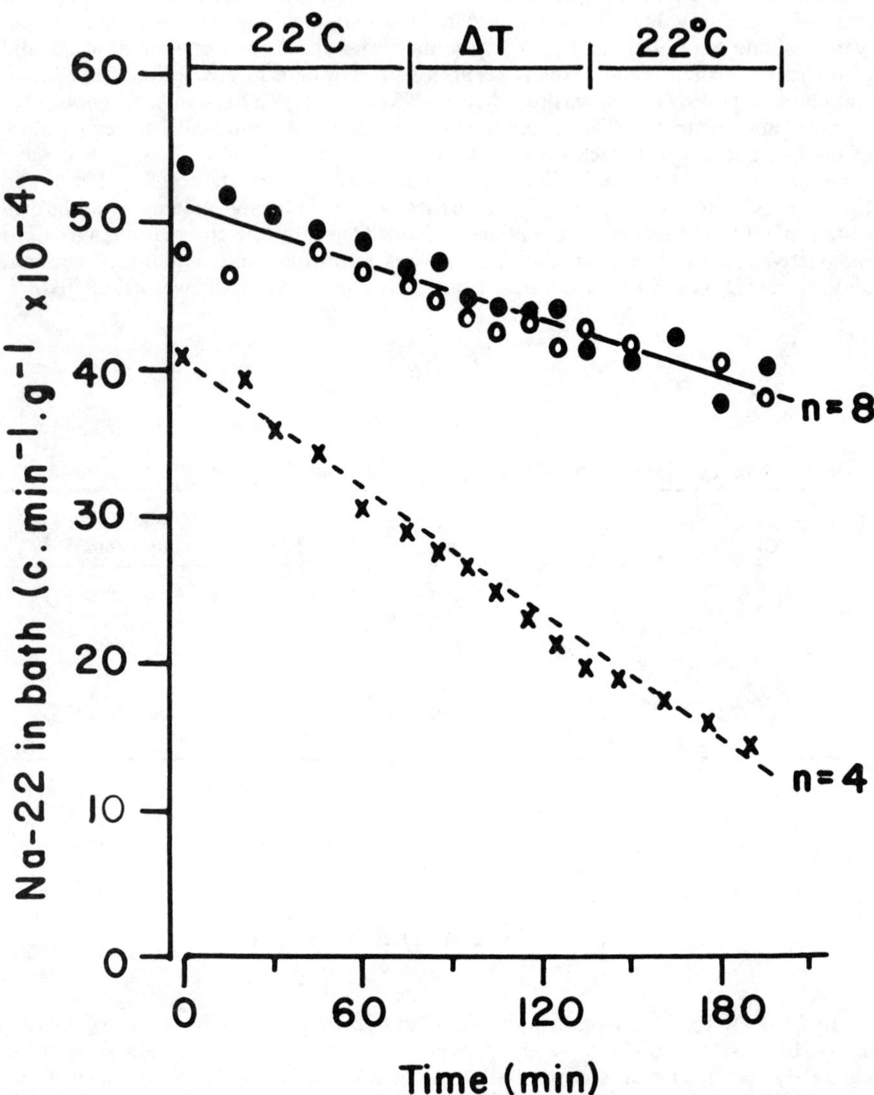

FIGURE 1. Effects of acute temperature changes of 7.0°C (*broken line*) and −7.0°C (*solid line*) on Na-22 loss from the experimental baths. $\bigcirc$ = small fish; $\bullet$ = large fish; x = large fish. Equations of regression lines are $y = 404562 - 1452x$ ($r^2 = -0.99$) and $y = 510133 - 646x$ ($r^2 = -0.933$), respectively, where y = Na-22 and x = time.

a fish to which the goldfish is closely related taxonomically. The fact that sodium influx is apparently unaffected by a 7.0°C change in temperature is therefore quite surprising. Water fluxes across the gill lamellae have been shown to be temperature-sensitive,[4,6] and this has been associated with alterations in lamellar blood flow in response to altered oxygen demand.[7] Sodium uptake, in contrast, occurs on the gill filament, and is not subject to similar changes in blood flow. In view of the fact that the sodium-transporting cells are directly exposed to the changes in temperature, it is difficult to understand why sodium uptake is unaffected.

REFERENCES

1. PROSSER, C. L. 1973. Comparative Animal Physiology, 3rd ed. W. B. Saunders. Philadelphia, Pa.
2. MOFFITT, B. P. & L. I. CRAWSHAW. 1983. Effects of acute temperature changes on metabolism, heart rate and ventilation rate in carp (*Cyprinus carpio* L.). Physiol. Zool. **56:** 397-403.
3. MOFFITT, B. P. 1984. Effects of small, acute changes in temperature on calm carp. Ann. N.Y. Acad. Sci. **435:** 313-315.
4. MAETZ, J. & F. GARCIA-ROMEU. 1964. The mechanism of sodium and chloride uptake by the gills of a freshwater fish, *Carassius auratus.* J. Gen. Physiol. **47:** 1209-1227.
5. KERSTETTER, T. H., L. B. KIRSCHNER & D. D. RAFUSE. 1970. On the mechanisms of sodium ion transport by the irrigated gills of rainbow trout (*Salmo gairdneri*). J. Gen. Physiol. **56:** 342-359.
6. MOTAIS, R. & J. ISAIA. 1972. Temperature-dependence of permeability to water and sodium of the gill epithelium of the eel, *Anguilla anguilla.* J. Exp. Biol. **56:** 587-600.
7. STEEN, J. B. & A. KRUYSSE. 1964. The respiratory function of teleostean gills. Comp. Biochem. Physiol. **12:** 127-142.

Blood-Brain Barrier Permeability during Hypocapnia in Halothane-Anesthetized Monkeys

CHARLES J. HANNAN, JR.,[a] THOMAS M. KETTLER,[a]
ALAN A. ARTRU,[b] AND ROBERT ARONSTAM,[c]

[a]*Department of Clinical Investigation*
Madigan Army Medical Center
Tacoma, Washington 98431-5454

[b]*Anesthesiology Department*
University of Washington
Seattle, Washington 98195

[c]*Medical College of Georgia*
Augusta, Georgia 30912

Halothane anesthesia has been demonstrated to increase the diffusion permeability of glucose[1] and cocaine[2] across the blood-brain barrier (BBB). There is also evidence that hypocapnia increases BBB permeability to glucose and amino acids.[3] Hypocapnia is frequently employed during anesthesia for neurosurgery because hypocapnia causes cerebral vasoconstriction, thereby reducing intracranial pressure while the cranium is closed and improving surgical exposure when the cranium is opened. However, when hypocapnia and halothane anesthesia are combined, the increase of BBB permeability caused by halothane anesthesia may be additive with the increase of BBB permeability caused by hypocapnia. This study was designed to examine the BBB permeability to endogenous albumin, cortisol, and an intravenous radioactive tracer technetium (technetium-99m) during prolonged halothane anesthesia combined with normocapnia or hypocapnia. These tracers were selected to permit evaluation of a spectrum of possible permeability changes due to large pores or vesicles (albumin) and small hydrophilic (technetium-99m) or lipophilic diffusional routes (cortisol).

Five monkeys (*Macaca nemestrina*) were anesthesized with halothane (0.5%) and N_2O (66%) for five hours. Ventilatory rates were adjusted to achieve a range of P_aCO_2 values between normocapnia ($P_aCO_2 \sim 35$ torr) and hypocapnia ($P_aCO_2 \sim 22$ torr). A spinal catheter was inserted into the cisterna magna and cerebrospinal fluid (CSF) was continuously withdrawn and pumped through an ice bath at 2 μl/min/kg. CSF was pooled for each 30 min. A 2.5-ml venous blood sample was also obtained each 30 min. Arterial blood gases were determined intermittently during the experiment. Arterial blood pressure, end-tidal CO_2, and temperature were continuously recorded. The technetium-99m was counted immediately in a gamma counter and the remainder of sample was stored at $-70°C$ before analysis of albumin by radialimmunodiffusion and cortisol by radioimmunoassay.

The two monkeys with the lowest P_aCO_2 exhibited CSF/plasma albumin ratios which exceeded normal values (approximately 8×10^3, based on human studies)[4]

(FIG. 1) indicating increased BBB permeability to albumin. In these animals high CSF/plasma ratios also were observed for Tc^{99m} (TABLE 1). Both the albumin and Tc^{99m} ratios were higher in the two most hypocapnic animals than in the other three animals at each of the eight sampling times. The CSF/plasma ratio of cortisol was increased in hypocapnic animals only during the last hour (210- and 240-min samples). Because cortisol is bound in the plasma by a specific globulin (CBG), passage of cortisol into the CSF may be affected by alterations in CBG binding as well as by changes in BBB permeability. Albumin is larger than CBG so changes due to protein transfer should appear at the same time as albumin ratio changes. Alkalinization of the plasma, as occurred during hypocapnia, increases the CBG affinity for cortisol[5] and so would produce an effect opposite to what was observed. As plasma cortisol is gradually increased during the experiment, it is possible that the binding capacity (approximately 45 μg/dl in *Macaca nemestrina*) of CBG for cortisol was exceeded in the hypocapnic animals. This could only be verified if CBG concentrations were measured, and they were not.

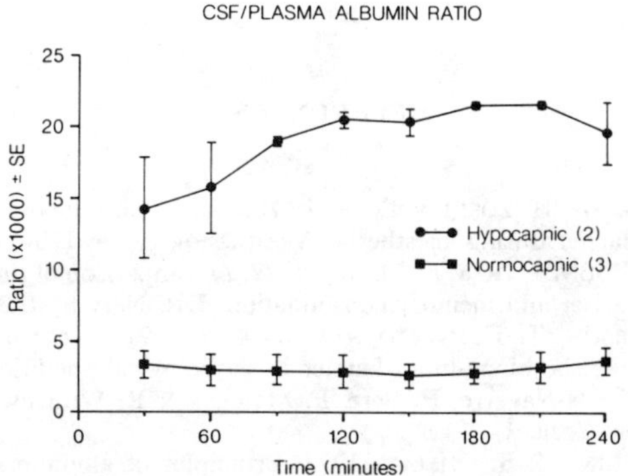

FIGURE 1. Ratio of CSF/plasma ($\times$1000) albumin in hypocapnic ($P_aCO_2 < 25$) and normocapnic ($P_aCO_2 > 25$) monkeys. Mean changes over 4 hours of anesthesia with 0.5% halothane and 66% nitrous oxide.

One means by which halothane may increase permeability of the BBB is through halothane action on cyclic nucleotides. A positive relationship between cyclic-GMP and cyclic-AMP concentration in cerebral capillary endothelial cells and pinocytotic vesicle abundance has been established.[6] There also has been reported a positive dose-response relationship between halothane and cyclic-GMP production in brain.[7] Halothane-stimulated cyclic GMP production was evident with as low as 0.5% of this agent.[7]

Previous reports of halothane-induced increase of BBB permeability employed halothane concentrations of > 1.5%, but the duration of exposure was brief.[1,2] In the present study 5 hours of 0.5% halothane with normocapnia or modest hypocapnia caused no alteration of BBB permeability. In previous studies hypocapnia increased BBB permeability only when P_aCO_2 was reduced to < 18 torr.[3] In the present studies 0.5% halothane combined with $P_aCO_2 = 21$-24 torr resulted in increased BBB permeability. A synergistic increase in BBB permeability may be produced by halothane and hypocapnia, but further studies in the same animal model will be necessary.

TABLE 1. Summary Data

| Monkey | | Weight | HDLC[a] | P_aCO_2 | P_aO_2 | | CSF/Plasma Ratios (mean ± SE)[b] | | |
No.	Group	(kg)	(mg/dl)	(torr)	(torr)	pH_a	Albumin	Tc[99m]	Cortisol
1	I	3.54	63	21	140	7.555	18 ± 4	128 ± 82	93 ± 27
2	I	5.10	52	24	127	7.499	20 ± 2	387 ± 96	118 ± 65
3	II	5.35	53	25	202	7.514	5 ± 1	18 ± 24	101 ± 10
4	II	5.28	15	27	109	7.417	2 ± 1	13 ± 11	74 ± 15
5	II	4.70	54	35	99	7.418	2 ± 1	20 ± 46	32 ± 7

NOTE: Group I = hypocapnic animals; Group II = normocapnic animals.
[a] High-density-lipoprotein cholesterol.
[b] Average of eight values. Samples were obtained each 30 min from hour 2 to hour 5 of the experiment.

REFERENCES

1. NEMOTO, E. M., S. W. STEZOSKI & D. MACMURDO. 1978. Glucose transport across the rat blood-brain barrier during anesthesia. Anesthesiology **49:** 170-176.
2. ANGEL, C., H. M. BOUNDS, JR. & A. PERRY. 1972. A comparison of the effects of halothane on blood-brain barrier and memory consolidation. Dis. Nerv Syst. **33:** 87-93.
3. SPATZ, M., F. BERSON, T. FUJIMOTO & I. KLATZO. 1976. Transport of nutrients and nonnutrients across the blood-brain barrier in pathological conditions. *In* The Cerebral Vessel Wall. J. Cevos-Navarro, E. Betz, F. Matakas & R. Wullenweber, Eds.: 225-232. Raven Press. New York.
4. TIBBLING, G., H. LINK & S. OHMAN. 1977. Principles of albumin and IgG analysis in neurological disorders. I. Establishment of reference values. Scan. J. Clin. Lab. Invest. **37:** 385-390.
5. WESTPHAL, U. 1971. Steroid-protein Interactions. Springer-Verlag. New York.
6. JOO, F., P. TEMESVARI & E. DUX. 1983. Regulation of the macromolecular transport in the brain microvessels: the role of cyclic GMP. Brain Res. **278:** 165-174.
7. DIVAKARAN P., B. M. RIGO & R. C. WIGGINS. 1980. Brain cyclic nucleotide and energy metabolite responses to subanesthetic and anesthetic concentrations of halothane. Experientia **36:** 655-656.

The Action of Atrial Natriuretic Factor on Glomerular Diameter *in Vitro*

WARREN L. ROSENBERG, CHRISTOPHER V. KEOGH,
BEVERLY A. HALL, AND LUIGI ALBANO

Department of Biology
Iona College
New Rochelle, New York 10801

Atrial natriuretic factor (ANF), a group of related atrial peptides possessing vaso-relaxant,[1,2] natriuretic[3,4] and diuretic[3,4] activities, most probably acts as a moderator of body-fluid balance. The factor is released from atrial myocytes in response to a volume expansion-induced stretch of atrial fibers. The natriuretic and diuretic activities are most probably the direct result of the ability of ANF to increase glomerular filtration rate (GFR).[5,6] The present study was designed to assess the effects of ANF on glomerular diameter, which is considered a determinant of GFR.[7]

ANF was isolated from rat atrial tissue according to modifications of the procedures of Pollock *et al.*[8] and deBold *et al.*[3] The natriuretic and diuretic activities of the extract were confirmed by injection into rats. All tests were run against controls of ventricular extract (VE) and phosphate-buffered saline solution (PBS). Glomeruli were isolated by passing rat renal cortical tissue through successively smaller brass sieves and collecting them in Medium 199. Purity was confirmed with phase-contrast microscopy (FIG. 1).

Atrial natriuretic factor-induced changes were determined by nephelometry[9] and confirmed by direct-video microscopy. ANF produced a significant increase in glomerular diameter that was immediate and of sustained duration (FIG. 2). Neither VE nor PBS produced similar results. The activity of ANF did not seem to be influenced by prior contraction of glomeruli with Angiotensin II (ANG-II); however, their response to AG-II was inhibited by prior exposure to ANF.

We hypothesize that the ANF-induced increase in glomerular diameter may reflect a relaxation in mesangial cell tonus and may account for ANF's ability to elevate GFR. This mechanism is probably analogous to ANF's relaxant effect on vascular smooth muscle.[1,2]

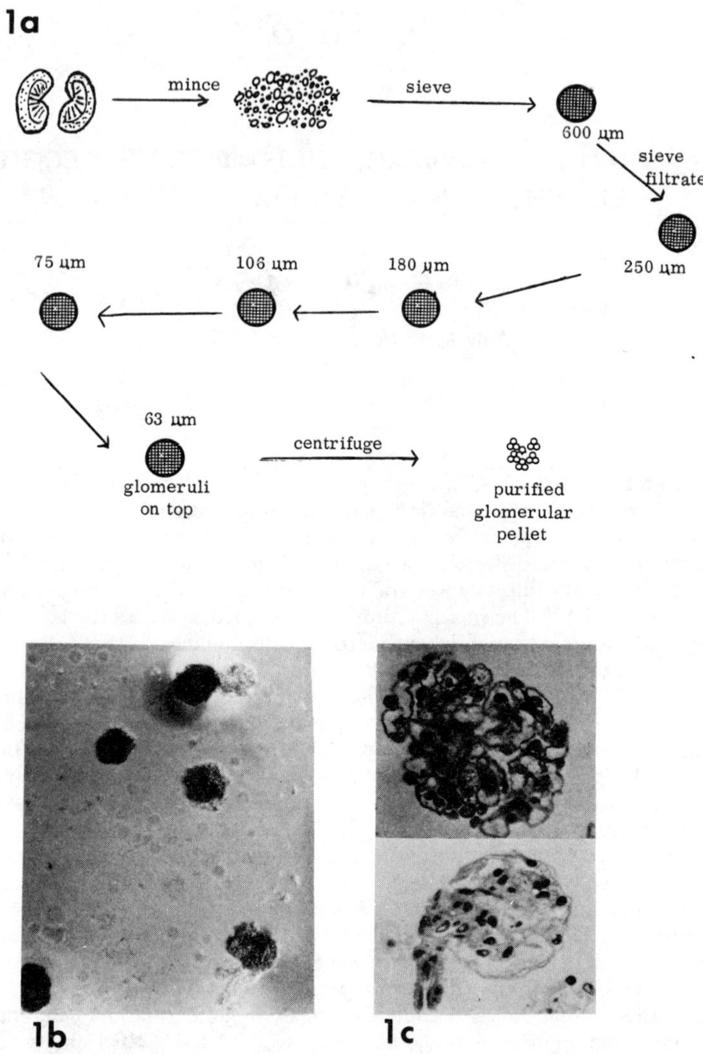

FIGURE 1(a). Protocol for the isolation of glomeruli from rat rental cortex. (b) and (c) Phase contrast and sectioned images of glomeruli obtained for assay of ANF activity.

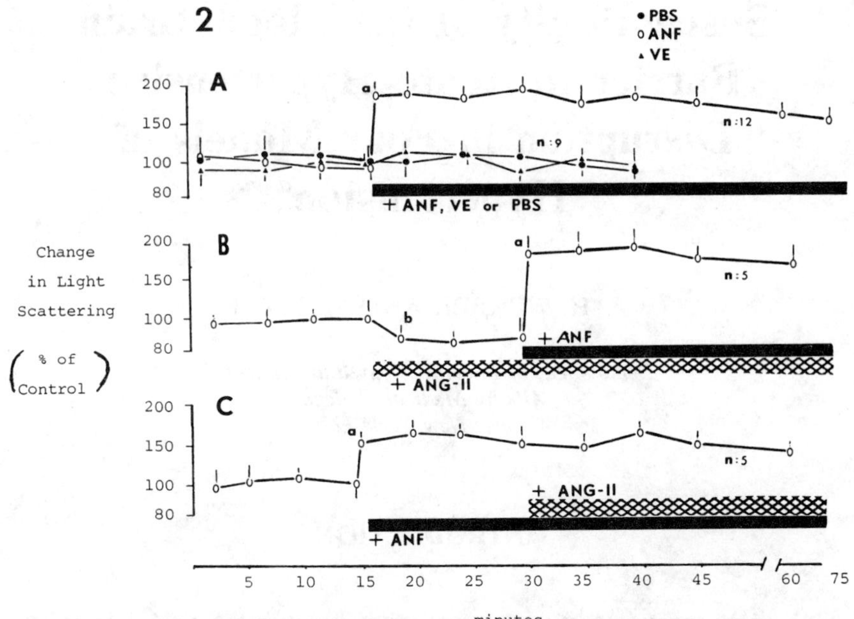

FIGURE 2. Glomerular responses to ANF, VE and PBS alone (**A**); to ANF subsequent to ANG-II (**B**); and to ANG-II subsequent to ANF (**C**). Abbreviations are described in text. Data are $\overline{X} \pm$ SEM. (**a**) Data from this point on differ from time $= 0$ ($p < .005$). (**b**) Data from this point on differ from time $= 0$ ($p < .02$).

REFERENCES

1. DETH, R. C., K. WONG, S. FUKOZAWA, R. ROCCO, J. L. SMART, C. J. LYNCH & R. AWAD. 1982. Fed. Proc. **1:** 983.
2. OSHIMA, T., M. G. CURRIE, D. M. GELLER & P. NEEDLEMAN. 1984. Circ. Res. **54:** 612.
3. deBOLD, A. J., H. B. BORENSTEIN, A. T. VERESS & H. SONNENBERG. 1981. Life Sci. **28:** 89.
4. BRIGGS, J. P., B. STEIPE, G. SCHUBERT & J. SCHNEERMAN. 1982. Pflügers Arch. **395:** 271.
5. CAMARGO, M., H. KLEINERT, S. A. ATLAS, J. SEALEY, J. H. LARAGH & T. MAACK. 1984. Am. J. Physiol. **246:** F447.
6. HUANG, C-L., J. LEWICKI, L. JOHNSON & M. COGAN. 1985. J. Clin. Invest. **75:** 769.
7. DWORKIN, L., I. ICHIKAWA & B. M. BRENNER. 1983. Am. J. Physiol. **244:** f95.
8. POLLOCK, D. M., M. MULLINS & R. O. BANKS. 1983. Renal Physiol. **6:** 295.
9. WESTENFELDER, C. & R. BARANOWSKI. 1983. Kid. Int. **23:** 248.

Susceptibility of the Blood-Brain Barrier to Acute Hypertensive Disruption in Four Models of Hypertension[a]

A. H. WERBER AND M. FITCH

Department of Pharmacology and Toxicology
Albany Medical College
Albany, New York 12208

INTRODUCTION

It has been suggested that acute hypertensive disruption of the blood-brain barrier (BBB) is the result of the change in tension on arterioles during the increase in intraluminal pressure.[1] If this hypothesis is true, then one would observe less disruption of the BBB in hypertrophied cerebral arterioles. Because cerebral vessels undergo hypertrophy in spontaneously hypertensive rats (SHR), acute hypertensive disruption of the BBB was examined in this strain. As predicted by the hypothesis described above, there was less BBB disruption in SHR.[1,2] Because it is generally held that chronic hypertension per se produces generalized vascular hypertrophy,[3] it was expected that other models of chronic hypertension would display reduced acute hypertensive disruption of the BBB. Yet studies have shown that acute hypertensive disruption of the BBB was not reduced in two-kidney-one-clip Goldblatt hypertension.[4,5] Thus, the goal of this study was to determine which of the models previously studied had an anomalous response to acute hypertensive disruption of the blood-brain barrier.

METHODS

We studied four rat models of chronic hypertension [SHR, two-kidney-one-clip Goldblatt (2K1C), DOCA-NaCl (DOCA), and Dahl-S rats fed a high-salt diet (DS-HI)], and two groups of normotensive controls [Wistar-Kyoto (WKY) and Dahl-S fed a low-salt diet (DS-LO)]. We caused acute hypertension in some rats using

[a]This work was supported by a grant from the Northeast Affiliate of the New York Heart Association and Grant HL35769 from the National Institutes of Health.

bicuculline (1.2 mg/kg) and aortic occlusion. Rats without acute hypertension served as controls. Blood-brain barrier transport was quantitated using brain/blood ratio of [^{125}I]albumin (RISA) according to the methods of Mueller and Heistad.[2]

RESULTS

Acute hypertensive disruption was lower in SHR, DOCA-NaCl, and DS-HI rats, but not in 2K1C, than in normotensive controls (TABLE 1). Acute hypertensive disruption was higher in DS-LO than in WKY (TABLE 1). The brain/blood RISA ratio in control rats was significantly lower compared with that in rats with acute hypertension and not different among any of the groups. The differences in acute hypertensive disruption among groups could not be accounted for by the increase in blood pressure, rate of rise of blood pressure, clearance of RISA from the blood, blood gases or pH.

TABLE 1. Brain/Blood Ratio of RISA (%) in Anterior Forebrain of Rats with Acute Hypertension

	SHR	DOCA	WKY	2K1C	DS-HI	DS-LO
Mean	1.08	0.89	1.68	1.42	1.60	3.42
SEM	0.08[a]	0.25[a]	0.22	0.15	0.44[a]	0.30[b]
n	10	7	10	10	8	10

[a] Significantly different ($p < 0.05$) from normotensive control.
[b] Significantly different ($p < 0.05$) from WKY rats.

CONCLUSIONS

We conclude that: (1) chronic hypertension reduces susceptibility of the blood-brain barrier to acute hypertensive disruption; (2) 2K1C rats have an anomalous response to acute hypertensive disruption; (3) there may be a genetic influence on the susceptibility of the blood-brain barrier to acute hypertensive disruption since we found that the degree of acute hypertensive disruption differed between the two strains of normotensive rats.

REFERENCES

1. JOHANSSON, B. B. 1977. The cerebrovascular permeability to protein after bicuculline and amphetamine administration in spontaneously hypertensive rats. Evidence for increased

resistance to pressure-induced blood-brain barrier dysfunction. Acta Neurol. Scand. **56:** 397-404.

2. MUELLER, S. M. & D. D. HEISTAD. 1980. Effect of chronic hypertension on the blood-brain barrier. Hypertension **2:** 809-812.

3. FOLKOW, B., M. GUREVICH, M. HALLBACK, Y. LUNDGREN & L. WEISS. 1971. Haemodynamic consequences of regional hypotension in spontaneously hypertensive and normotensive rats. Acta Physiol. Scand. **83:** 532-541.

4. JOHANSSON, B. B. & L.-E. LINDER. 1980. The blood-brain barrier in renal hypertensive rats. Clin. Exp. Hyp. **2:** 983-993.

5. MUELLER, S. M. & F. C. LUFT. 1982. The blood-brain barrier in renovascular hypertension. Stroke **13:** 229-234.

Efficacy of Region-of-Interest Study to Quantitate Lung Injury in Nuclear Scans

DIBYENDU BANDYOPADHYAY AND
SHLOMO HOORY

*Division of Nuclear Medicine
Long Island Jewish Medical Center and
State University of New York at Stony Brook
New Hyde Park, New York 11042*

DIPAK K. DAS

*University of Connecticut School of Medicine,
Farmington, Connecticut 06032*

INTRODUCTION

The technique of nuclear imaging is one of the noninvasive modalities available to physicians for diagnosis of various disease processes. A small amount of appropriate radiopharmaceutical is injected into the subject and the dynamics of the tracer *in vivo* is followed by a gamma camera and computer. In our previous studies we have reported that the lung injury created by hyperoxic exposure in laboratory animals can be diagnosed noninvasively by this technique using ^{111}In-labeled heterologous leukocytes.[1,2] The injury was confirmed by different physicochemical tests as well as by sacrificing the animals, counting radioactivity in selected vital organs, and evaluating the histopathology. In the present investigation, we studied different regions of each organ in a nuclear scan using the computer and compared that to the actual biodistribution values of the respective organs obtained after sacrificing the animals. The object of the study was to assess whether region-of-interest (ROI) analyses present a "true" measure of the radionuclide uptake into the organs.

MATERIALS AND METHODS

As described previously, lung injuries were created in white New Zealand rabbits by exposing them to 100% oxygen for up to 96 hours.[2] Polymorphonuclear leukocytes (PMNs) were isolated in pure form from human blood using a standard technique.[3]

About 40 μCi of [^{111}In]oxine-labeled PMNs were injected into the rabbits through an ear vein and imaging was performed after 24 hours using a Siemens large-field-view gamma camera equipped with a medium-energy parallel-hole collimator. Acquired images were digitized in 128 × 128 matrices and processed in a Siemens MicroDELTA terminal networked to a VAX 11/750 computer. After imaging, animals were sacrificed and different organs were counted for radioactivity in a Picker Autowell counter.

For each organ in a digitized image, different regions of interest of any shapes were drawn using the joystick at the terminal. The program then automatically calculated several parameters including the average and maximum count-per-pixel for the area.

RESULTS AND DISCUSSION

TABLE 1 shows a typical analysis of ROIs in a 96-hour-exposed rabbit. The number of pixels in the region are an indicator of how small or how large was the area under study. Better statistics of the measurement could be achieved with larger area selection. This is quite evident from the counts in the lung and liver. Since the liver is a larger organ, an approximately ten-times-larger area was available for processing, resulting in less error in the count-per-pixel value. Maximum count-per-pixel is a constant number for the particular area under study. Any change of this value indicates a shift from the peak uptake area and hence the region should be redrawn.

Comparative evaluation of ROI and actual biodistribution data for control and experimental rabbits are presented in TABLE 2. Since there were substantial variations of data from animal to animal, comparison between the average values may not show true correlation of the two techniques. Hence only one animal in each group was compared for this purpose. Since the lung is the "target" organ, ratio of lung to two other organs (namely, liver and kidney) were compared. It is evident from the table, that the ROI method correlates well with the actual *in vivo* distribution of the radiopharmaceutical. TABLE 2 also confirms the lung injury in the 96-hour-exposed animal since the uptake of the radionuclide in the lung was three times greater in this rabbit as compared to control.

ROI technique is a routine choice in a clinical nuclear medicine laboratory for follow-up studies in patients. In that case the same subject serves as his or her own

TABLE 1. A Typical Analysis of Regions of Interest (ROI) in the Nuclear Scan of a Rabbit Exposed to 100% O_2 for 96 Hours[a]

Organ	No. of Pixels[b]	Total No. of Counts	Average Counts/ Pixel	Maximum Count per Pixel
Lung	54 ± 12	6752 ± 764	126 ± 18	636
Liver	402 ± 2	36982 ± 1116	91 ± 2	225
Kidney	43 ± 4	991 ± 146	22 ± 1	38

[a] Four different regions for each organ are generated and values are expressed as mean ± SD.

[b] All images are digitized in 128 × 128 matrices.

TABLE 2. Comparison of "Target-to-Nontarget" Ratios Obtained by ROI and Biodistribution Methods

| Target | Rabbit Exposed to 100% O_2 for 96 Hours | | Control Rabbit | |
Nontarget	ROI[a]	Biodistribution	ROI[a]	Biodistribution
Lung / Liver	1.38	2.26	0.28	0.31
Lung / Kidney	5.72	5.42	1.38	2.22

[a] Ratios of average counts-per-pixel for respective organs are shown.

control and the data are more comparable. In an experimental injury model, as opposed to a clinical situation, the efficacy of the ROI approach can be evaluated by actual biodistribution. Some degree of variation between the two methods is expected since there is always considerable time lag between the time of imaging and actual sacrifice, during which period the radiopharmaceutical might undergo redistribution. Also, if there is a focal uptake in an organ, that area appears more intensely in the scan, suppressing the image of other organs or other parts of the same organ. In that case, the region drawn only signifies the area of the highest uptake, but not the whole organ. Keeping in mind the limitations of comparison, we can conclude from the data presented above that the ROI technique is an invaluable noninvasive method in quantitating experimental lung injury.

REFERENCES

1. DAS, D. K., D. BANDYOPADHYAY, S. HOORY & H. STEINBERG. 1985. Ann. N.Y. Acad. Sci. **435:** 399-401.
2. DAS, D. K., D. BANDYOPADHYAY, S. HOORY & H. STEINBERG. 1986. Ann. N.Y. Acad. Sci. **463:** 270-273.
3. BAKER, W. J. & F. L. DATZ. 1984. J. Nucl. Med. Tech. **12:** 131-136.

A Rapid Method to Stabilize Biological Materials for Cardiovascular Surgery

JUAN C. CHACHQUES, BERNARD VASSEUR,
PATRICK PERIER, JACQUES BALANSA, SYLVAIN
CHAUVAUD, AND ALAIN CARPENTIER

Experimental Cardiovascular Laboratory
Hôpital Broussais
Paris, France 75014

The continuous search for new blood-compatible biomaterials for use in cardiovascular surgery has thus far been centered on biological autologous tissues such as pericardium.[1,2] Autologous pericardium has been used for the reconstruction of many different cardiovascular areas including the vena cava, atrial septum, ventricular septum, and right ventricular outflow tract as well as for valvular reconstruction.[3,4] Pericardial thickening, fibrosis, contraction and aneurysmal development in high-pressure circulatory areas are some of the drawbacks of unconditioned pericardium.[3,5] Stability of the tissue is a function of the number of cross-linkages between collagen molecules. To stabilize the biological materials, a glutaraldehyde solution was chosen in our laboratory[1] because of its dual properties: first, as a tanning agent, it gave the tissue increased stability by forming irreversible cross-linkages between collagen molecules; and second, it was found that it reduced markedly the antigenicity of the graft tissue. The goal of this work was to evaluate the optimal concentration and the minimal time required for a patch of autologous pericardium to be stabilized during surgery using a glutaraldehyde solution.

MATERIALS AND METHODS

In 20 goats (mean weight 56.2 kg) an 8 by 8 cm piece of pericardium was removed, stripped of fat, and placed in a 0.62% magnesium-buffered glutaraldehyde solution at room temperature and atmospheric pressure. A fragment of 1 cm^2 was isolated every minute during the first 20 minutes and then 3, 12, 24, 48, and 72 hours and 7, 14, 21, and 28 days after the glutaraldehyde contact with the pericardium. Fragments were rinsed, immersed and preserved in physiological saline solution for 2 days at 4°C. Stability of the tissue was tested with each pericardial segment by heat-shrinkage. The higher the temperature initiating retraction of the tissue, the stronger the cross-linkage. The degree of tissue fixation was investigated by scanning and transmission electron microscopy.

RESULTS AND DISCUSSION

Improvements in the chemical management of the glutaraldehyde process and, particularly, the evaluation of the optimal density of cross-linkages with tanned specimens allow us to determine that after 10 minutes of treatment of pericardial tissue with 0.62% glutaraldehyde the long-term stability of the tissue is insured (FIG. 1). There have been no structural or primary tissue failures at this stage of evaluation. The end-product of the tanning activity was studied with the electron microscope, which permitted a detailed and highly magnified examination of both surfaces of the pericardial patch. Preservation of cellular and extracellular components of connective tissue was adequate, and cellular elements showed minimal degrees of autolytic change.

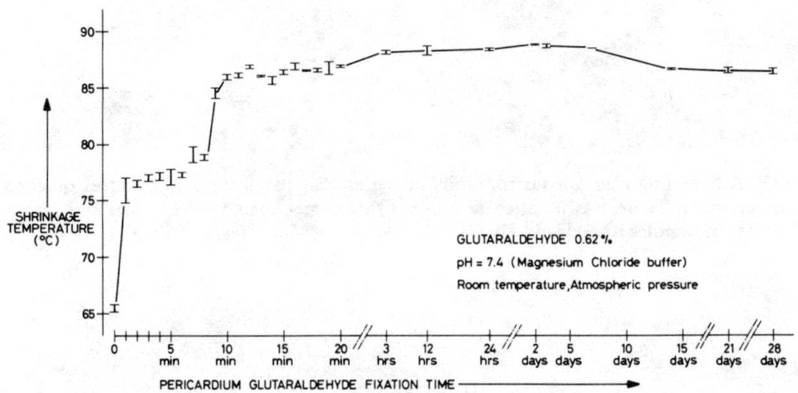

FIGURE 1. Effect of pericardial glutaraldehyde fixation time on shrinkage temperature tests.

The epipericardial surfaces of the patches were composed of bundles of collagen that coursed in various directions with a practically normal histologic structure. The mesothelial cells forming the serosa had a flattened shape and were connected to one another by intercellular junctions. The results showed that 10 minutes of treatment with a glutaraldehyde solution preserved the natural structure of the pericardial tissue. There was no alteration of cells or components of the tissue, and the crimp of the collagen bundles and elastic fibers was preserved (FIG. 2). Glutaraldehyde-tanned pericardial patches were investigated as a biological material to be used as a vascular substitute or for repair of cardiac defects. After treatment with glutaraldehyde solution, the quality of the tissue was improved as follows: there was adequate rigidity with elasticity; marked suppression of antigenicity; easy handling and preservation; high tolerance against internal pressure; and good disinfectant effect.[1,6]

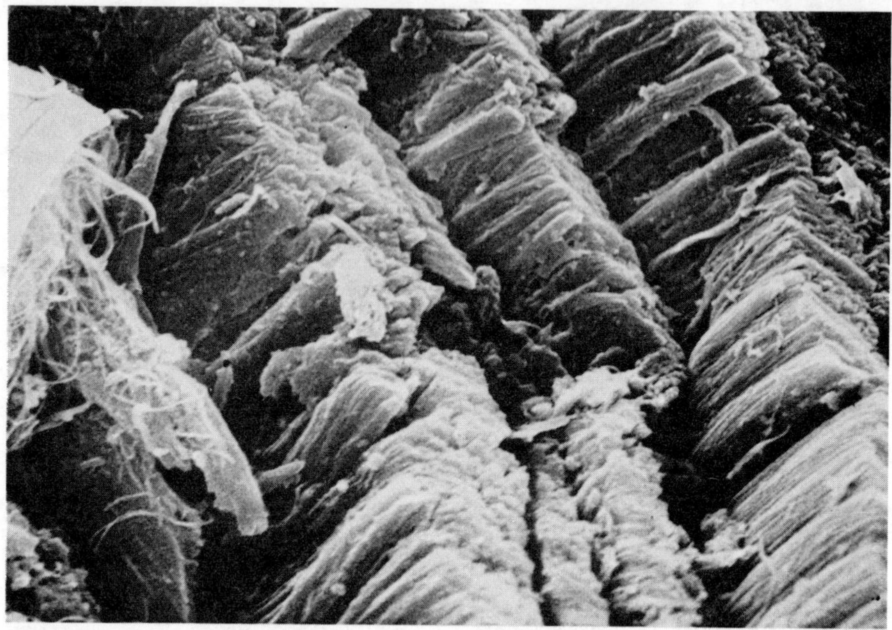

FIGURE 2. Scanning electron micrographs of pericardial patch with 10 minutes of glutaraldehyde treatment. There was no alteration of cells or components of the tissue, and the crimp of the collagen bundles and elastic fibers was preserved. (Magnification ×640.)

REFERENCES

1. CARPENTIER, A. *et al.* 1969. J. Thorac. Cardiovasc. Surg. **58:** 467.
2. ISHIHARA, T. *et al.* 1981. J. Thorac. Cardiovasc. Surg. **81:** 747.
3. HJELMS, E. *et al.* 1981. J. Thorac. Cardiovasc. Surg. **81:** 120.
4. CLARKE, C. P. *et al.* 1968. Thorax **23:** 111.
5. COBANOGLU, A. *et al.* 1984. J. Thorac. Cardiovasc. Surg. **87:** 371.
6. ONDA, Y. 1986. Artif. Organs **10:** 256.

Comparison of Myocardial Free Fatty Acids by Gas Chromatography and High-Performance Liquid Chromatography

G. CORDIS, R. PRASAD, H. OTANI, AND D. K. DAS

University of Connecticut School of Medicine
Farmington, Connecticut 06032

INTRODUCTION

The nanomole changes of myocardial free fatty acid concentrations during ischemia have been implicated in the changes in the biochemical and functional properties of the heart.[1] The careful analysis of free fatty acids (FFA) is essential to evaluate the mode of action of fatty acids. Therefore, in this paper we have modified and evaluated an established high-performance liquid chromatography (HPLC) technique[2] to obtain a reliable and a complete analysis of fatty acids, including polyunsaturates.

MATERIALS AND METHODS

Porcine myocardial biopsy specimens were obtained and homogenized, and lipids were extracted.[3] Ten nanomoles of heptadecanoic acid ($C_{17:0}$) were added to the homogenate as an internal standard. The lipid extracts were dried under nitrogen at 40°C. The lipid residues were dissolved in 2.0 ml of chloroform containing 0.01% BHT (w/v) and equally divided for subsequent analysis by HPLC and gas chromatography.

Ten nanomoles of pentadecanoic acid ($C_{15:0}$) were added to the samples for HPLC as a second internal standard. After the separation of neutral lipids from the phospholipids, FFA were derivatized to phenacyl esters and analyzed by HPLC.[3] An initial mobile phase of 80% acetonitrile was run for 50 min, followed by a 20-min linear gradient to 90% acetonitrile. Concentrations of FFA were calculated on the basis of the area of the internal standard $C_{17:0}$.

The FFAs were isolated by silica gel thin-layer chromatography (TLC) and converted to methyl esters as described by Culp *et al.*[4] and analyzed on 5% DEGS columns equipped with gas chromatograph HP-7620. The concentration of each fatty acid was calculated employing the area of the internal standard $C_{17:0}$.

TABLE 1. Retention Times (t_R) for Standard Phenacyl Esters and Methyl Esters of Free Fatty Acids[a]

Fatty Acid	t_R for Phenacyl Esters on HPLC (min)	t_R for Methyl Esters on GC (min)
$C_{12:0}$	11.62	—
$C_{14:0}$	22.45	—
$C_{15:0}$	31.70	—
$C_{16:0}$	45.03	4.00
$C_{17:0}$	61.86	5.50
$C_{18:0}$	73.08	7.20
$C_{18:1}$	48.66	8.50
$C_{18:2}$	28.29	11.00
$C_{18:3}$	18.12	15.00
$C_{20:4}$	24.12	26.00
$C_{20:5}$	15.75	35.00
$C_{22:6}$	19.95	74.00

[a] Standards for FFA were derivatized to phenacyl esters and run on HPLC, while standard methyl esters of FFA were run on GC as described in the MATERIALS AND METHODS section.

RESULTS AND DISCUSSION

The phenacyl esters of all the fatty acids, saturated, monounsaturated and polyunsaturated, which include $C_{18:2}$, $C_{18:3}$, $C_{20:4}$, $C_{20:5}$ and $C_{22:6}$ were totally separable in the HPLC method (TABLE 1). A stable baseline and homogenous baseline separation were obtained in the analysis of standard and myocardial samples. A comparison of the myocardial FFA between HPLC and the established gas chromatographic technique is given in TABLE 2. It can be seen that similar mole-percent of individual fatty acids and total free fatty acids was obtained from both of the methods. Thus, the

TABLE 2. Comparison of Porcine Myocardial Free Fatty Acids by High-Performance Liquid Chromatography and Gas Chromatography[a]

Free Fatty Acid	Biopsy 1 HPLC (nmol/g dry wt)	Biopsy 1 GC (nmol/g dry wt)	Biopsy 2 HPLC (nmol/g dry wt)	Biopsy 2 GC (nmol/g dry wt)
Arachadonic acid	85 (7.5)	60 (3.9)	110 (9.4)	34 (4.0)
Linoleic acid	102 (9.0)	195 (12.8)	103 (8.8)	79 (9.3)
Palmitic acid	366 (32.4)	476 (31.3)	437 (37.4)	294 (34.8)
Oleic acid	109 (9.6)	243 (16.0)	171 (14.6)	134 (15.8)
Stearic acid	468 (41.4)	546 (35.9)	349 (29.8)	305 (36.0)
Total free fatty acid	1130	1520	1170	846

[a] Myocardial lipids were extracted from porcine frozen biopsy specimens and equally divided for separation and derivatization of FFA. Phenacy esters of FFA were run on HPLC, and methyl esters of FFA were run on GC as described in MATERIAL AND METHODS. Values in parentheses indicate mole-percent.

HPLC method can be conveniently used to determine the amounts and acyl chain composition of biological samples, including myocardium. It should be pointed out here that this is the first report wherein the polyunsaturates like $C_{20:5}$ and $C_{22:6}$ have been shown to be completely separable from other fatty acids in an HPLC method.

REFERENCES

1. VAN DE VUSSE, G. J., TH. H. M. ROEMAN, F. W. PRINZEN, W. A. COUMANS & R. S. RENEMAN. 1982. Circ. Res. **50:** 538-546.
2. WOOD, R. & T. LEE. 1983. J. Chromatogr. **254:** 237-246.
3. DAS, D. K., R. M. ENGELMAN, D. FLANSAAS, H. OTANI, J. ROUSOU & R. H. BREYER. 1986. Basic Res. Cardiol. In press.
4. CULP, B. R., W. E. M. LANDS, B. R. LUCCHESI, B. PITT & J. ROMSON. 1980. Prostaglandin **20:** 1021-1031.

Metabolism and Turnover of Arachidonic Acid in Ischemic Myocardium

RANDALL JONES, HAJIME OTANI, GERALD
CORDIS, RENUKA PRASAD, SUBHAJIT DATTA,
AND DIPAK K. DAS

*University of Connecticut School of Medicine
Farmington, Connecticut 06032*

INTRODUCTION

Recent studies demonstrated accumulation of arachidonic acid secondary to the loss of membrane phospholipids in an ischemic myocardium.[1] Since the bulk of the arachidonate in mammalian heart is esterified in the fatty acyl chains of glycerophospholipids[2] and the free arachidonic acid level in heart should represent a balance between the liberation of the acid by hydrolysis and its re-esterification into complex lipids by acyltransferases,[3] we undertook a study to examine the fate of free arachidonic acid in the ischemic heart.

MATERIALS AND METHODS

Male Sprague-Dawley rats of about 300-gm body weight were anesthetized with intraperitoneal pentobarbital (6.5 mg/100 gm), and isolated and perfused heart was prepared according to the Langendorff technique.[4] The heart was perfused with buffer and, after the initial stabilization period, it was rendered ischemic for 30 min by stopping the perfusion. After the addition of [1-^{14}C] arachidonic acid (0.2 μM) to the perfusion circuit, perfusion was resumed for an additional period of 30 min. In control experiments, the same amount of radiolabeled arachidonic acid was added to the perfusion circuit and the heart was perfused for a further period of 30 min. After termination of the perfusion, the heart was immediately removed and extracted for lipids as described previously.[5] Neutral lipids and phospholipids were separated using thin-layer chromatography.[5] Individual spots for lipids, after visualization under iodine vapor, were compared with known standards, scraped off, and assayed for radioactivity by counting in a liquid scintillation counter.[5] Tissue levels of free arachidonic acid were estimated using high-pressure liquid chromatography. Individual phospholipids were assayed by phosphorus estimation.[5] A portion of the cardiac biopsy specimen was dried to a constant weight to determine the dry weight of the tissue. Statistical analysis was performed using Students' t test, and results are expressed as means $\pm$ SEM.

RESULTS AND DISCUSSION

Arachidonic acid was readily incorporated into membrane lipids during the 30 min of perfusion. As shown in TABLE 1, the major incorporation was associated with phospholipids (75.3%), whereas a smaller proportion was recovered into neutral lipids (24.7%). Amongst neutral lipids, isotopic incorporation was maximum into the triglyceride pool (84.8%). About 10.7% of the radioactivity remained in the total free fatty acid pool as free arachidonic acid.

TABLE 1. Effect of Ischemia on the Isotopic Arachidonic Acid Incorporation into Membrane Lipids of Isolated Rat Heart[a]

Lipids	Normal (cpm/g dry wt)	Ischemia (cpm/g dry wt)	$\frac{\text{Ischemia}}{\text{Normal}} \times 100$	
Total neutral lipids	10544 $\pm$ 1558	6195 $\pm$ 468	58.7	$p < .05$
Diglycerides	469 $\pm$ 107	370 $\pm$ 71	78.8	NS[b]
Triglycerides	8944 $\pm$ 1677	3928 $\pm$ 649	43.9	$p < .01$
Free arachidonic acid	1131 $\pm$ 265	1897 $\pm$ 412	167.7	$p < .05$
Total phospholipids	32041 $\pm$ 2889	17253 $\pm$ 3384	53.8	$p < .05$
Phosphatidylcholine	24321 $\pm$ 3182	11893 $\pm$ 2089	48.9	$p < .01$
Phosphatidylethanolamine	3032 $\pm$ 463	1635 $\pm$ 284	53.9	$p < .02$
Phosphatidylserine	803 $\pm$ 150	394 $\pm$ 81	49.1	$p < .05$
Phosphatidylinositol	2913 $\pm$ 479	2529 $\pm$ 376	86.8	NS
Lysophosphatidylcholine	54 $\pm$ 28	26 $\pm$ 8	47.5	NS
Sphingomyelin	263 $\pm$ 53	235 $\pm$ 41	89.5	NS
Cardiolipin	482 $\pm$ 116	397 $\pm$ 76	82.4	NS

[a] Value are mean $\pm$ SEM.
[b] Not significant.

Turnover of arachidonic acid was significantly reduced in the ischemic heart (TABLE 1). In the ischemic tissue, the incorporation of carbon-14 label into the neutral lipids was only 58.7% compared to that in normal control. Isotopic incorporation into tissue triglyceride and diglyceride was only 43.9% and 78.8%, respectively, compared to that in control hearts. Free arachidonic acid, on the other hand, increased to 167.7% in the ischemic heart. Ischemia also depressed the arachidonate incorporation into the membrane phospholipids. In the total phospholipid pool, the isotopic incorporation was only 53.8% of the control. Similar results were found for phosphatidylcholine (PC), phosphatidylethanolamine (PE) and phosphatidylserine (PS), whereas arachidonate incorporation was only slightly affected for phosphatidylinositol (PI), sphingomyelin and cardiolipin.

We also examined the effects of ischemia on the concentrations of membrane phospholipids (TABLE 2). No significant loss of membrane phospholipids was noticed

TABLE 2. Effect of Ischemia on the Membrane Phospholipid Concentrations of Isolated Rat Heart

Phospholipids	Normal[a]	Ischemia[a]	Loss of Membrane Phospholipids in Ischemic Tissue (% of control)	
	(μmol/g dry wt)			
Total phospholipids	231 ± 71	213 ± 61	92.2	NS[b]
Phosphatidylcholine	55 ± 4.5	54 ± 10	98.2	NS
Phosphatidylethanolamine	48 ± 5.8	46 ± 7.0	95.8	NS
Phosphatidylserine	11 ± 2.8	13 ± 3.1	118.2	NS
Phosphatidylinositol	32 ± 6.9	10 ± 1.3	31.3	$p <$ .01
Lysophosphatidylcholine	2.6 ± .1	7.6 ± 2.3	292.3	$p <$ .05
Sphingomyelin	14 ± 1.8	14.3 ± 17	102.1	NS
Cardiolipin	23 ± 4.8	36 ± 6.6	156.5	NS

[a] Values are mean ± SEM.
[b] Not significant.

except for PI, which suffered a loss of about 78.7%. Lysophosphatidylcholine (LPC) content was increased by 192.3% in the ischemic tissue.

Our results thus support the previous reports that arachidonic acid is liberated from the membrane phospholipids after ischemic insult,[1] and further indicate that the turnover of this acid into membrane lipids is significantly inhibited during ischemia, suggesting inhibited reacylation of lysolipids by acyltransferases.

REFERENCES

1. DAS, D. K., R. M. ENGELMAN, J. A. ROUSOU, R. H. BREYER, H. OTANI & S. LEMESHOW. 1986. Am. J. Physiol. **251:** H71–H79.
2. VAN DEN BOSCH, H. 1980. Biochim. Biophys. Acta **604:** 191–246.
3. LANDS, W. E. M. & B. SAMUELSSON. 1968. Biochim. Biophys. Acta **169:** 426–429.
4. WHITMER, J. T., J. A. IDELL-WENGER, M. J. ROVELTO & J. R. NEELY. 1978. J. Biol. Chem. **253:** 4305–4309.
5. DAS, D. K., R. M. ENGELMAN, D. FLANSAAS, H. OTANI, J. ROUSOU & R. H. BREYER. Basic Res. Cardiol. In press.

Isotachophoretic Analysis of Organic Acids in Human Cerebrospinal Fluid and Serum

PETER J. OEFNER,[a] PETER POHL,[b] AND GÜNTHER BONN[c]

Departments of Urology[a], Neurology[b] and Radiochemistry[c]
University of Innsbruck
Innsbruck, Austria

INTRODUCTION

The brain uses glucose as its only exogenous substrate and is totally dependent upon aerobic metabolism for its continued existence. Thus, cerebral metabolic abnormalities alter rapidly the levels of organic acids in the cerebral tissue and the cerebrospinal fluid.[1-3] Moreover, organic acids, which are fully ionized at physiological pH values, will diffuse only slowly across the blood-brain barrier as long as it is intact.[4-7] Hence, the prompt determination of organic acids in cerebrospinal fluid and in blood is of great value in assessing the cerebral acid-base status as well as the integrity of the blood-brain barrier.

Isotachophoresis is a powerful analytical method which is compatible with the basic requirements for screening procedures, namely, multicomponent information, rapid completion, reliability and low operational cost.[8-12] The present study evaluates its applicability for the analysis of organic acids in cerebrospinal fluid and serum.

MATERIALS AND METHODS

Sample Collection

Cerebrospinal fluid specimens were obtained by lumbar puncture. Simultaneously blood samples were taken from the cubital vein and deproteinized immediately by ultrafiltration, using the disposable micropartition system Centresart I (cut-off: 20,000 daltons; Sartorius GmbH, Göttingen, FRG), which had been filled with small glass beads to avoid coagulation. From 2ml of blood centrifuged with a swing-head rotor at $2000 \times g$ for 15 minutes at 4°C we obtained 200-300 μl of ultrafiltrate. The recovery rates for organic acids ranged from 98-100%. Cerebrospinal fluid samples were analyzed without any pretreatment.

Principle of Isotachophoresis

In isotachophoresis ions are separated according to their different net mobilities in a Teflon capillary tube kept at a constant temperature.[8–10,13] Samples are introduced at the interface between a leading electrolyte, the effective mobility of which is higher than that of any of the separands, and a terminating electrolyte, which has an effective mobility lower than that of any of the separands. The parameter which primarily influences the effective mobility is the degree of dissociation. Separation is achieved by means of voltage increment at a constant current. After an initial, moving-boundary separation process, an isotachophoretic steady-state configuration is obtained.[14] All separands are then arranged in contiguous zones, generally in the order of their effective mobilities. They migrate at an equal, constant speed. Since the concentration of the separand is constant within a zone, measurement of the zone-length provides quantitative information.

Isotachophoretic Conditions

The isotachophoretic analyses of cerebrospinal fluid and serum were carried out on an LKB 2127 Tachophor, equipped with a 250-mm Teflon capillary of an internal diameter of 0.5 mm. The leading electrolyte was 10 mM of hydrochloric acid adjusted to a pH value of 3.30 by the addition of β-alanine. Triton-X-100 was added to the leading electrolyte to reduce electroendosmosis. The terminating electrolyte was 10 mM of propionic acid. Separation were started at a current of 225 μA, which was gradually reduced to 75 μA shortly before the separated anions could be detected on account of their conductivity.

RESULTS AND DISCUSSION

The variation of pH, detection current, and leading and terminating electrolytes resulted in an operational system of high efficiency for the determination of anions commonly found in biological materials (FIG. 1).

FIGURE 2 shows the isotachophoretic analyses of cerebrospinal fluid and ultrafiltered serum. Oxalic acid, pyruvic acid, adenosine 5-triphosphate, phosphate, 2-ketoglutaric acid, citric acid, acetoacetic acid, lactic acid, 2-hydroxybutyric acid, glutamic acid, 3-hydroxybutyric acid and acetic acid could be detected. Their identity was confirmed by the injection of an additional small amount of each compound, which resulted in an increase in the length of the respective zones. Furthermore, the conductivity signals (step-heights) were characteristic of the detected compounds. The time of analysis ranged from 20-25 minutes.

Quantitative information from isotachophoretic analyses was obtained by measuring the zone lengths. The calibration curves for the detected anions revealed a detection limit of 1 nM and linearity over their ranges of biological concentration. The reproducibility of the measurements was checked by seven repeat determinations for which eight different samples of cerebrospinal fluid and serum were used. A

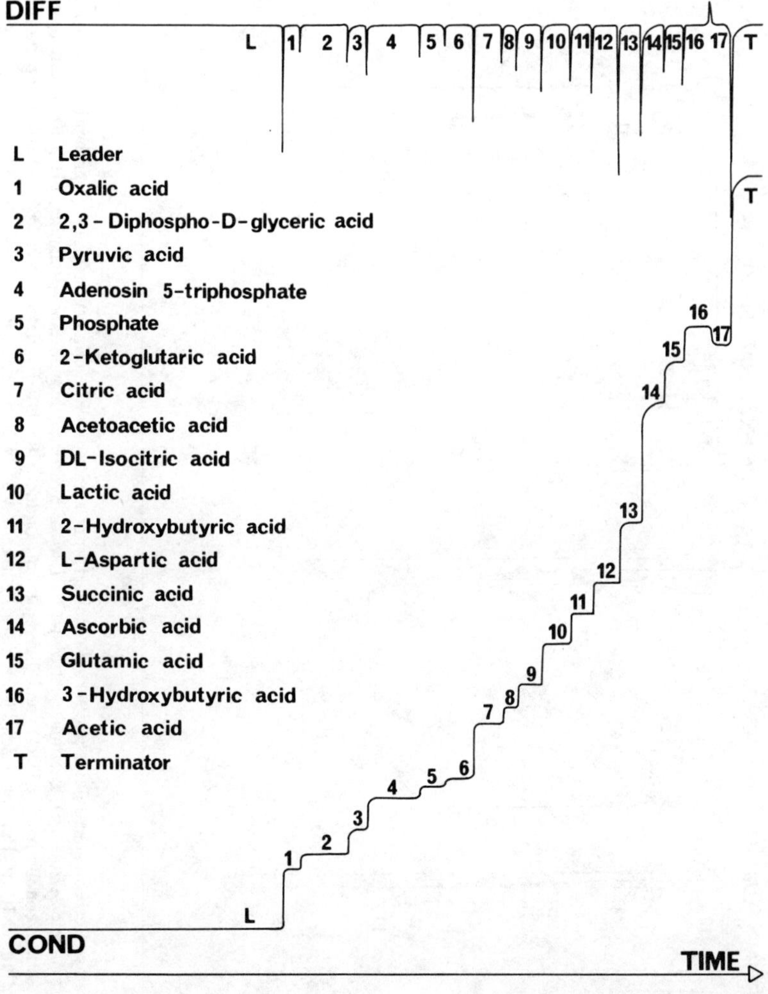

FIGURE 1. Isotachopherogram of a mixture of anions commonly found in biological materials. Leading electrolyte (L): 0.01 M of hydrochloric acid, and β-alanine, pH 3.30. Terminating electrolyte (T): 0.01 M of propionic acid. Detection: conductivity (COND) and differential of conductivity (DIFF); detection current: 75 μA; temperature: 20°C; chart speed: 0.2 mm/sec; sample volume: 2.5 μl (10 nM per zone).

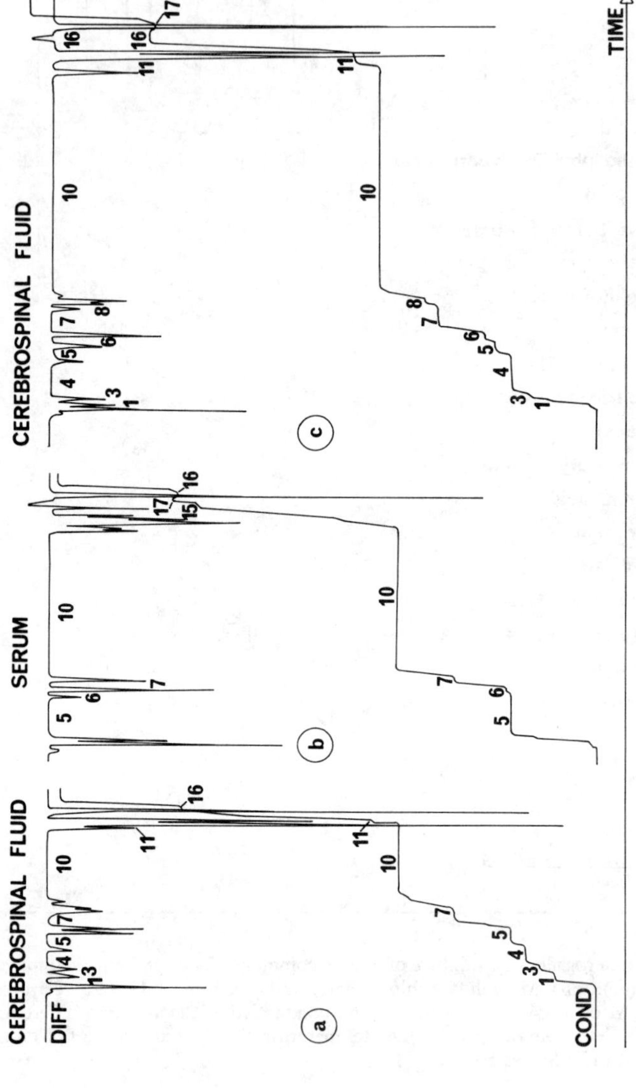

FIGURE 2. Isotachopherograms of cerebrospinal fluids and serum obtained from a patient with an intervertebral disc prolapse (**a** and **b**) and a patient who had just experienced an epileptic seizure (**c**). Leading electrolyte: 0.01 M of hydrochloric acid, and β-alanine, pH 3.30. Terminating electrolyte: 0.01 M of propionic acid. Detection: conductivity (COND) and differential of conductivity (DIFF); detection current: 75 μA; temperature: 20°C; chart speed: 0.5 mm/sec; sample volumes: 10 μl. For zone identification see FIGURE 1.

precision of $\pm 1.5\%$ was obtained, the main errors resulting from the injection of the samples and the measurement of the zone lengths. Moreover, a comparison of the enzymatic (lactate dehydrogenase) assay of lactate and the isotachophoretic results showed an excellent correlation ($r = 0.997$). Thus, it was not necessary to apply internal standards.

The analysis of clinical specimens, which had been obtained from 71 patients divided into several diagnostic groups, revealed significant increases in the levels of organic acids in cerebrospinal fluid mainly after epileptic seizures (FIG. 2c), brain injury, ischemic cerebral infarction and in the early course of meningitis. These findings are in accordance with previous studies.[15-23] However, the clinical implications of elevated levels of organic acids in cerebrospinal fluid and serum remain to be evaluated and are subject of current studies.

CONCLUSION

By allowing the accurate and reproducible analysis of a vast range of organic acids within minutes, isotachophoresis may serve as a valuable method in the diagnosis, prognosis and monitoring of pathologic cerebral conditions.

REFERENCES

1. SIESJÖ, B. 1978. Brain energy metabolism. John Wiley. Chichester.
2. SIESJÖ, B. 1982. Lactic acidosis in the brain: Occurrence, triggering mechanisms, and pathophysiological importance. *In* Metabolic Acidosis. R. Porter & G. Lawrenson, Eds.: 77-100. Ciba Foundation Symposium **87**. Pitman Books. London.
3. FISHMAN, R. A. 1980. Cerebrospinal fluid in diseases of the nervous system. W. B. Saunders. Philadelphia.
4. KLEIN, G. R. & N. S. OLSEN. 1947. Distribution of intravenously injected glutamate, lactate, pyruvate, and succinate between blood and brain. J. Biol. Chem. **167**: 1-5.
5. ALEXANDER, W. E., R. D. WORKMAN & C. J. LAMBERTSEN. 1962. Hyperthermia, lactic acid infusion, and the composition of arterial blood and cerebrospinal fluid. Am. J. Physiol. **202**: 1049-1054.
6. POSNER, J. B. & F. PLUM. 1967. Independence of blood and cerebrospinal fluid lactate. Arch. Neurol. **16**: 492-496.
7. PARDRIDGE, W. M. & W. H. OLDENORF. 1977. Transport of metabolic substrates through the blood-brain barrier. J. Neurochem. **28**: 5-12.
8. HAGLUND, H. 1970. Isotachophoresis. A principle for analytical and preparative separation of substances such as proteins, peptides, nucleotides, weak acids, metals. Sci. Tools **17**: 2-12.
9. ARLINGER, L. 1974. Analytical isotachophoresis. Resolution, detection limits and separation capacity in capillary columns. J. Chromatogr. **91**: 785-794.
10. EVERAERTS, F. M., J. L. BECKERS & TH. P. E. M. VERHEGGEN. 1976. Isotachophoresis. Theory, Instrumentation and Applications. Journal of Chromatography Library, Vol. 6. Elsevier. Amsterdam.
11. OEFNER, P. J., G. BONN & G. BARTSCH. 1985. Isotachophoretic analysis of organic acids in human seminal plasma. Fresenius Z. Anal. Chem. **320**: 175-178.
12. OEFNER, P.J., K. SCHEIBER, G. BONN, M. GOTWALD & G. BARTSCH. Isotachophoretic analysis of organic acids in human seminal plasma and prostatic fluid. Prostate. In press.

13. MIKKERS, F. & F. M. EVERAERTS. 1981. Theoretical and practical aspects of isotachophoresis. *In* Analytical Isotachophoresis. F. M. Everaerts, F. E. P. Mikkers & Th. P. E. M. Verheggen, Eds.: 1-17. Elsevier. Amsterdam.
14. VACIK, J. & V. FIDLER. 1981. Dynamics of the isotachophoretic separation. *In* Analytical Isotachophoresis. F. M. Everaerts, F. E. P. Mikkers & Th. P. E. M. Verheggen, Eds.: 19-24. Vol. 6. Elsevier. Amsterdam.
15. ZUPPING, R. 1970. Cerebral acid-base and gas metabolism in brain injury. J. Neurosurg. **33:** 498-505.
16. HAERER, A. F. 1971. Citrate and alpha-ketoglutarate in cerebrospinal fluid and blood. Neurology **21:** 1059-1065.
17. BROOKS, B. R. & R. D. ADAMS. 1975. Cerebrospinal fluid acid-base and lactate changes after seizures in unanesthetized man. Neurology **25:** 935-942.
18. SEITZ, H. D. 1976. Liquorveränderungen beim schweren Schädel-Hirn-Trauma und ihre therapeutische Bedeutung. Fortschr. Med. **94:** 2088-2094.
19. SIMPSON, H., A. H. HABEL & E. L. GEORGE. 1977. Cerebrospinal fluid acid-base status and lactate and pyruvate concentrations after convulsions of varied duration and aetiology in children. Arch. Dis. Child. **52:** 844-849.
20. BROOK, I., K. S. BRICKNELL, G. D. OVERTURF & S. M. FINEGOLD. 1978. Measurement of lactic acid in cerebrospinal fluid of patients with infections of the central nervous system. J. Infect. Dis. **137:** 384-390.
21. KNIGHT, J. A., S. M. DUDEK & R. E. HAYMOND. 1981. Early (chemical) diagnosis of bacterial meningitis-cerebrospinal fluid glucose, lactate, and lactate dehydrogenase compared. Clin. Chem. **27:** 1431-1434.
22. BUSSE, O. 1982. Zur Prognose des ischämischen Hirninfarkts. Fortschr. Med. **100:** 1197-1200.
23. JORDAN, G. W., B. STATLAND & C. HALSTED. 1983. CSF lactate in diseases of the CNS. Arch. Intern Med. **143:** 85-87.

Phorbol-Ester-Induced Alteration of α_1-Adrenergic Response in Rat Left Ventricular Papillary Muscles

HITOMI OTANI, HAJIME OTANI, AND
DIPAK K. DAS

University of Connecticut School of Medicine
Farmington, Connecticut 06032

INTRODUCTION

Tumor-promoting phorbol esters are known to alter α_1-adrenoceptor-mediated physiologic responses by stimulating the activity of protein kinase C in vascular smooth muscles and liver.[1-3] Phorbol diester receptors have also been identified and characterized in isolated rat cardiac myocytes.[4] Our recent study has suggested that α_1-adrenoceptor-mediated positive inotropic effect (PIE) may be caused by phosphoinositide (PI) breakdown, presumably through the action of inositol trisphosphate and protein kinase C activation.[5] In the present study, we investigated the effect of the phorbol ester, 12-O-tetradecanoylphorbol-13-acetate (TPA) on α_1-adrenoceptor-mediated PIE and PI breakdown in rat left ventricular papillary muscle.

MATERIALS AND METHODS

Isolated left ventricular papillary muscles from male Sprague-Dawley rats were treated with TPA or its vehicle, dimethylsulfoxide, for various intervals in 5 ml Tyrode's solution. The pulse-chase experiments for labeling of PI were performed by incubating these preparations with 20 μCi [^{3}H]inositol for 90 min followed by diluting free [^{3}H]inositol with unlabeled inositol. After exposure to phenylephrine in the presence of propranolol for 10 min, the labeling of PI and inositol phosphates (IP), as well as contractile force, was determined as described previously.[5]

RESULTS AND DISCUSSION

TPA produced a time- and dose-dependent potentiation followed by attenuation of PIE mediated by α_1-adrenoceptor stimulation in isolated rat left ventricular papillary muscle (FIG. 1). The TPA-induced potentiation of PIE became apparent within 1 hr at a concentration of 1 μM, whereas it required 2-3 hr at concentrations of 100 nM and 10 nM. However, TPA attenuated α_1-adrenoceptor-mediated PIE by 30-50% of the vehicle-treated control after 3-4 hr treatment. Thus, it was found that TPA-induced attenuation of PIE was preceded by its potentiation. We next examined whether the TPA alteration of PIE mediated by α_1-adrenoceptor stimulation was associated with changes in the PI response in these preparations (TABLE 1). In the present study, [^{3}H]IP formation was expressed as a percentage of the labeling of PI in order to normalize time-dependent decrease in net [^{3}H]IP formation resulting from reduced PI labeling. TPA had no effect on α_1-adrenoceptor-mediated IP formation when the preparations showed increases in PIE. This finding indicates that TPA-induced potentiation of PIE is independent of PI breakdown and that TPA has an additive effect on α_1-adrenoceptor-mediated PIE during relatively brief incubation periods. However, IP formation mediated by α_1-adrenoceptor stimulation was significantly attenuated compared to the vehicle-treated control after 3-4 hr TPA treatment. Therefore, TPA inhibition of contractile response and IP formation mediated by cardiac α_1-adrenoceptor stimulation is consistent with the hypothesis that the phorbol-ester-induced desensitization of α_1-adrenergic response is causally related to inhibition of PI breakdown.[6] Alternatively, it is possible that TPA and α_1-adrenoceptor stimulation acted through a common pathway in provoking PIE, thereby blunting the α_1-adrenoceptor-mediated contractile response. Further studies will be required to elucidate an unequivocal role of TPA in modulating cardiac contractile function.

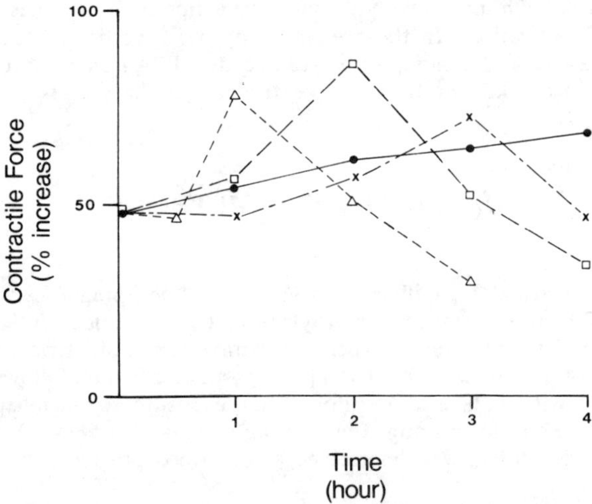

FIGURE 1. Effect of 12-O-tetradecanoylphorbol-13-acetate (TPA) on positive inotropic effect mediated by α_1-adrenoceptor stimulation. Isolated rat papillary muscles preincubated with TPA or its vehicle, dimethylsulfoxide, for various intervals were treated with 10 μM phenylephrine in the presence of 0.3 μM propranolol. The maximum contractile responses were compared to the basal contractile force measured immediately before the exposure to phenylephrine. $\bigcirc$, 0.1% dimethylsulfoxide; x, 10 nM TPA; $\square$, 100 nM TPA; $\triangle$, 1 μM TPA. Each symbol represents a mean value of three to six experiments.

TABLE 1. Effect of 12-*O*-Tetradecanoylphorbol-13-Acetate (TPA) on Inositol Phosphate (IP) Formation Mediated by α_1-Adrenoceptor Stimulation

Treatment	1	2	3	4	Time (hours)
	[³H]IP formation (% of ³H-labeled phosphoinositides)				
0.1% dimethylsulfoxide	7.41 ± 0.58	7.02 ± 0.33	6.96 ± 0.76	6.52 ± 0.49	
10 nM TPA			6.99 ± 1.32	5.53 ± 0.60	
100 nM TPA		7.96 ± 0.43		4.15 ± 0.41[a]	
1 μM TPA	7.13 ± 0.81	5.60 ± 0.26[a]	4.72 ± 0.28[a]		

NOTE: Isolated rat left ventricular papillary muscles were preincubated with TPA or its vehicle, dimethylsulfoxide, and [³H]inositol as described in the MATERIALS AND METHODS section. These preparations were exposed to 10 μM phenylephrine in the presence of 0.3 μM propranolol for 10 min. [³H]IP formation was expressed as percent of the labeling of phosphoinositides. The results are shown as mean ± SEM of four to six experiments.

[a] $p < 0.05$ compared to 0.1% dimethylsulfoxide.

REFERENCES

1. DANTHULURI, N. R. & R. C. DETH. 1984. Biochem. Biophys. Res. Commun. **125:** 1103–1109.
2. COLUCCI, W. S., M. A. GIMBRONE, JR. & R. W. ALEXANDER. 1986. Circ. Res. **58:** 393–398.
3. LYNCH, C. J., R. CHAREST, S. B. BOCCKINO, J. H. EXTON & P. F. BLACKMORE. 1985. J. Biol. Chem. **260:** 2844–2851.
4. LIMAS, C. J. 1985. Arch. Biochem. Biophys. **228:** 300–304.
5. OTANI, H., H. OTANI & D. K. DAS. 1986. Biochem. Biophys. Res. Commun. **136:** 863–869.
6. COLUCCI, W. S. & R. W. ALEXANDER. 1986. Proc. Natl. Acad. Sci. USA **83:** 1743–1746.

The Effects of Sympathomimetic Stimulation on Pulmonary Vascular Responsiveness

R. J. PORCELLI, P. R. VILLANO, AND
W. A. MAHONEY[a]

Veterans Administration Medical Center
Northport, New York 11768

One of the many unique characteristics of the pulmonary circulation is that the response to one pressor agent, such as epinephrine or acute hypoxia, subsequently reduces or reverses the pulmonary pressor responses to the catecholamines, histamine, acute hypoxia and hypercapnic acidosis.[1,2] One explanation for these observations is that the stimuli that interact either compete for, or share, a common contractile pathway. The use of that pathway by one substance limits the response to the others. The use of epinephrine to study this phenomenon has an important drawback, in that epinephrine has been shown to constrict or dilate the pulmonary circulation, depending on a myriad of preconditions. Not the least of such conditions is the ability of this amine to stimulate both α- and β-adrenergic activity. Acute hypoxia, in addition to altering vasoactivity, may change smooth muscle cell energetics such that interpretation of these data may also be very complex. The present study, therefore, was designed to examine this phenomenon of prior responsiveness on pulmonary vasoactivity by using sympathomimetic stimulation with norepinephrine and phenylephrine, agents that are reported to be without such variable vasoactivity. Furthermore, we studied a number of different vasoactive agents to define clearly the extent to which such stimulation alters subsequent pulmonary vasoactivity.

The experiments described used the isolated blood-perfused lung *in situ*. Once three to five control responses to a given agent were recorded, three infusions of 20 μg of either norepinephrine or phenylephrine were given at 5-minute intervals. After these infusions and after vascular resistances had returned to control levels, the first agent was readministered at 15-minute intervals for $1\frac{1}{2}$ hours and these values were compared to the controls. Only one agent was used in each animal: 20 μg of histamine, serotonin (5HT), epinephrine, isoproterenol or angiotensin II (AII), as well as alveolar hypoxia (8% O_2) and hypercapnic acidosis (10% CO_2; pH $=$ 7.10). Since each animal served as its own control, the control responses were compared to the post-sympathomimetic responses by a paired Student's t test and the level of significance was set at $p < 0.05$.

The present study demonstrates that, unlike the results with epinephrine and acute hypoxia in this model, norepinephrine and phenylephrine were somewhat specific for reducing the pulmonary pressor responses to the catecholamines, but not to acute hypoxia (FIGS. 1 and 2). Furthermore, this antagonism could not be attributed to

---

[a] Deceased.

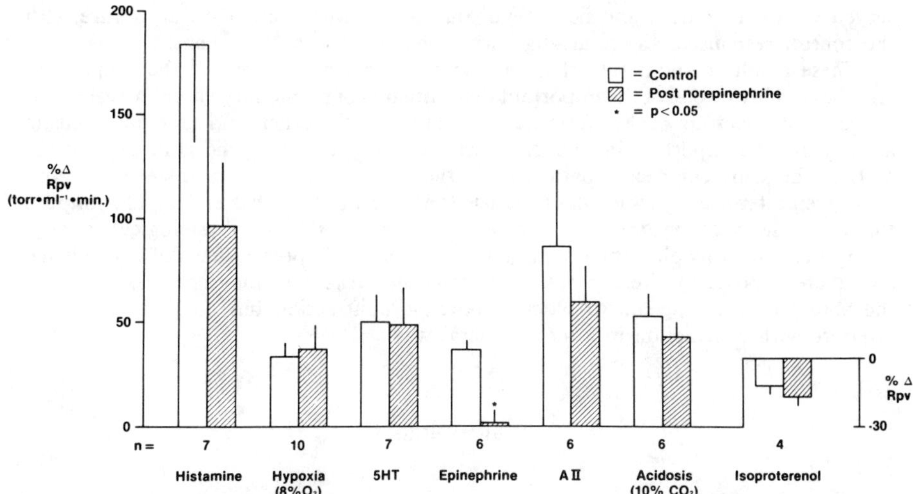

FIGURE 1. The effects of norepinephrine pretreatment on pulmonary vascular responsiveness. % Δ Rpv = percent change in pulmonary vascular resistance. AII = angiotensin II.

the pressor action of either norepinephrine or phenylephrine since: (a) the pressor responses to each agent were the same throughout each study, and (b) the baseline vascular resistances after the infusions of each agent (0.48 ± 0.05 and 0.72 ± 0.05 torr/ml per min, respectively) did not significantly differ from the same values during control conditions (0.46 ± 0.04 and 0.62 ± 0.05 torr/ml per min, respectively). The present observations with norepinephrine and phenylephrine are similar to those ob-

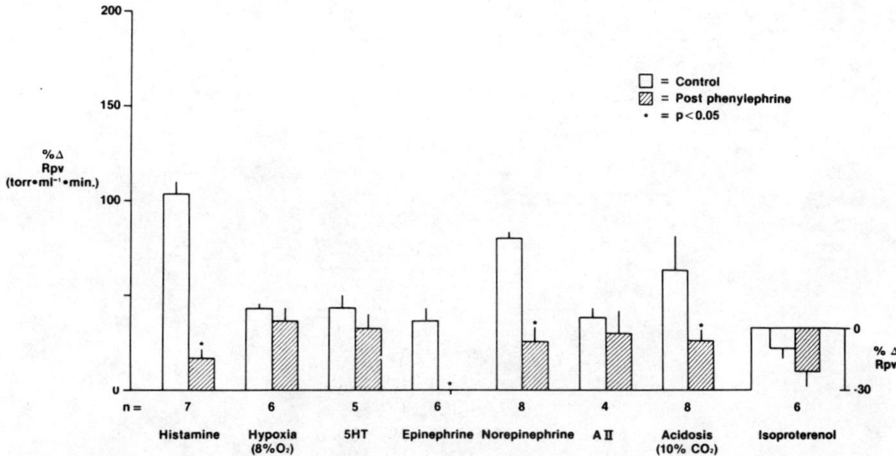

FIGURE 2. The effects of phenylephrine pretreatment on pulmonary vascular responsiveness. Abbreviations are the same as in FIGURE 1.

served with epinephrine and acute hypoxia, in that they were acute in nature with the control responsiveness returning within one hour of the treatment.

These results demonstrate that the sympathomimetic history of the pulmonary circulation may be one of the important determinants of pulmonary vascular reactivity. Indeed, one agent may act as a partial antagonist for the other, and such antagonistic activity may be important in such cases where two agents share the same receptor(s) and/or the same contractile pathway. Furthermore, such antagonism seems to continue even after the physiological response (vasoconstriction) has ended. This implies that the presence of the amine at the receptor site, rather than its physiological activity, is important for this phenomenon. Lastly, the more widespread effects of epinephrine and acute hypoxia in altering pulmonary vascular reactivity may not rely solely on the ability of these agents to block adrenergic contraction but on their ability to interfere with contraction in a more general way.

REFERENCES

1. PORCELLI, R. J. & E. H. BERGOFSKY. 1973. J. Appl. Physiol. **34:** 483-488.
2. PORCELLI, R. J., M. J. BERGMAN & M. V. CUTAIA. 1984. Ann. N.Y. Acad. Sci. **435:** 478-480.

Inhibitory Effects of Various Structural Analogues of Methionine on the Cellular Proliferation of Lewis Lung Carcinoma and B16-F10 Melanoma

ANTHONY DeFELICE,[a] MARTIN ROSENBERG,[b]
JONATHAN R. MATIAS,[a] AND
NORMAN ORENTREICH[a,b]

[a]Orentreich Foundation for the
Advancement of Science, Inc.
Biomedical Research Station
Cold Spring-on-Hudson, New York 10516

[b]NYU School of Medicine
Department of Dermatology
New York, New York 10016

Methionine, an essential amino acid, functions as a methyl donor for numerous biosynthetic processes which require transmethylation reactions. It has recently been shown that the rate of transmethylation proceeds faster in many types of cancer cells in comparison to normal cells.[1] Other investigators studying the methionine adenosyltransferase (MAT) reactions in extracts of tumor and normal tissues have suggested the use of inhibitors of the MAT reaction as a means of inhibiting cancer growth.[2]

We explored this possibility by supplementing the normal growth media of Lewis lung carcinoma (LLC) and B16-F10 melanoma cells with various structural analogues of methionine. The data in FIGURE 1 (see inset) show that the addition of L-ethionine at a concentration of 500 μg/ml inhibited the proliferation of LLC cells grown in a media which contained 38.5 μg/ml of methionine. This inhibition occurred at an ethionine:methionine ratio of 13:1. Viability of the ethionine-treated cells was high (85-90%). The absence of vacuolization, the lack of cell detachment from the substrate, and the maintenance of the typical cellular morphologic pattern suggest that ethionine may act by inhibiting cellular proliferation rather than through a direct toxic effect. A dose-response study on L-ethionine showed that the antiproliferative effects were evident even at a concentration of 250 μg/ml for LLC cells. In contrast, a similar degree of growth inhibition of B16-F10 melanoma cells occurred at a much higher concentration (FIG. 1).

Methionine is unique because it is the only amino acid that the body can use in either the L- or D-form. Thus, it is surprising to find that the L-isomer was 100% more potent than D-ethionine. Using preparations of the MAT enzymes, it has been previously demonstrated that +2-amino-bicyclo-(2.1.1)-hexane-2-carboxylic acid (compound A), L-2-amino-4-hexynoic acid (compound B), and cycloleucine were significantly more effective as inhibitors than ethionine.[2] However, when tested on intact cells, we observed that ethionine produced greater antiproliferative action than these analogues (FIG. 2).

Numerous papers have already dealt with the toxic and carcinogenic effects of ethionine when administered to laboratory animals at high doses. Recently, we have shown that supplementation of the diet with low doses of ethionine (0.2%-0.3%) can reduce the rate of cancer growth in tumor-bearing rats and mice.[3] The present *in vitro* data, in concert with our recent *in vivo* observations, suggest that inhibitors of the MAT reaction may be considered as another means of modulating the proliferative capacity of cancer cells.

ACKNOWLEDGMENTS

We would like to express our thanks to Dr. J. Sufrin (Department of Physiology and Biophysics, Washington University School of Medicine, St. Louis, MO) for providing samples of the methionine analogues, and to Ms. Carole Taub for her technical assistance.

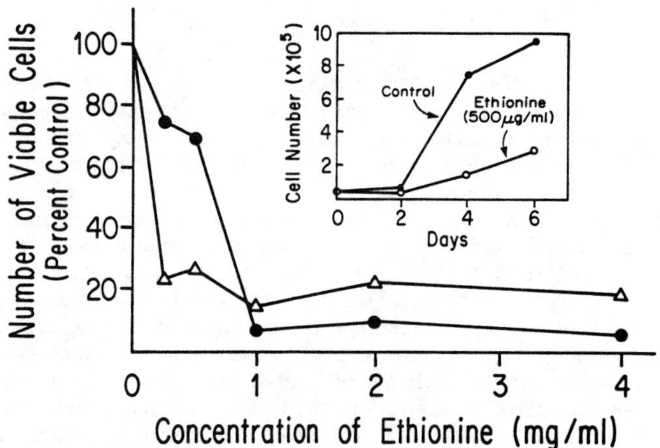

FIGURE 1. The inhibitory effects of L-ethionine on the proliferation of Lewis lung carcinoma (LLC) and B16-F10 melanoma. Cells were seeded at a density of 10,000 cells per well using 96-well microtiter plates. LLC cells were grown in RPMI media, while B16-F10 cells were grown in Minimal Essential media. Both were supplemented with 1% penicillin/streptomycin and 10% fetal calf serum. *Inset* shows the growth pattern of cells grown in ethionine-supplemented and normal control media. Main diagram shows the effect on LLC and B16-F10 cells grown for four days at various concentrations of ethionine (△, LLC cells; ●, B16-F10 cells).

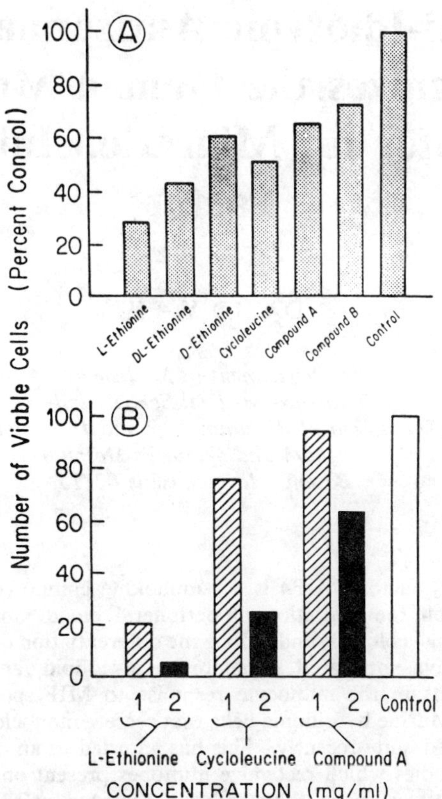

FIGURE 2. Comparison of the inhibitory effects of various analogues of methionine on the proliferation of (**A**) LLC and (**B**) B16-F10 cells. See FIGURE 1 for details.

REFERENCES

1. HOFFMAN, R. M. 1985. Anticancer Res. **5:** 1–30.
2. LOMBARDINI, J. B. & J. R. SUFRIN. 1983. Biochem. Pharmacol. **32:** 489–495.
3. MATIAS, J. R., A. CHIN, V. MALLOY & N. ORENTREICH. 1985. Anticancer Res. **5:** 593.

Anti-Idiotypic Antiserum that Recognizes the Human Monocyte Receptor for Migration Inhibitory Factor[a]

DAVID Y. LIU[b]

Department of Medicine
Harvard Medical School, and
Department of Rheumatology and Immunology
Brigham and Women's Hospital
Boston, Massachusetts 02115

Migration inhibitory factor (MIF) is a lymphokine composed of several molecular species which inhibit the migration of peripheral blood monocytes, stimulate the growth of macrophage colonies, and induce the differentiation of promyelocytic cells.[1,2] Previous studies have implicated a membrane glycolipid[3] and a glycoprotein[4,5] as essential components in the monocyte response to MIF, possibly in a membrane receptor complex. Murine hybridoma cells that secrete monoclonal antibodies to MIF have been established quite recently.[6] This has afforded us an opportunity to produce anti-idiotypic antibodies which recognize idiotopes present on anti-MIF monoclonal antibodies and its mirror image, the membrane receptor for MIF. Application of this procedure will, it is hoped, result in the successful production of a biological and molecular probe with which to investigate the interaction between MIF and its cell membrane receptor.

In order to generate an anti-idiotypic response, we injected syngeneic BALB/c mice with a mixture of two hybridoma cells which secrete idiotypic anti-human T cell hybridoma MIF monoclonal antibodies (Id).[6] After the initial intravenous injection of 1×10^6 cells, the mice were boosted three times every other week for six weeks. Several days after the third injection the mice were bled from the tail vein and the serum collected for subsequent analysis of anti-idiotypic antibody production by using the microdroplet agarose assay to assess the human blood monocyte response to MIF.

It was important to first determine the ability of the potential anti-idiotypic antiserum to bind to anti-MIF idiotopes by examining the effect of the binding of anti-idiotypic antiserum on the ability of idiotypic anti-MIF antibodies to neutralize MIF activity in the migration bioassay. As demonstrated in FIGURE 1, anti-MIF binds and completely neutralizes MIF activity. However, when anti-Id antiserum #5 was prein-

[a]This work was supported by Grants PCM-8208262 and DCB-8510469 from the National Science Foundation and Grants AI22532 and CA40550 from the National Institutes of Health.

[b]Present address: Cetus Corporation, Department of Immunology, 3400 West Bayshore Road, Palo Alto, CA 94303.

cubated with anti-MIF antibody and then added to the monocytes in the presence of MIF, complete restoration of MIF bioactivity occurred. That is, upon binding of anti-MIF (Id) by anti-Id, the ability of anti-MIF to bind and neutralize MIF was disrupted such that full MIF activity could be expressed. In confirmation of this observation, dot blot analysis revealed that antiserum #5 was able to bind ^{125}I-labeled anti-MIF

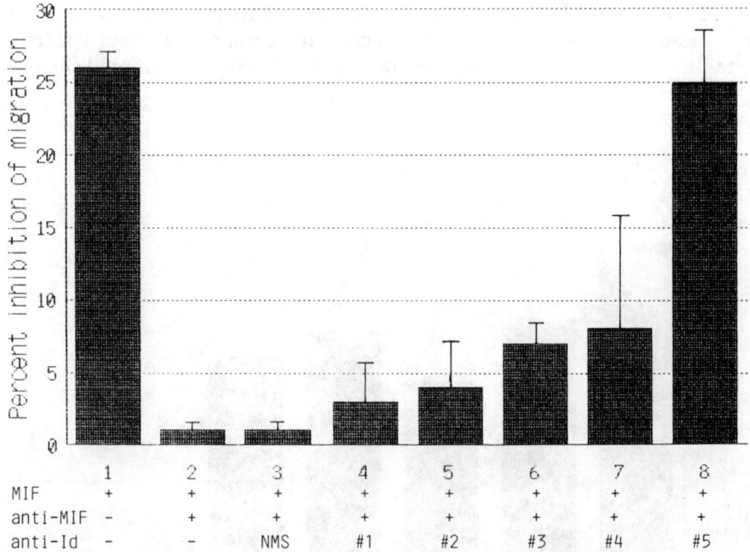

FIGURE 1. Anti-Id inhibits neutralization of MIF by anti-MIF. Supernatants enriched for MIF activity were prepared from a human T-cell hybridoma line that was stimulated for three days with 10 μg/ml concanavalin A.[6] The agarose microdroplet assay was used to assess MIF response of peripheral blood monocytes.[4] Murine anti-MIF hybridoma cells were propagated and supernatants harvested as previously described.[6] These supernatants were then subjected to a 45% ammonium salt precipitation to isolate the immunoglobulin fraction. To test for the ability of anti-Id sera to affect the neutralization of MIF activity by anti-MIF monoclonal antibody, we preincubated anti-Id sera (0.5 μl) or normal BALB/c mouse serum (NMS) with anti-MIF preparation (5 μg) for 1 hr at 37°C. This mixture was then added to an agarose droplet containing 10^6 mononuclear cells bathed in RPMI tissue culture medium (2 mM HEPES, 50 U/ml penicillin and 50 μg/ml streptomycin) containing 10-20 λ of 100-fold-concentrated MIF preparation. The monocyte response to MIF, denoted by percent inhibition of migration (*ordinate*), was assessed after a 16-hr incubation. Each *bar* and *bracket* represent the mean and standard error of the mean of three experiments. Using the Student's *t* test, it was found that $p < .008$ for all samples when compared to control (bar 1) except for anti-Id sera #4 (bar 7; $p = .08$) and #5 (bar 8; $p = .71$). Note that only anti-Id serum #5 was able to fully restore MIF activity in the presence of anti-MIF antibody.

antibodies (Liu, unpublished results). Interestingly, neither normal mouse serum nor any of the other anti-Id sera (#1-#4) had any significant effect on neutralization. These sera had no direct effect on normal cell migration in the absence of MIF.

It has been demonstrated that some anti-idiotypic antibodies may be quite useful in their ability to recognize the membrane receptor for their respective biological ligand.[7-13] In light of this possibility, we examined the potential of each anti-Id antiserum to directly block the monocyte response to MIF. As shown in FIGURE 2, when each anti-Id antiserum was preincubated with monocytes, only sera #4 and #5 were able to completely block the cellular response to MIF. This effect was noticeable with as little as 0.3 μl antiserum for every million cells. Normal mouse serum and anti-Id sera #1-#3 did not markedly affect the MIF response. Antibody serum #4 was able to directly block the cellular response although in FIGURE 1 it was shown that serum #4 had a marginal effect on the neutralization of MIF bioactivity by anti-MIF. These data would suggest that anti-idiotypic serum #5 recognizes the monocyte receptor for MIF. In addition, antiserum #5 was active in binding to protein lysates from a monocyte-like cell line as determined by an ELISA analysis (Liu, unpublished results).

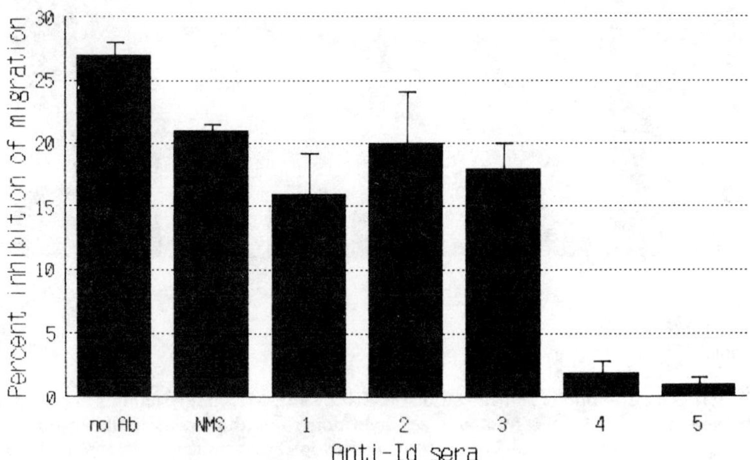

FIGURE 2. Anti-Id blocks monocyte response to MIF. Anti-idiotypic antisera (1-2 λ) from the five different mice (denoted #1-#5 on abscissa) were preincubated with monocytes for 30 min at 37°C before testing the monocyte response to MIF, denoted by percent inhibition of migration (*ordinate*). Each *bar* and *bracket* represent the mean and standard error of the mean of three experiments. When compared to the medium (M) control (MIF in the absence of any antibody) the *p* values were less than 0.05 for all samples except for serum #2. Note that only anti-Id sera #4 and #5 were able to completely block the monocyte response to MIF.

Taken in its entirety, the results of this study would suggest that anti-idiotypic antibodies can be elicited by immunization with anti-MIF hybridoma cells. These antibodies, found in the serum of mouse #5 in particular, were able to bind idiotypic anti-MIF antibodies, block neutralization of MIF activity by anti-MIF, directly inhibit the monocyte response to MIF, and bind to lysates from U-937 cells previously stimulated with phorbol diester. This potential anti-receptor activity will have important applications in the study of lymphokine interactions with the cell surface. To facilitate this, we will attempt to establish murine hybridoma cells which secrete monoclonal anti-idiotypic antibodies.

ACKNOWLEDGMENTS

I would like to thank Dr. Weishui Y. Weiser for the kind gift of mouse hybridoma cells that secrete anti-MIF antibodies and human T hybridoma cells that secrete MIF, Ms. Joy Y. Ree for excellent technical assistance, and Ms. Barbara Wolk and Ms. Joan Murphy for assisting in the preparation of this manuscript.

REFERENCES

1. WEISER, W. Y., D. K. GREINEDER, H. G. REMOLD & J. R. DAVID. 1981. Studies on human migration inhibitory factor: Characterization of three molecular species. J. Immunol. **126:** 1958-1962.
2. KAWAGUCHI, T., D. W. GOLDE, A. MEDNIS, N. BERSCH & H. G. REMOLD. 1986. T cell-derived migration inhibitory factor and colony stimulating factor share common structural elements. Blood **67:** 1619-1623.
3. LIU, D. Y., S. YU, H. G. REMOLD & J. R. DAVID. 1985. Macrophage glycolipid receptors for human migration inhibitory factor (MIF): Differentiated HL-60 cells exhibit MIF responsiveness and express surface glycolipids which both bind MIF and convert non-responsive cells to responsiveness. Cell Immunol. **90:** 605-613.
4. LIU, D. Y. & R. F. TODD, III. 1986. A monoclonal antibody specific for a monocyte-macrophage membrane component blocks the human monocyte response to human migration inhibitory factor. J. Immunol. **137:** 448-455.
5. TODD, R. F. III, P. A. ALVAREZ, D. A. BROTT & D. Y. LIU. 1985. Bacterial lipopolysaccharide, phorbol myristate acetate, and myramly dipeptide stimulate the expression of a human monocyte surface antigen, Mo3e. J. Immunol. **135:** 3869-3877.
6. WEISER, W. Y., H. G. REMOLD & J. R. DAVID. 1985. Generation of human hybridomas producing migration inhibitory factor (MIF) and of murine hybridomas secreting monoclonal antibodies to human MIF. Cell. Immunol. **90:** 167-178.
7. SEGE, K. & P. A. PETERSON. 1978. Use of anti-idiotypic antibodies as cell-surface receptor probes. Proc. Natl. Acad. Sci. USA **75:** 2443-2447.
8. MARASCO, W. A. & E. L. BECKER. 1982. Anti-idiotype as antibody against the formyl peptide chemotaxis receptor of the neutrophil. J. Immunol. **128:** 963-968.
9. LAMBRIS, J. D. & G. D. ROSS. 1982. Characterization of the lymphocyte membrane receptor for factor H (β1H-globulin) with an antibody to anti-factor H idiotype. J. Exp. Med. **155:** 1400-1411.
10. ISLAM, M. N., B. M. PEPPER, R. BRIONES-URBINA & N. R. FARID. 1983. Biological activity of anti-thyrotropin anti-idiotypic antibody. Eur. J. Immunol. **13:** 57-63.
11. VENTER, J. C., J. A. BERZOFSKY, J. LINDSTROM, S. JACOBS, C. M. FRASER, L. D. KOHN, W. J. SCHNEIDER, G. L. GREENE, A. D. STROSSBERG & B. F. ERLANGER. 1984. Monoclonal and anti-idiotypic antibodies as probes for receptor structure and function. Fed. Proc. **43:** 2532-2539.
12. OSHEROFF, P. L., T. CHIANG & D. MANOUSOS. 1985. Interferon-like activity in an anti-interferon anti-idiotypic hybridoma antibody. J. Immunol. **135:** 306-313.
13. CO, M. S., G. N. GAULTON, A. TOMINAGA, C. J. HOMCY, B. N. FIELDS & M. I. GREENE. 1985. Structural similarities between the mammalian β-adrenergic and reovirus type 3 receptors. Proc. Natl. Acad. Sci. USA **82:** 5315-5318.

Assay of Low Levels of Serum Alpha-Fetoprotein

S. MANIMEKALAI, G. S. SUNDARAM,
D. STRUMMER, D. BERMAN, J. L. WOLFE,
AND P. J. GOLDSTEIN

*Sinai Hospital of Baltimore
Baltimore, Maryland 21215*

Alpha-fetoprotein (AFP), a specific fetal serum globulin consisting of a single polypeptide chain with a molecular weight of 68,000, is produced by the fetal liver, yolk sac, and gastrointestinal tract. AFP is found in amniotic fluid; with increasing gestational age, it reaches a peak level at 12-14 weeks, declining thereafter.[1] As more AFP is produced, more AFP diffuses into maternal circulation, resulting in a continuous rise in maternal serum.

Serum AFP assay is a screening test to aid in detection of open neural type defects and other birth defects.[2] Elevated levels of AFP may be indicative of fetal maldevelopment, open neural tube defect, amphalocele, congenital nephrosis, esophageal or duodenal atresia, impending spontaneous abortion, multiple fetuses, fetal death, and

TABLE 1. Alpha-Fetoprotein Assay: Comparison of Assay Parameters

Sample No.	HYB			AM			DPC		
	MEAN (IU/ml)	SD	%CV	MEAN (IU/ml)	SD	%CV	MEAN (IU/ml)	SD	%CV
A: INTRA-ASSAY VARIATION									
1	13.0	0.5	4	8.0	0.5	6	17.6	0.6	3
2	60.3	2.2	4	51.9	1.2	2	74.3	3.6	5
3	134.4	4.0	3	123.0	5.4	4	157.5	5.7	4
B: INTER-ASSAY VARIATION									
1	12.5	1.5	12	7.1	0.8	11	17.2	1.3	8
2	58.7	6.1	10	50.2	3.4	7	72.7	5.1	7
3	126.3	7.1	6	120.4	5.2	4	155.1	11.2	7
C: ESTIMATION IN LOWER RANGE (CONCENTRATIONS, IU/ML)									
	16.5	8.8		10.3	6.1		20.0	8.8	

NOTE: HYB, Hybritech; AM, Amersham; DPC, Diagnostic Products Corporation; SD, standard deviation; %CV, percent coefficient of variation.

TABLE 2. Alpha-Fetoprotein Assay: Comparison of Assay Parameters

	Derived Value (IU/ml)		
Expected Value (IU/ml)	HYB Mean + SD	AM Mean + SD	DPC Mean + SD
A: Recovery			
75	84.1 + 8.2	75.4 + 6.4	96.6 + 9.2
50	47.3 + 2.8	48.1 + 4.2	63.3 + 4.6
30	29.3 + 3.9	28.2 + 3.1	38.0 + 3.3
Percent Variation			
75	112	101	129
50	95	96	126
30	98	94	127
B: Paralellism			
Dilution		(IU/ml)	
Undiluated	200	200	200
1:2	96.3 + 5.1	98.1 + 6.4	104.5 + 7.3
1:4	47.3 + 2.8	48.1 + 4.2	63.3 + 4.6
Percent Variation			
1:2	96	98	105
1:4	95	96	126

C: Comparison

	HYB vs. AM		HYB vs. DPC		AM vs. DPC	
Mean	50.2	44.2	50.3	60.0	44.2	60.1
SD	36.7	33.1	36.9	41.7	33.5	41.8
r		0.971		0.977		0.989
M		0.876		1.100		1.240
Y		0.219		4.600		5.540
n		90		89		88

NOTE: R, correlation coefficient; M, slope; Y, intercept; *n*, number of samples; SD, standard deviation; HYB, Hybritech; AM, Amersham; DPC, Diagnostic Products Corporation.

miscalculated gestational period. Elevated levels have been found in many patients suffering from hepatoma and teratoma of the testis and ovary, cirrhosis of the liver, viral hepatitis, tyrosinosis, and ataxia telangiectasia. Low levels have been reported in cases of fetal growth retardation and Down's syndrome. These pathologic conditions need a sensitive technique to differentiate the clinical situation at all levels. Radioimmunoassay (RIA) kits for AFP are generally designed to perform in the 20 to 400 ng/ml range of AFP in the sera. However, accurate and precise assay of low levels of serum AFP is of clinical importance in monitoring pregnant patients at risk for Down's syndrome and patients with some type of neoplasm. In the present study, we have compared RIA kits from Amersham (AM) and Diagnostic Products corporation (DPC) with an enzyme immunoassay (EIA) kit from Hybritech Corporation (HYB) for their performance using pooled patients' sera ranging from 0-200 IU/ml. The RIA procedure is based on competitive binding principles, while the Tandem-E AFP EIA assay is based on solid-phase sandwich technique using monoclonal antibody raised against two sites of the antigen (AFP) molecule. The percent intra- and interassay coefficient of variations (% CV) determined in our laboratory and shown in TABLE 1 are within the limits claimed by the manufacturers. When pooled sera with AFP concentrations of 0-20 IU/ml were analyzed, AM protocol consistently estimated lower concentrations than did HYB and DPC protocols, suggesting that HYB and DPC kit formulations are more sensitive for differentiating lower concentrations encountered in Down's syndrome (TABLE 1C).

A known amount of standard was added to the zero standard of AM kit and the percent recovery of AFP by each assay protocol was determined; TABLE 2A shows that DPC protocol has a greater percentage of recovery compared to AM and HYB protocols. Pooled patients' sera with a known high AFP was diluted two and four times with zero standard and analyzed for linearity of assay (TABLE 2B). DPC showed a higher percent value compared to HYB and AM in 1:4 dilution. All the three kits correlated very well, although varying in their absolute concentration when samples with different concentration were analyzed (TABLE 2C). Considering the EIA method's greater need for attention to details and use of dedicated instruments than the RIA technique, the fact that radioactive isotopes are not used and precise values are obtained even in the low ranges, we rank HYB protocol at the top for overall performance followed by AM and DPC protocols.

REFERENCES

1. MASSEYEFF, R., J. GILLI, B. KREBS, ET AL. 1975. Ann. N. Y. Acad. Sci. **259:** 17-28.
2. ———— 1977. Report of U.K. Collaborative study on alpha-fetoprotein in relation to neural tube defects. Lancet **1:** 1323-1332.

Cytoplasmic Filaments in Cultured Cells Used for Autoantibody Detection

AUDREY S. OCHAL,[a,b] LYNN B. KEIL,[b] AND
VINCENT A. DeBARI[b,c]

[a]Department of Biology
Montclair State College
Upper Montclair, New Jersey 07043

[b]The Renal Laboratory
Department of Medicine
Seton Hall University
Graduate School of Medicine
St. Joseph's Hospital and Medical Center
Paterson, New Jersey 07503

The cytoskeleton is a major component of all vertebrate cells and is important for maintenance of cellular integrity, shape and motility. Three types of fibers have been found to compose the cytoskeletal elements (CSE), the microtubules, microfilaments, and intermediate filaments. The microtubules are 20-26 nm in diameter and contain tubulin as their major structural protein. Microfilaments are 5-6 nm in diameter and are composed primarily of actin. Intermediate filaments are 7-11 nm in diameter and, regardless of the source from which they are isolated, share very similar ultrastructure and morphologic patterns. There are five major classes of intermediate filaments; cytokeratin, desmin, vimentin, neurofilaments, and glial fibrillary acidic protein (GFAP).[1] Antibodies to CSE have been detected in a number of autoimmune, neoplastic and infectious diseases.[2] The fluorescent antinuclear antibody test (FANA), using cultured cells, has proved to be a valuable technique in the evaluation of autoimmune diseases. HEp-2 and KB, human epithelioid cell lines, are commonly used as substrates for this test.[3] The observation of cytoplasmic filament staining is not uncommon during the course of laboratory examination of FANA from clinical specimens. The purpose of the work described herein is to determine, using CSE-specific antibodies, which CSE antigens are present in commercially available HEp-2 and KB cells.

[c]To whom all correspondence should be addressed.

MATERIALS AND METHODS

Materials

Cell lines were purchased in kit form: KB cells from Electro-Nucleonics Laboratories, Inc. (Bethesda, MD) and HEp-2 cells from Kallestad Laboratories (Chaska, MN). Antisera, including fluroescein conjugates, were obtained from Miles Scientific (Napierville, IL), Clark Laboratories, Inc. (Jamestown, NY) and ICN Immuno-Biologicals (Lisle, IL).

Methods

Cytoskeletal elements were demonstrated by indirect immunofluorescence, and KB and HEp-2 cells were incubated (30 min at 25°C) with untreated or heat-denatured (95° for 15 min) antisera followed by three washes (5 min each) with PBS. The cells were then incubated with the appropriate FITC-conjugated antibody for 30 min at 25°C. The washing procedure was then repeated. Slides were observed with the optical system previously described.[4]

RESULTS AND DISCUSSION

The data in TABLE 1 summarize the observed specificities for the antibody probes to CSE antigens in KB and HEp-2 cells. With the exception of α-actinin, detectable in KB but not in HEp-2, the qualitative CSE profiles presented by these lines are quite similar. From a quantitative standpoint, the KB cells react at a slightly higher titer than HEp-2, a phenomenon that we have also observed with antinuclear antibodies.[3] This may be more the result of fixation techniques used in the production of the pre-coated slides than an inherent difference in the amount of CSE present. A discrepancy exists with regard to the presence of desmin. A monoclonal anti-desmin does not react with either cell line, although a polyclonal antiserum reacts with both. Desmin occurs primarily in muscle[1] and, thus, would not be expected in cell lines of epithelial origin. The structural homology of the intermediate filaments has been found to result in some degree of cross-reactivity.[1] The fact that the monoclonal antibody to desmin does not react argues strongly in favor of the hypothesis that desmin is not present and that the polyclonal anti-desmin reactivity is artifactual. Considerable differences are observed in the titers of the anti-vimentin antibodies used in this study. These differences are not, in fact, limited to the titers, as demonstrated in TABLE 1, but extend to other observations made in the laboratory (unpublished data). It is clear from these observations that the epitopic specificities of the two anti-vimentin antibodies are different; that they both react reinforces the finding of this CSE in epithelioid cells maintained in culture.[5] There has also been a previous report of the presence of vimentin in HEp-2,[5] although it has not heretofore been demonstrated in KB.

The immunohistologic distributions of some of the CSE proteins are illustrated by the fields shown in FIGURE 1. In general, the extensive cytoplasmic spreading seen with HEp-2 allows better visualization of the intracellular distribution of CSE as compared to KB. FIGURES 1i, j, k and l, all representing control maneuvers, clearly illustrate the minimal contribution of the fluoresceinated conjugate to the observed staining patterns. FIGURES 1a and 1b, representing an anti-microfilament antibody and FIGURES 1c and 1d, representing an anti-microtubular antibody, demonstrate a characteristic cotton-like, diaphanous appearance of these CSE. Occasional thick filaments can be seen with anti-tubulin, but the density of the staining pattern does not allow this observation to be made in many cases. FIGURES 1e, f, g, and h, representing

TABLE 1. CSE Antigen Specificities Observed in Cultured Cell Lines

		Reciprocal Titer	
Probe	Source/Clone	KB	HEp-2
Polyclonal antibodies			
Anti-actin	Rabbit	640	160
Anti-myosin	Rabbit	1280	640
Anti-tubulin	Rabbit	2560	1280
Anti-α-actinin	Rabbit	800	Negative
Anti-desmin	Rabbit	1280	160
Anti-keratin	Guinea pig	5120	2560
Monoclonal antibodies			
Anti-vinculin	VIN-11-5	Negative	Negative
Anti-desmin	DE-B-5	Negative	Negative
Anti-cytokeratin 18	R-2	> 50,000	40,960
Anti-vimentin	#9	80	40
Anti-vimentin	#V-9	5,000	5,000
Anti-GFAP	N.A.	Negative	Negative
Anti-neurofilament HMW	NE 14	Negative	Negative
Anti-neurofilament MMW	NN 18	Negative	Negative

intermediate filament antibodies, demonstrate the more distinct, discrete fibers observed with these antibodies. These two intermediate filament antibodies can be distinguished from each other by the characteristic perinuclear aggregation of vimentin fibers.[6] Nuclear staining of KB cells was observed with antibodies to actin (FIG. 1a), myosin, tubulin (FIG. 1c), α-actinin, polyclonal anti-desmin, and polyclonal anti-keratin. The nuclei of HEp-2 cells were stained by antibodies to actin (FIG. 1b), myosin and tubulin (FIG. 1d).

In conclusion, we have begun to elucidate the CSE antigens present in cultured cells applied to autoimmune disease testing. Further study may lead to an increased ability to ascertain the specificity of autoantibodies found in patients with this group of diseases.

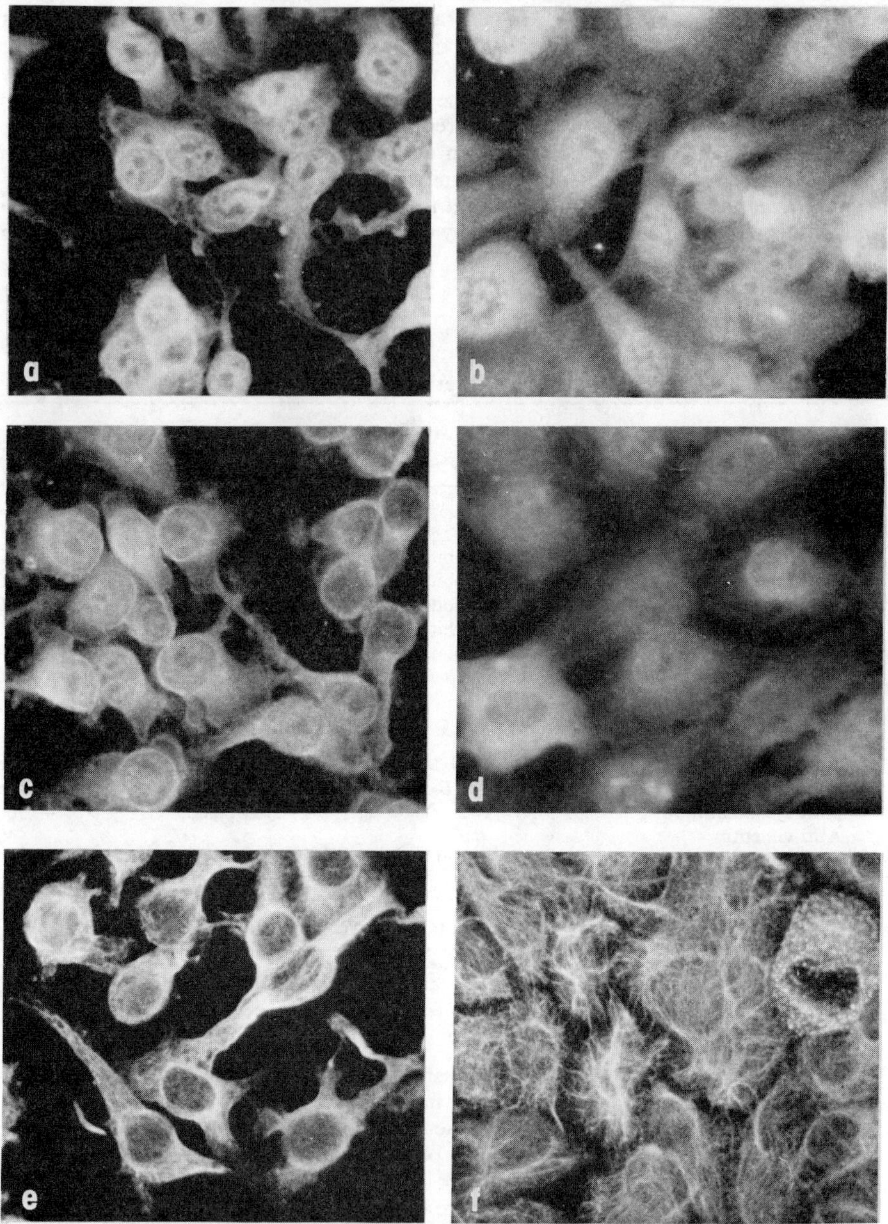

FIGURE 1a-f. Immunohistology of CSE in KB and HEp-2 cell. **a,** anti-actin on KB; **b,** anti-actin on HEp-2; **c,** anti-tubulin on KB; **d,** anti-tubulin of HEp-2; **e,** anti-cytokeratin 18 on KB; **f,** anti-cytokeratin 18 on HEp-2; Original magnification ×400; reduced by 30%.

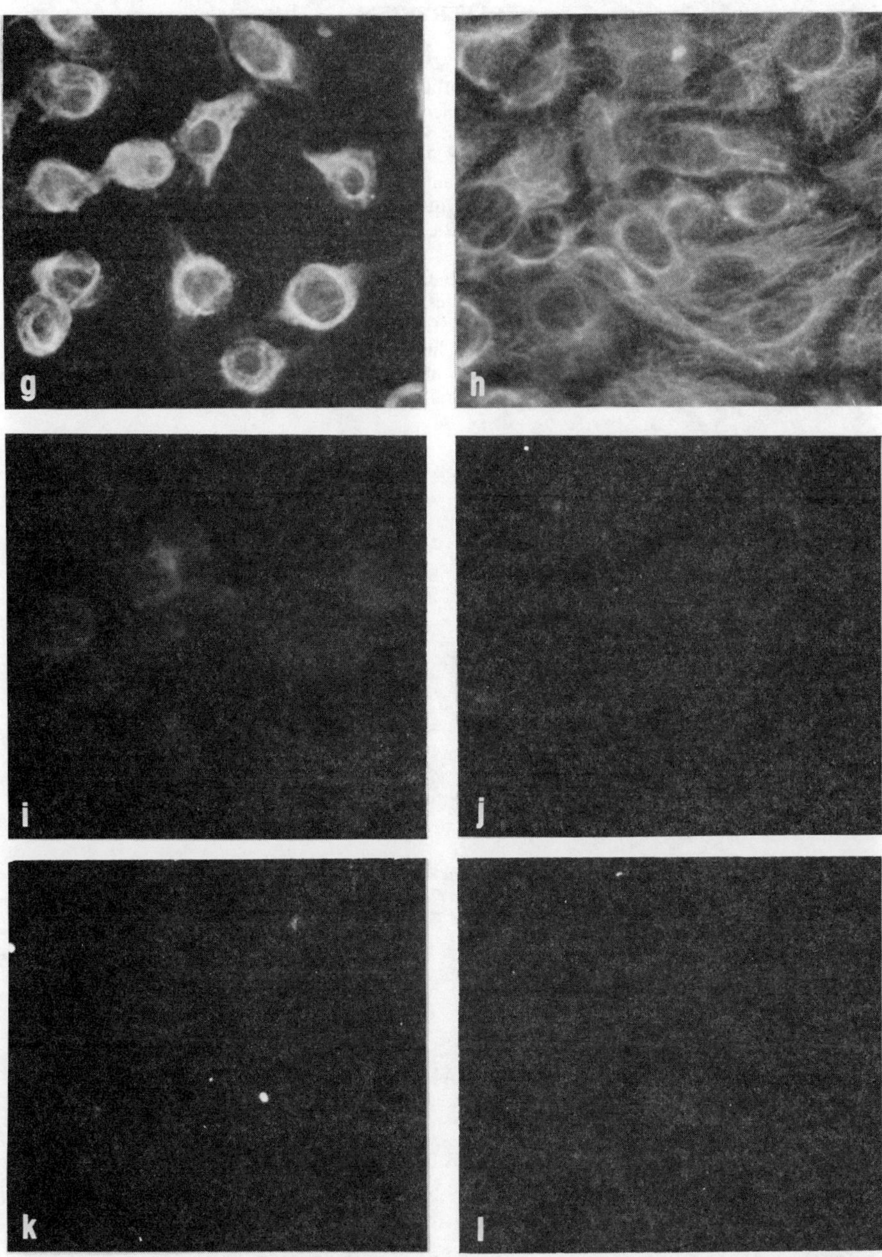

FIGURE 1g-l. g, anti-vimentin (clone #9) on KB; **h,** anti-vimentin on HEp-2; **i,** PBS control for FITC conjugated antiserum on KB; **j,** PBS control on HEp-2; **k,** denatured antibody control for KB (illustrated with anti keratin) and **l,** denatured antibody control for HEp-2 (illustrated with anti-actin). Original magnification ×400; reduced by 30%.

REFERENCES

1. FRANKE, W. W., E. SCHMID, D. L. SCHILLER, S. WINHER, E. D. JARASCH & R. MOLL. 1982. Differentiation-related patterns of expression of proteins of intermediate-size filaments in tissues and cultured cells. Cold Spring Harbor Symp. Quant. Biol. **46:** 431-453.
2. KURKI, P., H. DENK, B. W. JACKSON & K. ILLMENSEE. 1984. The detection of human antibodies against cytoskeletal components. J. Immunol. Meth. **67:** 209-223.
3. KEIL, L. B., V. A. DEBARI & M. A. NEEDLE. 1984. KB cells for antinuclear antibody determination: Comparison with HEp-2 cells and the Crithidia luciliae assay. Diagnostic Immunol. **2:** 213-218.
4. DEBARI, V. A. & M. A. NEEDLE. 1981. Cation complexed DNA. A novel class of immunoadsorbents for anti-DNA antibodies. J. Immunol. Meth. **40:** 89-94.
5. SENECAL, J. L., N. F. ROTHFIELD & J. M. OLIVER. 1982. Immunoglobulin M autoantibody to vimentin intermediate filaments. J. Clin. Invest. **69:** 716-721.
6. BLOSE, S. H. & S. CHACKO. 1976. Rings of intermediate (100-A) filament bundles in the perinuclear region of vascular endothelial cells. J. Cell. Biol. **70:** 459.

The Role of 2-(S)-*n*-Butyl-(1-Phenyl- Hydrazino-Carbonyl)-Hexanoic Acid in the Anti-inflammatory Process

HENRICH H. PARADIES

Märkische Fachschule
Department of Chemical Engineering and Biotechnology
FB Physical Technique 7
Frauenstuhlweg 31
D-5860 Iserlohn, Federal Republic of Germany

KARL E. SCHULTE

Institut für Pharmazeutische Chemie
Westfälische Wilhelms-Universität
4400 Münster, Federal Republic of Germany

The noncyclic form of 1-phenyl-4-*n*-butyl-pyrazolidine-(3,5)-dione and its disposition, anti-inflammatory activity and pharmakokinetics have been described by Schulte *et al.*[1-3]

Utilizing a cell-free system we studied the formation of 5-HETE and di-HETE including LTB_4 by incubating the 10,000 $\times$ g supernatant with [^{14}C]arachidonic acid and in the presence and absence of Ca^{2+} (Mg^{2+}) and 2-(S)(-)-*n*-butyl-(1-phenyl-hydrazino-carbonyl)-hexanoic acid (FIG. 1) (2-(S)(-)BMPH.[4,5] When the 10,000 $\times$ g supernatant was incubated with [^{14}C]arachidonic acid (AA) alone, only small amounts of lipoxygenase products were synthesized. However, addition of Ca^{2+} markedly increased the formation of hydroxy acids, namely 5-HETE and di-HETE according to MS analysis. The di-HETE band contained LTB_4 and other isomers also, which is dose-dependent. But after addition of Ca^{2+} and 2-(S)(-)BMPH there was little or no 5-HETE and 5,12-di-HETE. However, after a 15-min incubation no LTB_4 or its isomers were detected. Instead the substances 1, 2 and 3 were formed and their structure elucidated by MS, UV and NMR spectroscopy including chemical synthesis (FIG. 2). When a 5000 $\times$ g supernatant from broken platelets was incubated with [^{14}C] AA with or without Ca^{2+}, no effect of Ca^{2+} (Mg^{2+}) on the AA metabolism was observed, but a potentiation of 5-HETE and 5,12-HETE biosynthesis by a Ca^{2+}

ABBREVIATIONS: 5-HETE: 5-D-hydroxy-6, 8, 11, 14-hydroperoxy-eicosatetraenoic acid; di-HETE: 5, 12-dihydroxy-6, 8, 11, 14-hydroperoxy-eicosatetraenoic acid; LTB_4: (5S, 12R)-dihydroxy-6, 14-*Cis*-8, 10-transeicosatetraenoic acid; A 23187: calimycin; GSH: glutathione.

ionophore A 23187 for PMN leukocytes was detected. Following addition of 10 μM 2-(S)(-)BMPH, no 5-HETE or di-5,12-HETE were detected, but the products 1 and 2 were discovered and their structure determined by MS, UV spectroscopy and enzymatic degradation.

The steady-state kinetics of the conjunction of glutathione with LTA$_4$ in the presence of 2-(S)(-)BMPH were studied with three forms of human glutathione-S-transferase. The glutathione concentration was fixed to 5 mM, since this concentration

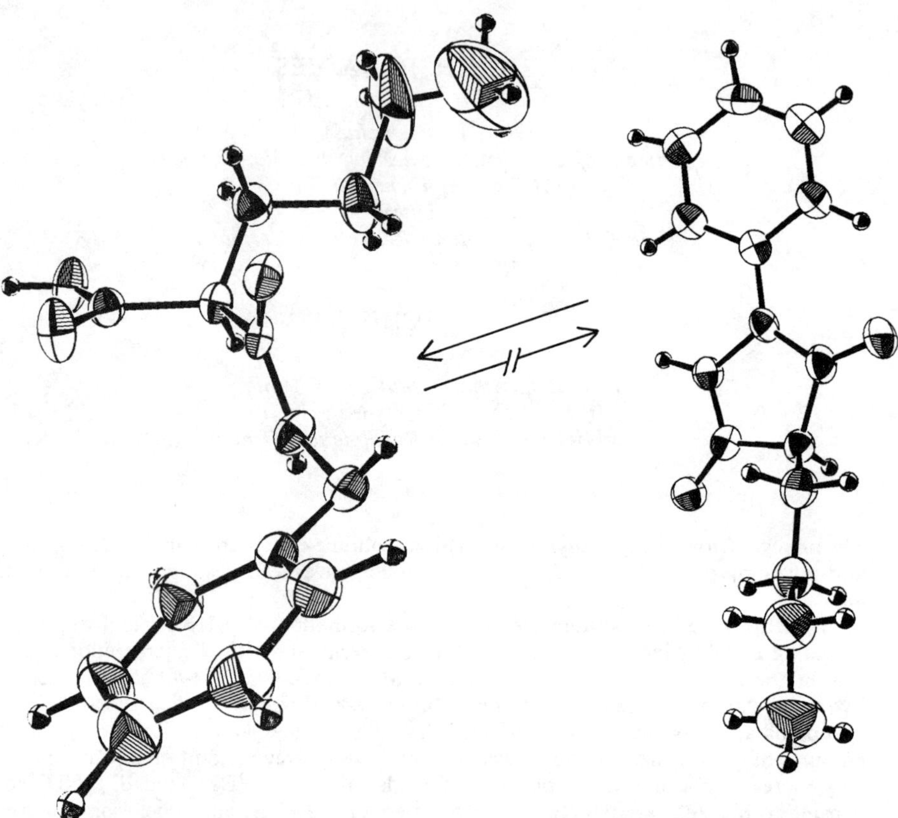

FIGURE 1. Conformation of 2-(S)(-)-*n*-butyl-(phenyl-hydrazino-carbonyl)-hexanoic acid (*right*) and the cyclic form of 1-phenyl-4*n*-butyl-pyrazolidine-(3,5)-dione (*left*).

is representative of intracellular glutathione concentration. TABLE 1 gives the K_m and V_{max} values determined by nonlinear regression analysis as well as specific activities determined in the standard assay, and the data for the different forms of BMPH are shown in TABLE 2. The compound 1 is mainly formed by the glutathione-S-transferase and GSH from the (α-ϵ) type. Appropriate derivatives of 2-(S)(-)BMPH reveal the same pattern: reduction of cystei-containing leukotriene due to reaction of 2-(S)(-)BMPH with the epoxy group of LTA$_4$. The compound 3 is formed enzymatically

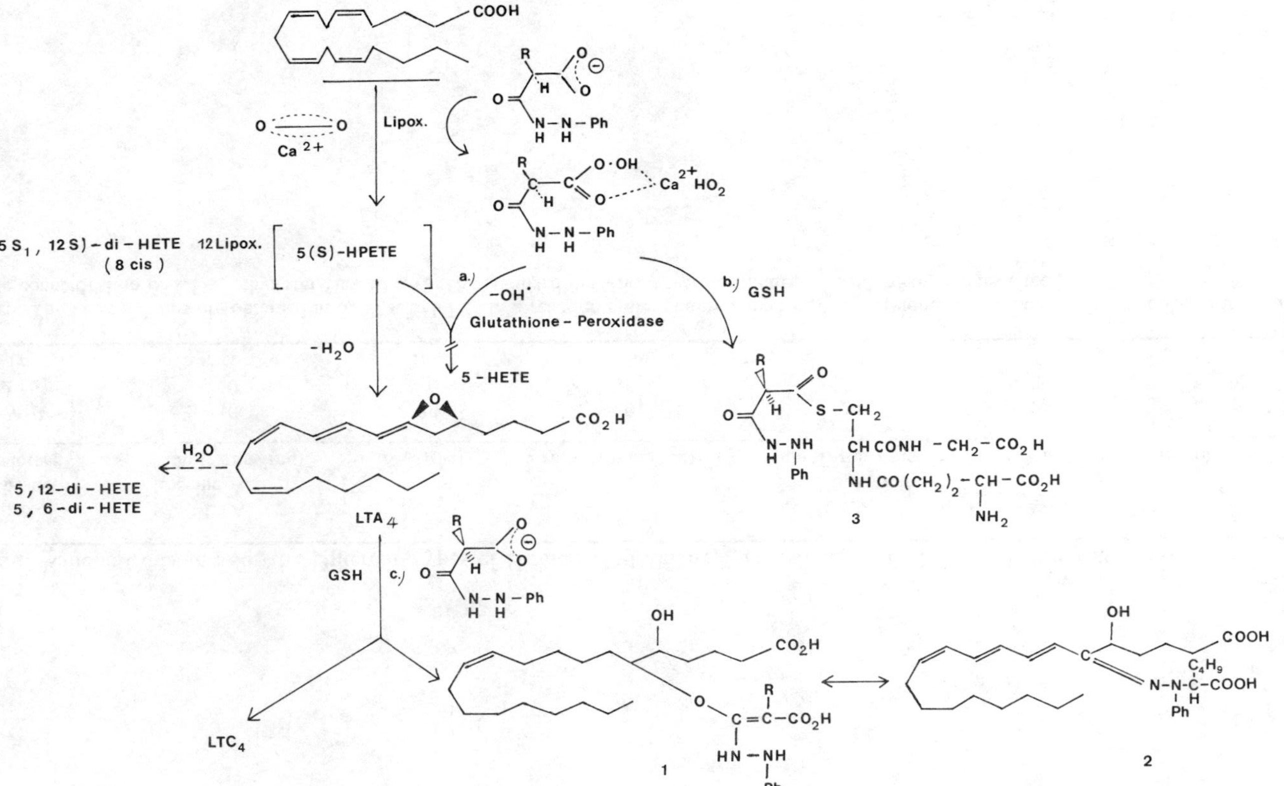

FIGURE 2. Reaction of 2-(S)(-)-*n*-butyl-(phenyl-hydrazino-carbonyl)-hexanoic acid within the 5-lipoxygenase pathway resulting in the enzymatic formation of substances 1, 2 and 3.

TABLE 1. Kinetic Constants for the Different Types of Human Glutathione-S-Transferase with LTA_4 Ethylester as Substrate

Nature of Glutathione-S-Transferase	LTA_4 Specific Activity (nmol/min/mg)	K_m (μM)	V_{max} (nmol/min/mg)	LTA_4 Specific Activity (nmol/min/mg)	K_m (μM)	V_{max} (nmol/min/mg)
Basic (α, β, γ)	10	120	60	15	20	80
Neutral (μ)	60	190	195	250	20	790
Acidic (π)	1	210	25	30	25	85

NOTE: The kinetic data were obtained at 32°C at pH 7.25 in a reaction system containing 5 mM glutathione. The concentration of LTA_4 was 30 μM, and the concentration of LTA_4 ethylester was 10 μM. The standard deviations were estimated as 25% of the values cited.

TABLE 2. Kinetic Constants for the Different Types of Human Glutathione-S-Transferase with Mofebutazone and 2-(S)(−)BMPH as Substrates

Nature of Glutathione-S-Transferase	Specific Inhibitory Activity of Mofebutazone (μmol/min/mg)	K_i (mM)	V_{max_i} (μmol/min/mg)	Specific Inhibitory Activity of 2(S) (−) BMPH (μmol/min/mg)	K_i (μM)	V_{max} (nmol/min/mg)
Basic (α, β, γ)	100	90	100	10	40	100
Neutral (μ)	120	120	30	25	15	700
Acidic (π)	15	135	45	2	10	45

NOTE: The kinetic data were obtained at 32°C at pH 7.25 in a reaction system containing 5 mM glutathione, a concentration of 30 μM of LTA$_4$ or 10 μM of LTA$_4$ ethylester. The concentration of mofebutazone [the cyclic form of 2-(S)(−)BMPH] was 60 μM, and the concentration of 2-(S) (−) BMPH was 10 μM. Under these conditions the Lineweaver-Burk plots are all linear, and the inhibition of 2-(S) (−) BMPH can be completely eliminated by high levels of LTA$_4$ ethylester.

by means of GSH with the μ-form of glutathione-S-transferase at pH 7.5, but at the expense of compound 1. The formation of compound 3 is apparently subject to a redox equilibrium which is H^+-dependent. Furthermore, compound 2 cannot be converted to LTD_4 by means of the λ-glutamyl-transferase.[6]

The present data reveal that the human glutathione transferase is active in the conjugation of LTA_4 with 2-(S)(-)BMPH, apart from GSH. In view of the high intracellular concentration of these enzymes, this capacity is high *in vivo*. Further studies are required to clarify whether the reactions in which LTA_4 as well as PGH_2 (and its enzymatic reaction with the thromboxane-synthetase) are involved with respect to 2-(S)(-)BMPH catalyzed by (*a*) cytosolic glutathione transferase or (*b*) membrane-bound thromboxane-synthetase are due to (1) catalysis of the conjugation of epoxides[6,7] in general or (2) to specific physiological appearance. The half-life time for a decrease of cellular glutathione is approximately 2.5 hours *in vivo*, and that of the turnover of 2-(S)(-)BMPH with LTA_4 is approximately 90 minutes.[5] It is interesting to note that RBL cells (rat basophil leukemia cells) which were preincubated with 2-(S)(-)BMPH and with mofebutazone for up to 20 hours showed no change in the formation of cyclo-oxygenase products, but rather manifested a three-fold increase in the formation of HETEs with a simultaneous *complete* inhibition of the thioether compounds of leukotrienes. An influence on the viability or the morphology of these RBL cells could not be ascertained.

2-(S)(-)BMPH and compounds derived from 2-(S)(-)BMPH reveal the same pattern: reduction of cysteine-containing leukotrienes due to reaction of 2-(S)(-)BMPH with the epoxy group of LTA_4 thus reducing the edema and smooth-muscle-stimulating properties, including the reduction of allergic and anaphylactic reactions.

SUMMARY

The noncyclic form of 1-phenyl-4*n*-butyl-3,5-dioxo-pyrazolidine and the active forms of 2-*n*-butyl-(1-phenyl-hydrazino-carbonyl)-hexanoic acid and its derivates [2-(S)(-)BMPH] act directly on the biosynthesis of leukotrienes, LT_4, LTD_5 and LTB_4. Whereas 2-(S)(-)BMPH inhibits the glutathione S-transferase by a noncompetitive inhibition due to the formation of an *O*-ether, rather than a thioether, the inhibition of the γ-glutamyl-transferase is reached exclusively by 2-(S)(-)BMPH at 10 μM, dependent on the substituted phenylring and the size and nature of the ester of the hexanoic acid. 2-(S)(-BMPH and its derivatives act as a false substrate for this enzyme system. The inhibition is irreversible owing to formation of a thioether compound between 2-(S)(-)BMPH and glutathione in the presence of GSH. The absolute configuration of some 2-(S)(-)BMPH compounds has been elucidated by means of X-ray analysis. Appropriate derivatives of stereoisomers of 2-(S)($\pm$)BMPH reveal the same pattern: reduction of cystei-containing leukotrienes due to reaction of 2-(S)(-)BMPH or its derivates with the epoxy group of leukotriene A_4 thus reducing the edema and smooth-muscle-stimulating properties, including the reduction of allergic and anaphylactic reactions.

REFERENCES

1. BASS, V.-M., R. MRONGOVIUS & K. E. SCHULTE. 1980. Eur. J. Drug Metab. Pharmacokin. **5:** 201-206.
2. MRONGOVIUS, R., V.-M. BASS & K. E. SCHULTE. 1982. Eur. J. Med. Chem. **17:** 275-279.
3. PARADIES, H. H. 1987. J. Pharm. Sci. **76:** 920-930.
4. KASSEM, M. A. & K. E. SCHULTE. 1984. Eur. J. Drug Metabol. Pharmacokin. **9:** 223-227.
5. PARADIES, H. H., H. H. FLÄMIG & K. E. SCHULTE. 1985. Deutsche Apotheker-Zeitung **125:** 749-753.
6. PARADIES, H. H. 1986. Presented at 1986 meeting of the American Chemical Society Anaheim, California. Abstract 79, Section of Medicinal Chemistry.
7. PARADIES, H. H., H. H. FLÄMIG & K. E. SCHULTE. 1986. Deutsche Apotheker-Zeitung **126:** 477-483.
8. PARADIES, H. H. 1986. Eur. J. Med. Chem. In press.
9. MANNERVIK, B. 1985. Adv. Enzymol. **57:** 357-417.
10. BOYLAND, E. & K. WILLIAM. 1965. Biochem. J. **94:** 190-197.

Production of Anti-DNA Antibodies in Normal Mice after Immunization with Anti-Idiotypic Antibodies

T. M. PHILLIPS,[a] S. C. FRANTZ,[a] AND
T. V. HOLOHAN[b]

[a]Immunochemistry Laboratory
The George Washington University Medical Center
Washington, DC 20037

[b]United States Public Health Service
Food and Drug Administration
Rockville, Maryland 20857

The role of idiotypic (Id) antibodies in immune regulation is well established[1,2] and the presence of these antibodies has been shown to have both enhancing and suppressive effects on both cellular and humoral immune responses.[2] Therefore, it has been hypothesized, but not proven experimentally, that idiotypic antibodies elicit the production of autoantibodies, even in the absence of detectable levels of antigen.

To investigate this hypothesis, we have immunized normal Balb/C mice with defined fragments of murine DNA and isolated the idiotypic antibody responses to these fragments by affinity chromatography. These Id antibodies were complexed with naturally occurring rheumatoid factor (RF) as previously described[3] and used to induce the formation of anti-Id antibodies. The anti-Id antibodies were isolated by high-performance immunoaffinity chromatography,[4] using the original Id as the immobilized ligand. These anti-Ids were again complexed with RF and injected into normal 6-month-old Balb/C male mice.

The animals were monitored weekly for the presence of anti-DNA antibodies by the Jerne plaque assay and by indirect immunofluorescence using Hep-2 cells as the substrate. The presence of antibody-secreting cells was detected at Day 42 by the plaque assay, and this was confirmed by the presence of homogeneous antinuclear antibodies and detected by immunoprecipitation of the original antigen (FIG. 1). These results were seen in 25 of 40 animals which were injected with the complexed anti-Id, and circulating immune complexes containing DNA and anti-DNA antibodies were detected in six of the 25 reactive animals. Kidney deposition of these complexes, leading to immune complex glomerulonephritis, was seen in three of the six complex-positive animals. These results are summarized in TABLE 1.

Although these results are preliminary, we believe that we have shown that it is possible to elicit detectable autoantibodies in normal, young animals, by the injection of enhancing idiotypic antibodies complexed to RF, which is a commonly detected anti-antibody found in a variety of autoimmune-associated disease states. From our findings we conclude that RF is an important immune regulatory agent which can induce the production of idiotypic and anti-idiotypic antibodies by presenting the appropriate antibody, as an antigen, to the immune system.

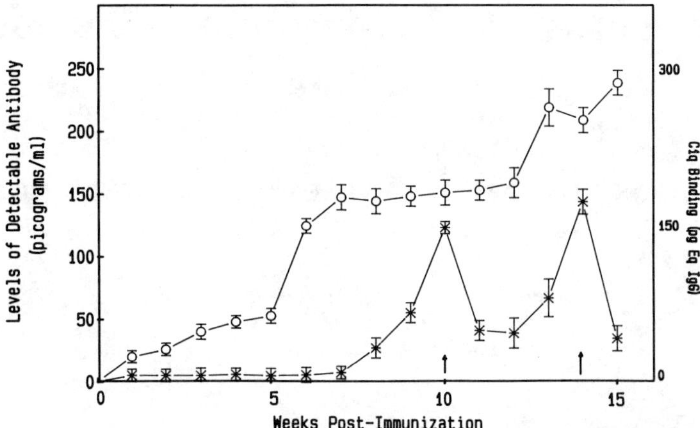

FIGURE 1. Development of anti-DNA antibodies in normal animals after injection of anti-idiotypic antibodies complexed to rheumatoid factor. Anti-DNA antibodies detected by immunoprecipitation of original antigen (*open circles*) are compared to the presence of DNA/anti-DNA immune complexes (*asterisks*). The *arrows* indicate the time points when positive immune deposits were detected in the kidneys.

TABLE 1. Summary of Laboratory Findings

Laboratory Variable	No. of Positive Animals
Anti-DNA antibodies	25/40
DNA immune complexes	6/25
Glomerular involvement	3/6

REFERENCES

1. URBAIN, J. *et al.* 1983. Ann. N.Y. Acad. Sci. **418:** 1.
2. MCNAMARA, M. *et al.* 1983. Ann. N.Y. Acad. Sci. **418:** 65.
3. PHILLIPS, T. M. *et al.* 1984. Ann. N.Y. Acad. Sci. **435:** 475.
4. PHILLIPS, T. M. *et al.* 1985. Liquid Chromatog. **3:** 962.

Serum Alpha-Fetoprotein in Tumor Patients

S. G. SUNDARAM, S. MANIMEKALAI, S. UNNI, AND
P. J. GOLDSTEIN

Obstetrics and Gynecology Research Divisions
Sinai Hospital of Baltimore
Baltimore, Maryland 21215

The desirable characteristics of a serum tumor marker are as follows: The marker should (1) precede and predict recurrences before a tumor becomes clinically detectable; (2) positively correlate with tumor volume and extent; (3) reflect the current status of the disease; (4) change as the status of the tumor changes over time; and (5) predict a high or low risk of recurrence. No one marker known at present fulfills all of these criteria. Serum alpha-fetoprotein (SAFP), the major fetal protein similar to albumin and synthesized by fetal liver and yolk sac, is present only in trace amounts in humans from the second year of life. However, SAFP has been found to be elevated in some types of tumors.[1] Since elevated levels have been demonstrated in the sera of patients with benign tumors[2] and in pregnant women, SAFP cannot be used for tumor screening, but only for tumor monitoring. A knowledge of SAFP levels can be useful in the care of a patient or potential patient with cancer to aid in the differential diagnosis and tumor staging, to aid in therapy by determining tumor burden, and to aid in the detection of tumor recurrence.

SAFP can be quantitated by several techniques. At present, radioimmunoassay (RIA) is the method of choice for most laboratories. However, the RIA kits are designed to measure SAFP in the 20-400 ng/ml range and generally perform with poor precision and sensitivity in the 0-20 ng/ml range. If SAFP concentration is to be used for tumor monitoring, then the chosen assay method should be capable of measuring low concentrations of SAFP. In this study, we evaluated an immunoenzymetric assay (EIA) kit from Hybritech, Inc., San Diego, California (H) and compared it with an RIA tumor marker kit from Diagnostics Products Corporation, Los Angeles, California and a routinely used RIA AFP assay kit from Amersham Corporation, Arlington, Illinois, for their efficacy in determining SAFP in tumor patients' sera and in reproducibly measuring low levels of SAFP. Sera from normal men and women ($n = 250$), from patients ($n = 88$) with eight different types of malignant tumors, from patients ($n = 26$) with benign fibroid tumors, as well as pooled control sera were analyzed in replicate by the different kit protocols as specified by the manufacturer. Micromedic autogamma counter/MACC data reduction system was used for RIA methods and Hybritech Photon Immunoassay Analyzer was employed for the EIA method.

The essential features of the three kits are compared in TABLE 1. Ninety-five percent of normal subjects had undetectable SAFP concentrations; only in 5% were basal levels of less than 5 ng/ml found. Assay variables such as recovery, parallelism, and inter- and intra-assay variations were within limits claimed by the kit manufac-

"

TABLE 1. Alpha-Fetoprotein Assay: Comparison of Kit Features

	AM	DPC	HYB
Assay type	RIA	RIA	EIA
Separation	PEG	PEG + AB	Two-site solid phase
Intended use	Serum, plasma and amniotic fluid AFP	Serum AFP for managing testicular tumor	Serum AFP for managing testicular tumor
Antibody	Rabbit	Not given	Mouse monoclonal
FDA-Approved	Yes	No	Yes
Calibration factor (ng to IU)	1.0/1.09	0.83/1.0	1.13/1.0
Calibration range (ng/ml)	20–400	3.6–363	200
Sample volume (μL)	100	100	20
Total incubation time (hr)	21 + 3	3.25	4.5
Incubation temperature (C)	15–30	15–30	37
Sensitivity (ng/ml)	2	3.6	2

NOTE: AM, Amersham Corporation; DPC, Diagnostics Products Corporation; HYB, Hybritech, Inc.; RIA, Radioimmunoassay; EIA, Enzymeimmunometric assay; PEG, polyethylene glycol; AB, antibody.

turers. Concentrations by all the three kit protocols correlated well, but the H EIA method gave a relatively higher value, possibly due to the greater affinity of its monoclonal antibody directed against two different antigenic sites of AFP molecule. TABLE 2 shows that while SAFP was detectable in 28 to 67% of tumor sera when analyzed by the RIA method, it was detectable in 67 to 92% of the same sera when analyzed by the EIA method. However, both methods detected AFP in similar number (31 and 27%) of sera of patients with noncancerous fibroid tumors.

TABLE 2. Alpha-Fetoprotein Concentration in Different Tumor Sera

Tumor Type (No. of Samples)	Number of Samples in Which AFP Was Detectable (% Detectable)		AFP Concentration (ng/ml) Mean + SEM	
	by RIA	by EIA	by RIA	by EIA
Bladder (7)	3 (43)	6 (86)	1.0 + 0.8	15.2 + 4.3
Breast (21)	13 (62)	17 (81)	1.5 + 0.5	15.2 + 2.0
Colon (25)	7 (28)	22 (88)	2.4 + 1.6	14.6 + 1.3
Lung (12)	5 (42)	11 (92)	0.6 + 0.3	12.1 + 1.7
Lymphoma (3)	1 (33)	2 (67)	10.0 + 1.5	9.6 + 4.4
Multiple myeloma (3)	2 (67)	2 (67)	1.3 + 1.2	6.3 + 2.1
Ovarian (4)	1 (25)	3 (75)	0.9 + 0.4	15.0 + 5.9
Prostate (9)	5 (56)	7 (78)	2.7 + 1.5	17.3 + 3.5
Renal (4)	1 (25)	3 (75)	0.21 + 0.05	13.2 + 3.1
Benign fibroid (26)	8 (31)	7 (27)	4.4 + 2.6	19.4 + 5.5

NOTE: Number in parenthesis in column 1 represents the number of tumor sera analyzed for AFP; number in parenthesis in columns 2 and 3 represents the percent of samples in which alpha-fetoprotein can be assayed. The EIA method refers to Hybritech EIA protocol and the RIA method refers to Amersham RIA protocol. (Results by DPC protocol do not vary significantly from those by Amersham protocol.)

The data suggest that if SAFP measurement is used to monitor tumor development or regression, a technique that is capable of reproducibly measuring low levels of SAFP should be used; the data also suggest that SAFP measurement may be of value in tumor types other than testicular tumor.

REFERENCES

1. WALDMANN, T. S. & K. R. McINTIRE. 1974. Cancer 34: 1510.
2. ENDO, Y., K. KANAI, T. ODA et al. 1975. Ann. N.Y. Acad. Sci. 259: 234.

Studies on the Relationship between Plasma Proteins and Amyloid Fibrils Found in Alzheimer's Disease

LAWRENCE K. DUFFY,[a] MARCIA BERMAN-PODLISNY,
CHAUDRI G. RASOOL,[b] RAYMOND L. WALSH,
AND DENNIS J. SELKOE

Department of Neurology (Neuroscience)
Harvard Medical School, and
Center for Neurologic Diseases
Brigham and Women's Hospital
Boston, Massachusetts 02115

[b]*CNS Section*
Lederle Laboratories
Pearl River, New York 10965

It has been observed for some time that antibodies to some serum proteins crossreact with amyloid deposits found in human brain.[1,2] We have recently observed the cross-reaction of antibodies made to human central nervous system (CNS) amyloid with several unidentified plasma proteins. More than other types of amyloid fibrils, CNS amyloid has proven to be extremely insoluble and therefore resistant to biochemical analysis.[3] The extent of amyloid deposition varies among senile plaques in humans and has been used as a criterion for the age or "maturity" of a plaque. The renewed realization that many of the CNS amyloid deposits occurring in humans or lower primates lie in close proximity to blood vessels and even occur in meningeal arteries outside the brain parenchyma lends support to the hypothesis[1,5] that the amyloid may originate from a protein precursor synthesized outside of the nervous system, carried to it by blood, and processed into insoluble amyloid fibrils by local cells such as endothelial or perithelial cells of the vessels or microglial cells in the brain. However, more work is needed before a "serum-derived amyloid" model for age-related CNS amyloid can be accepted.

Enriched fractions of amyloid cores can be prepared from the brain tissue of persons with Alzheimer's disease ("Alzheimer brains") using fluorescence-activated particle sorting.[3,4] The amino acid sequence of the meningeal amyloid has been determined by Glenner and Wong[5] and more recently by us. With the exception of the glutamic versus glutamine difference at position 11, we obtained an identical sequence from amyloid isolated from meningeal vessels of Alzheimer brains. We have used various brain amyloid preparations to prepare the following panel of polyclonal antisera that recognize the amyloid cores and vascular amyloid. The immunogens used were

---

[a] Address for correspondence: L.K. Duffy, Center for Neurologic Diseases, Brigham and Women's Hospital, Boston, Massachusetts 02115.

(1) native amyloid cores, (2) HPLC-purified amyloid protein, (3) amyloid from congophilic angiopathy, and (4) a synthetic peptide representing the NH_2-terminal 28 residues of the Alzheimer's disease amyloid fibril. Using these antibodies, we observed several crossreacting proteins on Western blots of human plasma. We present here our initial studies on the characterization of these proteins.

Several plasma proteins which crossreacted with the anti-amyloid antibodies on the Western blots had approximate molecular weights (M_r) of 32 kDa, 47-49 kDa, and 87 kDa. We also observed faint staining of 3-4 bands between 120 and 200 kDa. Isoelectric focusing followed by immunoprecipitation indicated that the native immunoreactive proteins had isoelectric points in the range of 4.0-6.0 (FIG. 1). Normal preimmune rabbit serum did not precipitate these same plasma proteins. Dye affinity chromatography using Cibacron Blue F3GA (Matrix Blue A) was used to separate the plasma proteins into two fractions, an unbound fraction, I, and a bound fraction which could be eluted with 2M NaCl. The 47-kDa protein appeared in fraction I while the 87-kDa protein was in fraction II. In order to remove the albumin from the 87-kDa fraction, a 40% NH_4SO_4 precipitation of the plasma was used followed by the Matrix Blue A chromatography (FIG. 2). When this fraction was immunostained with one of the anti-amyloid core antisera, after it had been absorbed with amyloid cores, the 87-kDa band no longer stained. When antibodies were immunopurified on the 87-kDa band of the Western blots, these antibodies stained the senile plaque cores and vascular amyloid in Alzheimer disease brain sections.

These data suggest that amyloid cores and the Alzheimer's disease vascular amyloid share determinants with certain normal human plasma proteins, particularly with two proteins with M_r of 47 kDa and 87 kDa and these determinants are preserved even after SDS-extraction of the fibrils. We have also observed that these proteins can be recognized by the lectin Con A. Whether these proteins are covalently bound (i.e., crosslinked) to the amyloid fibrils or are the actual precursors of the amyloid fibrils

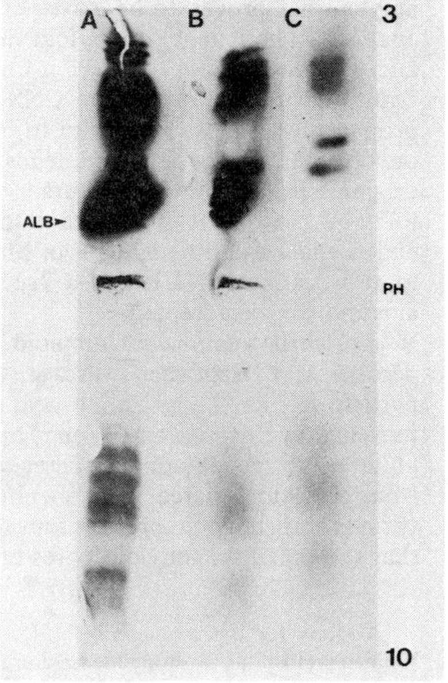

FIGURE 1. Immunofixation of amyloid-related serum proteins (A) human serum, Coomassie blue staining pattern; (B) human serum, fixation with anti-amyloid protein antibody; (C) human serum, fixation with anti-native amyloid core antibody.

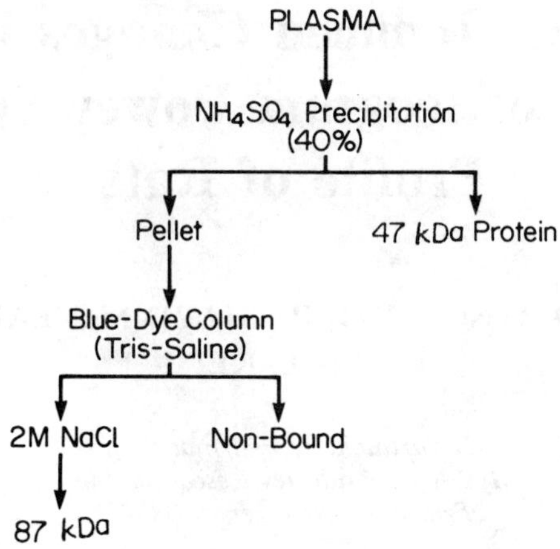

FIGURE 2. Purification of two proteins with crossreaction of anti-amyloid antibodies.

remains to be seen. Further characterization of these polyclonal antibodies to the Alzheimer's disease amyloid as well as the production of monoclonal antibodies could eventually be of diagnostic value through identification of the precursor protein in plasma or through imaging of the amyloid deposits in the brain.

REFERENCES

1. GLENNER, G. G. 1984. *In* Biological Aspects of Alzheimer's Disease. R. Katzman, Ed. Banbury Report **15:** 137-144. Cold Spring Harbor, N.Y.
2. GOUST, J. M., M. M. MANGUM & J. M. POWERS. 1984. J. Neuropath. Exp. Neurol. **43:** 481-486.
3. SELKOE, D. J., C. R. ABRAHAM, M. B. PODLISNY & L. K. DUFFY. 1986. J. Neurochem. **46:** 1820-1834.
4. ROHER, A., D. WOLFE, M. PALUTKE & D. KUKURUGA. 1986. Proc. Natl. Acad. Sci. USA **83:** 2662-2666.
5. GLENNER, G. G. & C. W. WONG. 1984. Biochem. Biophys. Res. Commun. **120:** 885-890.

Hypoxia-Induced Changes in the Electrocorticogram Power Spectral Profile of Rats

RONALD R. NOTVEST, RICHARD McNEAL, AND
KEVIN L. KEIM

Department of Pharmacology
Ayerst Laboratories Research, Inc.
Princeton, New Jersey 08543

In animal studies, acute exposure to hypoxia results in cerebral depression.[1,2] The purpose of this study was to quantify the effects of hypoxia on the central nervous system by a physiologic measure. The electrocorticogram (ECoG) was chosen as the functional measure because it is sensitive to various compromising conditions[2-4] it can be continuously monitored, and it is easily quantified by power spectral frequency analysis.

Male Sprague-Dawley rats with cortical electrodes were placed in an air-tight plastic chamber, and a 10-minute ECoG baseline was recorded under normoxic conditions (i.e., 21% O_2, 79% N_2). The gas mixture was then changed to produce an environment of either 3.5% hypoxia (3.5% O_2, 96.5% N_2; $n = 6$) or 6% hypoxia (6% O_2, 94% N_2; $n = 6$). A 10-minute ECoG was recorded during the hypoxia. On-line frequency analysis (Nicolet Pathfinder) was used to quantify the ECoG.

Changes in the power spectral profile of the ECoG were apparent within the first minute of 3.5% hypoxia (FIG. 1). The initial ECoG response was a transient increase in power in the delta frequency band without significant changes in the theta, alpha, and beta frequency bands. By the fourth minute of hypoxia, power was reduced in all four frequency bands, with the reduction being significant for theta, alpha, and beta activity. The secondary response persisted throughout the remaining duration of the hypoxic episode. The response to 6% hypoxia (not shown) was similar, but the transient response was prolonged (i.e., 5 minutes), and the magnitude of the secondary response was reduced.

The initial transient response is a synchronized ECoG commonly observed during states of low arousal; the secondary response or steady-state ECoG is indicative of a state of cerebral depression. These data demonstrate that power spectral ECoG analysis of rats can quantify the onset and severity of hypoxia-induced brain impairment. This model may be useful for pharmacologic studies designed to identify or characterize potential therapeutics with antihypoxidotic properties.

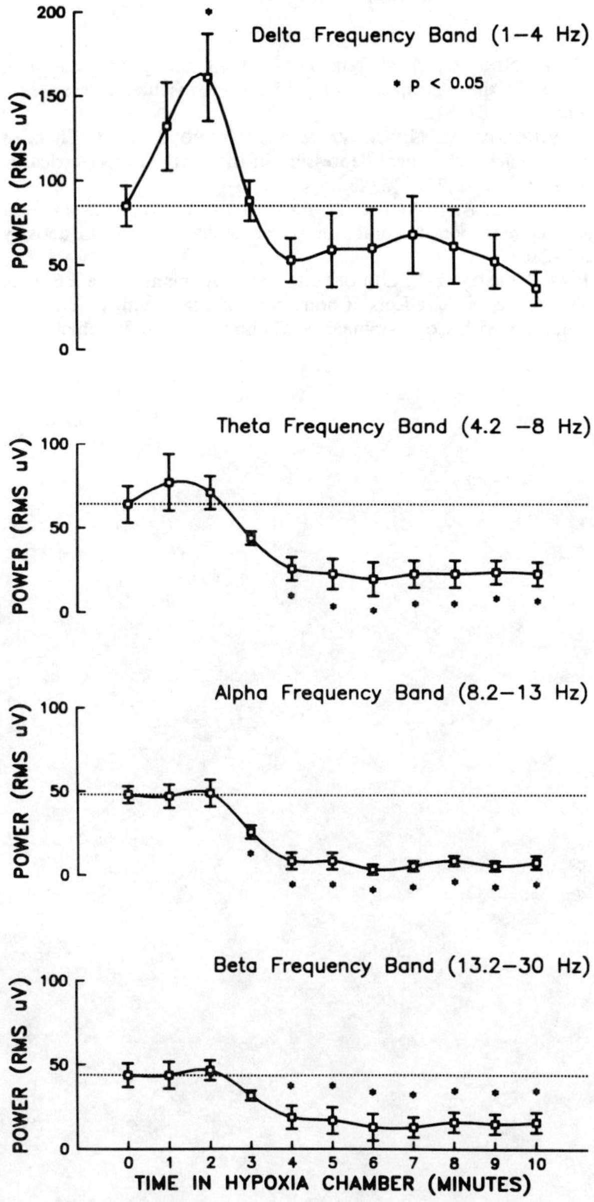

FIGURE 1. Time course of the mean ($\pm$ SEM) electrocorticographic response to hypoxia (3.5% O_2, 96.5% N_2) as recorded for six rats. All comparisons were made relative to baseline. A Student's t test (two-tail) was used to determine significance at the $p < 0.05$ level.

REFERENCES

1. MILANOVA, D., R. NIKOLOV & M. NIKOLOVA. 1983. Study on the antihypoxic effect of some drugs used in the pharmacotherapy of cerebrovascular disease. Meth. Find. Exp. Clin. Pharmacol. **5:** 607-612.
2. DIMOV, S., R. NIKOLOV, M. NIKOLOVA & S. MOYANOVA. 1983. Effect of piracetam in some models of general and local depression of the cortical bioelectrical activity in cats. Arch. Int. Pharmacodyn. **262:** 13-23.
3. FEISE, G., K. KOGURE, R. BUSTO, P. SCHEINBERG & O. M. REINMATH. 1976. Effect of insulin hypoglycemia upon cerebral energy metabolism and EEG activity in rat. Brain Res. **126:** 263-280.
4. APORTI, F., J. M. NELSON & L. GOLDSTEIN. 1982. A quantitative EEG study at cortical and subcortical levels of the effects of brain cortical phosphatidyl serine (BC-PS) in rats, and of its interaction with scopolamine. Res. Commun. Psych. Psychiat. Behav. **7:** 131-143.

In Vitro Androgen Receptor Binding Affinity and *in Vivo* Inhibitory Activity of 5α-Pregnane-3, 20-Dione

SAUDHAMINI PARTHASARATHY, ANDREA CHIN,
VIRGINIA MALLOY, AND JONATHAN MATIAS

Biomedical Research Station
Orentreich Foundation for the Advancement of Science, Inc.
Cold Spring-on-Hudson, New York 10516

Aside from having hormonal effects on the female reproductive system, progesterone (P) also acts as an antiandrogen by inhibiting the transformation of testosterone (T) to its active metabolite, 5α-dihydrotestosterone (DHT), via competition with testosterone for the active site of the 5α-reductase enzyme.[1,2] This enzyme also efficiently converts progesterone to 5α-dihydroprogesterone (DHP, 5α-pregnane-3,20-dione)[3] which theoretically can compete with DHT for binding to the androgen receptor protein.

Wright *et al.*[4] previously showed that DHP had an antiandrogenic activity which was equivalent to that of cyproterone acetate (CA), a potent androgen receptor inhibitor. Because of the continuing interest in the antiandrogenic effect of progesterone and its analogues and metabolites for treatment of prostate cancer, we have re-evaluated the receptor binding affinity of DHP *in vitro* and inhibitory activity *in vivo*.

Using the rat ventral prostate receptor assay, we observed a 50% displacement of [³H]DHT with CA at a ten-fold higher concentration. For DHP to produce a 50% displacement, an 800-fold excess was required (FIG. 1). DHP was also ineffective as an antiandrogen *in vivo*. When injected subcutaneously in the Copenhagen rat, we were unable to demonstrate any inhibitory effect on the size of the prostate or seminal vesicles of normal, adult males (TABLE 1). Although the growth of the Dunning R3327H prostate adenocarcinoma was slower in the DHP-treated animals, this difference was not statistically significant owing to the variability of the growth rate of the tumors. By comparison, administration of CA under the same regimen has been shown to produce a dramatic inhibition of tumor growth.[5] There has been speculation that the effect of progesterone may be modulated via DHP. In view of the clear absence of antiandrogenic effect of DHP, the inhibition produced by progesterone on the tumor and seminal vesicles is more likely due to its 5α-reductase blocking action or its effects on the pituitary-gonadal axis rather than via further metabolism to DHP.

Our receptor binding data contradict those reported previously.[4] Although methodologic differences may be cited as one explanation, the exact reason for the divergence remains unclear. Our present *in vivo* data support the *in vitro* findings which show DHP's lack of significant antiandrogenic effects in the prostate, prostate tumor and seminal vesicles. However, further examination of the biological effect of this steroid

239

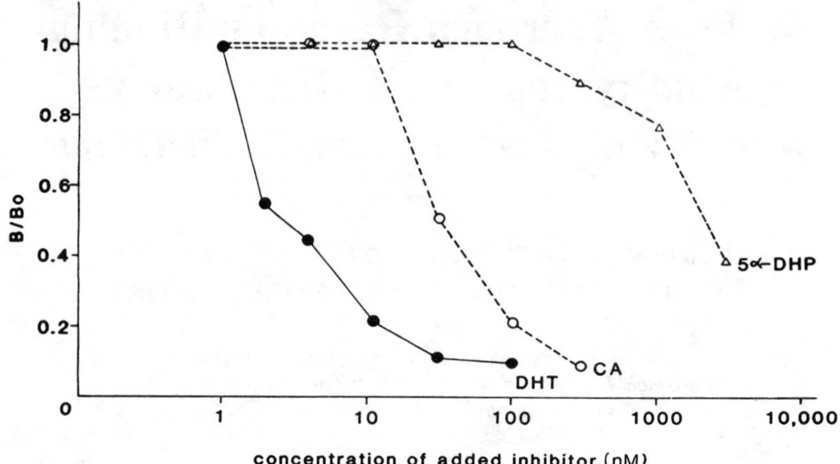

FIGURE 1. Comparison of the relative binding characteristics of DHT, cyproterone acetate (CA) and 5α-pregnane-3, 20-dione on the rat ventral prostate. Inhibition of receptor binding was measured by a modification of the method of Zava *et al.*[6] Diisopropyl-fluorophosphate was used as a protease inhibitor to stabilize the incubation medium. Samples were incubated with 1 nM [3H]DHT and varying concentrations of unlabeled steroids. (B = concentration in cpm of bound [3H]DHT in the presence of competitor and Bo = concentration of bound [3H]DHT in the absence of competitor). All data were corrected for nonspecific binding.

is warranted since an increasing body of information shows that progesterone and its metabolites have other actions, such as anti-inflammatory and anaesthetic effects, which are separate from their reproductive roles.

TABLE 1. Antiandrogenic Effect of Progesterone and DHP on the R3327H Prostate Adenocarcinoma and Prostate and Seminal Vesicles of the Copenhagen Rat

Experiment[a]	(n)	Prostate Tumor (g)	Seminal Vesicles (% body weight)	Prostate (% body weight)
Control	8	5.2 ± 1.0	0.31 ± 0.02	0.035 ± 0.004
Progesterone	8	0.6 ± 0.2[b]	0.17 ± 0.01[b]	0.041 ± 0.010[c]
DHP	8	3.6 ± 0.7[c]	0.26 ± 0.02[c]	0.060 ± 0.030[c]

[a] The steroids were suspended in an aqueous vehicle and injected s.c. on alternate days to the test animals at a dose of 10 mg/kg body weight. After 4 months of treatment, the animals were sacrificed and the weights of tissues were determined.

[b] $p < 0.001$ when compared to control by Student's *t* test.

[c] Not significant at 0.05 level of probability when compared to the control.

REFERENCES

1. VOIGT, W., E. FERNANDEZ & S. L. HSIA. 1970. J. Biol. Chem. **245:** 5594-5599.
2. VERMORKEN, A. J. M., C. M. A. A. GOOS & H. M. T. ROELOP. 1980. Br. J. Dermatol. **102:** 455-460.
3. MAUVAIS-JARVIS, P., N. BAUDOT & J. P. BERCOVICI. 1969. J. Clin. Endocrinol. **29:** 1580-1585.
4. WRIGHT, F., M. D. KIRCHHOFFER & P. MAUVAIS-JARVIS. 1978. J. Steroid Biochem. **10:** 419-422.
5. MALLOY, V., A. CHIN, J. MATIAS & N. ORENTREICH. 1987. Ann. N. Y. Acad. Sci. **494:** 310-312.
6. ZAVA, D. T., B. LANDRUM, K. B. HORWITZ & W. L. McGUIRE. 1979. Endocrinology **104:** 1007-1012.

Hematopoietic Cell Transplantation in Murine Globoid Cell Leukodystrophy (the Twitcher Mouse)[a]

Effects on Galactosylceramidase and Psychosine Levels in the Nervous System

ANDREW M. YEAGER,[b,c] TAKAO ICHIOKA,[d,e,f] AND
YASUO KISHIMOTO[d,e]

[b]Oncology Center and
Departments of [d]Neurology and [c]Pediatrics
The Johns Hopkins University School of Medicine
Baltimore, Maryland 21205

[e]The John F. Kennedy Institute
Baltimore, Maryland 21205

The therapeutic effects of allogeneic bone marrow transplantation (BMT) in human lysosomal storage diseases with central nervous system (CNS) involvement are not established; it has been suggested that the blood-brain barrier (BBB) may prevent entry of donor-derived enzyme and/or mononuclear cells into the CNS. Using an animal model of human galactosylceramidase deficiency (Krabbe disease; globoid cell leukodystrophy [GCL]), the "twitcher" mouse, we have previously shown that transplantation of normal congenic marrow and spleen cells (hematopoietic cell transplantation; HCT) into 10-day-old irradiated (900 rad) twitcher mice prolongs survival by two- to three-fold, and is associated with some remyelination of sciatic nerves.[1] To determine the extent of hydrolase replacement and degradation of substrate at selected times after HCT in murine GCL, we examined levels of galactosylceramidase and psychosine (the major toxic metabolite in GCL[2,3]) in neural and non-neural tissues of twitcher mice and normal littermates that underwent HCT at 10 days of age. Galactosylceramidase activity was determined with tritiated galactosylceramide as the substrate,[4] and psychosine was determined by a new highly sensitive technique that used N-acetylation with ^{3}H-labeled acetic anhydride, benzoylation, and TLC densitometry.[5]

[a]The work was supported in part by Grants NS 13559 and 13569 (Y.K.), K08 HD00535 and R01 NS24097 (A.M.Y.) from the National Institutes of Health, Grant BNS 8314337 from the National Science Foundation (Y.K.), and Basil O'Connor Starter Research Grant 5-485 from the March of Dimes Birth Defects Foundation (A.M.Y.).

[f]Present address: Department of Pediatrics, University of Tokushima School of Medicine, Tokushima 770, Japan.

TABLE 1. Galactosylceramidase Activity after Hematopoietic Cell Transplantation (HCT)

	Age at Sacrifice (days)	Galactosylceramidase Activity (nmol/hr/mg protein)[a]			
Status		Clipped Tails at age 7 Days	Brain	Kidney	Spleen
Twitcher					
Non-HCT	40 (n = 10)	0.034 ± 0.002	0.0186 ± 0.007	0.0176 ± 0.005	0.002 ± 0.0001
HCT	50 (n = 4)	0.09 ± 0.07	0.075 ± 0.018	0.023 ± 0.012	0.21 ± 0.07
	70 (n = 4)	0.05 ± 0.05	0.090 ± 0.061	0.146 ± 0.083	0.31 ± 0.01
	100 (n = 4)	0.06 ± 0.02	0.341 ± 0.173	0.30 ± 0.10	1.90 ± 1.25
Control					
Non-HCT	40 (n = 10)	0.74 ± 0.047	0.223 ± 0.04	1.005 ± 0.117	0.70 ± 0.02
HCT	50 (n = 3)	0.79 ± 0.07	0.28 ± 0.10	0.34 ± 0.06	0.69 ± 0.15
	70 (n = 3)	0.98 ± 0.08	0.59 ± 0.48	1.20 ± 0.12	0.74 ± 0.07
	100 (n = 5)	0.87 ± 0.05	0.39 ± 0.18	1.62 ± 0.65	1.17 ± 0.02

[a] Mean values ± 1 standard error are shown. Animals received HCT at age 10 days.

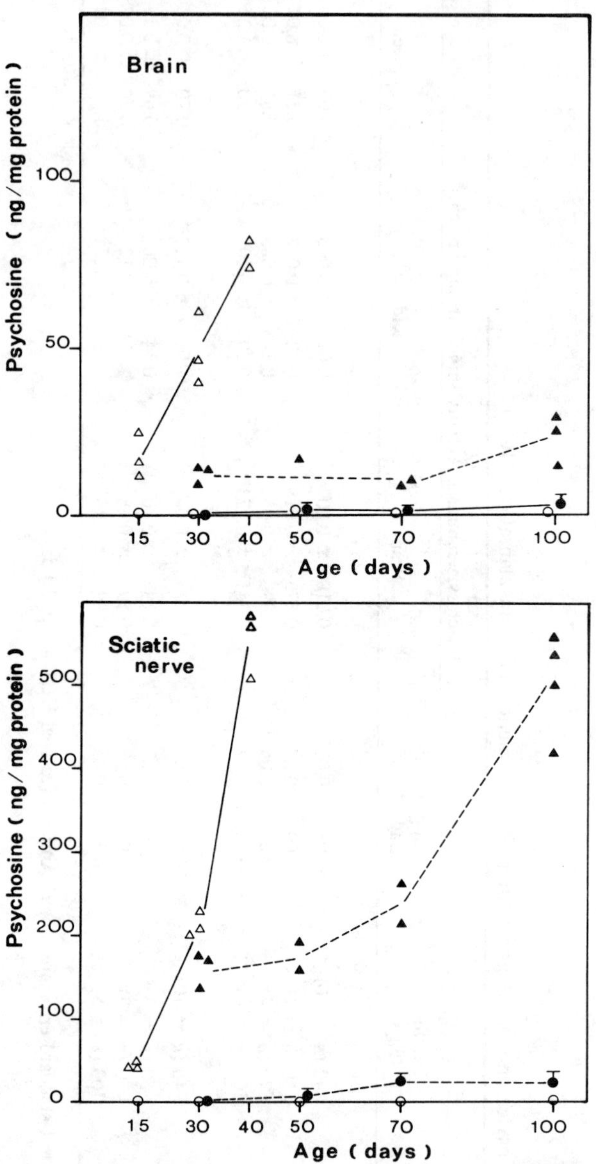

FIGURE 1. Changes in psychosine concentrations in brains (*upper panel*) and sciatic nerves (*lower panel*) of twitcher and control mice as a function of the age of the animals. Twitcher mice with or without hematopoietic cell transplantation (HCT) are indicated by *solid* or *open triangles*, respectively. Enzymatically normal animals with or without HCT are indicated by *solid* or *open circles*, respectively.

Levels of galactosylceramidase in tissues of untreated twitcher mice were consistently low (< 8% of control values), but gradually reached 15-25% of control by 40 to 60 days after HCT (TABLE 1). By 90 days after HCT, galactosylceramidase activity in the brains of twitcher mice was in the normal range (TABLE 1), even though no specific techniques were used to induce or maintain permeability of the BBB. Compared to values in normal littermates, psychosine levels in brains and sciatic nerves, respectively, of untreated twitcher mice were elevated by 44- and 90-fold in 30-day-old mice and by 69- and 232-fold in 40-day-old animals (FIG. 1). Although still substantially elevated relative to values in control normal littermates, psychosine levels in the brains were reduced to 30-35% of those seen in untreated twitchers by 20 days after HCT (i.e., 30-day-old animals) and remained stable throughout 90 days after transplant. In contrast, psychosine content was similar in the sciatic nerves of both untreated and HCT-treated 30-day-old twitcher mice and increased gradually in the transplanted animals, so that the levels in nerves from 100-day-old HCT-treated twitchers were similar to those seen in 40-day-old untreated animals (FIG. 1).

Our studies indicate that HCT in this murine sphingolipid storage disease provides both enzyme replacement and favorable alterations in the major toxic metabolite in the CNS. It has been shown that globoid cells in the brains of twitcher mice are of monocytemacrophage origin;[6] it is not known whether the observed alterations in galactosylceramidase and psychosine after HCT are attributable to the formation of globoid cells from donor-derived monocytes that cross the BBB in twitcher mice, or whether the pre-transplant irradiation induces permeability of the BBB to donor cells. Alternatively, perturbation of the BBB may occur as a consequence of the natural history of murine GCL. Since psychosine is highly toxic to oligodendroglia and remains elevated in the CNS after HCT is performed in 10-day-old twitcher mice, HCT in younger (e.g., 2- to 5-day-old) animals may prevent accumulation of psychosine in the CNS and lead to further prolongation of survival and prevention of neurodegeneration in murine GCL.

REFERENCES

1. YEAGER, A. M., S. BRENNAN, C. TIFFANY, H. W. MOSER & G. W. SANTOS. 1984. Science **225:** 1052-1054.
2. MIYATAKE, T. & K. SUZUKI. 1972. Biochem. Biophys. Res. Commun. **48:** 538-543.
3. IGISU, H. & K. SUZUKI. 1984. Science **224:** 753-755.
4. SUZUKI, K. 1978. Methods Enzymol. **50**(C): 456-488.
5. ICHIOKA, T., Y. KISHIMOTO & A. YEAGER. 1987. Anal. Biochem. **166:** 178.
6. KOBAYASHI, A., M. KATAYAMA, E. BOURQUE, K. SUZUKI & K. SUZUKI. 1986. Devel. Brain Res. **20:** 49-54.

Characterization of the Interrelationship of the Secretory Activities of Gastric Tissue[a, b]

CHRISTOPHER CASCIANO[c] AND ALFRED BENNUN

Department of Biological Sciences
Rutgers University
Newark, New Jersey 07102

The understanding of the process of gastric cytoprotection requires the elucidation of the mechanism of bicarbonate (HCO_3^-) secretion. This includes the identification of the source(s) of HCO_3^- and mechanism(s) that regulate the secretion. These studies have been hindered by the means chosen to prevent interference of acid secretion on the assay for HCO_3^-.[1,2] The methods previously used for the removal of this acid interference, which employed isolated stomach preparations, include the following: the use of histamine antagonists,[3,4] which incompletely inhibit basal acid secretion; treatment of the tissue with nonphysiological doses of SCN^-, which combines with H^+ in the mucosal solution to form HSCN, which then diffuses back to the cell cytosol to dissipate the proton gradient;[5,6] and extrapolation from experiments conducted on isolated antral mucosa, which is devoid of parietal cells and thus free of acid interference.[3]

Sch 28080 [2-methyl-8-(phenylmethoxy)imidazo[1,2-a]-pyridine-3-acetonitrile] (MPIPA), a potent inhibitor of the H^+/K^+ ATPase pump in the parietal cell,[7] was found to have both antisecretory and cytoprotective activity.[8] It was also shown to stimulate HCO_3^- secretion in the isolated guinea-pig gastric mucosa.[9] These pharmacologic properties of MPIPA afford it a potentially valuable role for furthering studies of HCO_3^- secretion by the fundic mucosa. Therefore, it was desirable to characterize the mechanism through which MPIPA inhibits acid and stimulates HCO_3^- secretions in the isolated fundic mucosa of the rat. Accordingly, slices from this tissue were incubated in a HCO_3^-/CO_2-free bathing media, and the effect of the drug was examined in relationship to intracellular and extracellular agonists of acid secretion.

[a]This work was aided by grants from the United States Public Health Service (RR 07059) and Schering Corporation.

[b]Publication No. 32 of the Department of Biological Sciences, Rutgers University, Newark, New Jersey.

[c]Address for correspondence: RD4, Box 295, Newton, New Jersey 07860.

METHODS

The animals were stunned by a blow to the head and killed by cervical dislocation. The stomachs were quickly excised, opened along the lesser curvature and rinsed with bathing media at room temperature. The composition of the bathing media was (mM): Na^+ 143, K^+ 5.9, Ca^{+2} 2.5, Mg^{+2} 1.2, Cl^- 149, SO_4^{-2} 1.2, dextrose 16. The fundic mucosae were then cut into four equal sections and placed into test tubes, covered by parafilm to prevent evaporation. Each tube contained 1 ml of nonbuffered bathing media, pH adjusted to 7.00, which was gassed with 100% O_2 and incubated for 2 hours in a water bath maintained at 36°C.

At the end of the incubation period, the tissue was removed and samples from the (at all times capped) bathing media were used to determine HCO_3^- levels by enzymatic assay.[10] Additional samples of the bathing media were taken for the determination of Cl^-, Na^+, and K^+ using a System E4A Electrolyte Analyzer (Beckman Instruments, Inc.). Blanks of bathing media plus additions, containing no tissue, were similarly incubated and assayed.

RESULTS AND DISCUSSION

Using an enzymatic assay to measure HCO_3^-, we were able to demonstrate that MPIPA increased HCO_3^- secretion in a dose-dependent (10^{-6}M - 10^{-3}M) manner. The doses selected were greater than those necessary for inhibition of acid secretion.[9] FIGURE 1 shows that in the absence of MPIPA, histamine (10^{-5}M), methacholine (10^{-4}M), and dbcAMP (1 mM) plus theophylline (1 mM) stimulate HCO_3^- secretion. These levels were reduced in the presence of 10^{-6}M MPIPA and subsequently stimulated in a dose-dependent manner by the drug. MPIPA-stimulated HCO_3^-

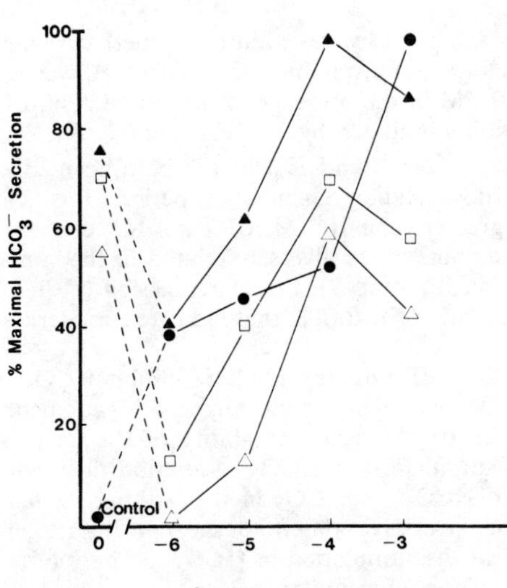

FIGURE 1. Dose-response of HCO_3^- secretion by rat fundic mucosa. Tissue slices were incubated for 2 hours in the absence (controls, initiating ---) and presence of 10^{-6}M to 10^{-3}M MPIPA. ●, MPIPA alone; ▲, + 10^{-5}M histamine; □, + 10^{-4}M methacoline; △, + 1 mM dbcAMP and 1 mM theophylline. To average any topologic difference in secretory activity, alternating slices from three rat fundic mucosae were divided into four equal sections and incubated separately. Hence, each point in the curves represents the mean of 12 independent determinations. Maximal HCO_3^- secretion at 100% corresponds to 1.095 ± .138 (SEM) nEq HCO_3^-/mg tissue/hr.

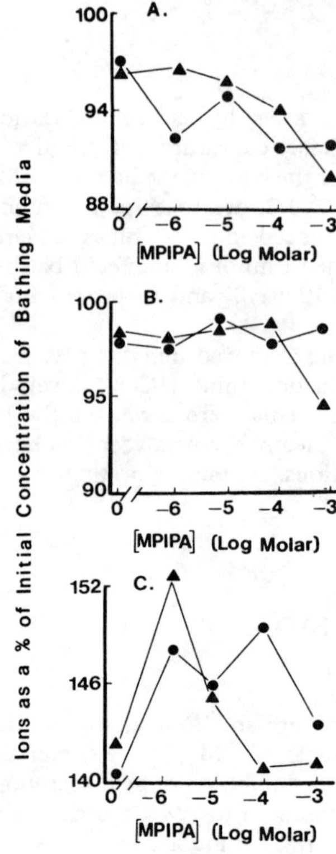

FIGURE 2. Effect of MPIPA on Cl^-, Na^+, and K^+ fluxes in gastric fundic mucosa. Slices of tissue were incubated in the absence (controls) and presence of 10^{-6}M to 10^{-3}M MPIPA and ion fluxes assayed: (A) Cl^-; (B) Na^+; (C) K^+ concentrations as a percentage of the initial concentration of the bathing media. Concentrations at 100% correspond to the initial ion concentrations of the bathing media as indicated in the METHODS section. Ion concentrations were determined after incubation with: ●, MPIPA alone; ▲, + 10^{-5}M histamine. Each point in the curves represents the mean of 12 determinations sampled as described for FIGURE 1. Linear regression analysis yielded correlation coefficients (r) between HCO_3^- secretion-response curves in FIGURE 1 (●,▲) and their respective Cl^- and K^+ fluxes (FIGURE 2, same symbols) with values for the response to 10^{-6}M to 10^{-3}M MPIPA of: FIGURE 1 versus FIGURE 2A for ●, −0.382; for ▲, −0.6323; and FIGURE 1 versus FIGURE 2C for ●, −0.7539; for ▲, −0.9452.

secretion at 10^{-6}M appeared to be slightly inhibited by the addition of methacholine and dbcAMP plus theophylline, while the addition of histamine had no effect. However, MPIPA-stimulated HCO_3^- secretion at 10^{-4}M in the presence of histamine resulted in a two-fold increase over HCO_3^- secretion stimulated by MPIPA alone.

The postincubation concentrations of Cl^-, Na^+, and K^+ in the bathing media were also determined and compared with those of the preincubation period (FIG. 2). It was observed that at doses of MPIPA greater than 10^{-6}M, Cl^- and K^+ concentrations, particularly in the presence of histamine, were inversely related to the stimulation of HCO_3^- secretion (FIGURE 1 versus FIGURE 2). Linear regression of these results yielded r values, as reported in the legend to FIGURE 2, that indicate a moderate to high degree of inverse correlation.

In conclusion: (1) Methacholine and dbcAMP plus theophylline inhibit HCO_3^- secretion stimulated by MPIPA at 10^{-6}M, while histamine appears to augment HCO_3^- secretion stimulated by MPIPA at 10^{-4}M. (2) The ability of the gastric fundic mucosa to respond to MPIPA for stimulation of HCO_3^- secretion does not appear to be dependent on the presence of HCO_3^- nor CO_2 in the bathing media. (3) Cl^- and K^+ ion fluxes appear to be inversely related to the appearance of measurable HCO_3^- secretion, indicating that the stimulation of HCO_3^- secretion by MPIPA may be coupled to either or both of these ion fluxes.

It would be expected that as the H^+/K^+-ATPase pump within the apical membrane of the parietal cell is inhibited, a concomitant inhibition in associated HCO_3^- extrusion from the basolateral portion of the same cell would also be inhibited. This would then allow a MPIPA-dependent stimulation of a nonparietal cell HCO_3^- secretion to be measurable.

REFERENCES

1. ALLEN, A. & A. GARNER. 1980. Gut **21:** 249–262.
2. KONTUREK, S. J., J. BILSKI, J. TASLER & J. KANIA. 1984. Mechanisms of Mucosal Protection in the Upper Gastro-intestinal Tract. A. Allen *et al.,* Eds. Raven Press. New York, NY.
3. FLEMSTROM, G. & A. GARNER. 1981. Clin. Sci. **60:** 427–433.
4. CORUZZI, G., M. ADAMI & G. BERTACCINI. 1984. Agents and Actions **4**(3/4): 516–521.
5. REENSTRA, W. W. & J. G. FORTE. 1986. Am. J. Physiol. **250:** G76–G84.
6. WALLMARK, B., B. JARESTEN, H. LARSSON, B. RYBERG, A. BRANDSTROM & E. FELLENIUS. 1983. Am. J. Physiol. **245:** G64–G71.
7. BEIL, W., I. HACKBARTH & K-FR. SEWING. 1986. Br. J. Pharmacol. **88:** 19–23.
8. LONG, J. F., P. J. S. CHIU, M. J. DERELANKO & M. STEINBERG. 1983. J. Pharmacol. Exp. Ther. **226:** 121–125.
9. CHIU, P. J. S., C. CASCIANO, G. TETZLOFF, J. F. LONG & A. BARNETT. 1983. J. Pharmacol. Exp. Ther. **226:** 114–120.
10. FORRESTER, R. L., L. J. WATAJI, D. A. SILVERMAN & J. PIERRE. 1976. Clin. Chem. **22:** 243.

Mucin Secretion by Duodenal Mucosa in Response to Arachidonic Acid[a]

M. KOSMALA, S. R. CARTER, J. BILSKI,
A. SLOMIANY, AND B. L. SLOMIANY[b]

Gastroenterology Research Laboratory
Department of Medicine
New York Medical College
Valhalla, New York 10595

INTRODUCTION

Mucus gel covering the epithelial surfaces of the gastrointestinal tract constitutes the first line of mucosal defense against mechanical, chemical, enzymatic and bacterial insults.[1,2] The continuous renewal and resilient nature of the gel provides an ideal milieu that confines the reaction between HCO_3^- and H^+ entering the gel in such a way that a neutral pH is maintained on the mucosal side. The process of HCO_3^- secretion involves the prostaglandin mechanism, and the alkaline output of the mucosa varies, depending upon the location along the gastrointestinal tract, with the highest being in the duodenum.[3] We report here on the content and composition of mucus glycoprotein elaborated by canine proximal and distal duodenum in response to luminal application of prostaglandin precursor, arachidonic acid.

MATERIALS AND METHODS

The studies were conducted with ten mongrel dogs. Five animals were used for the preparation of proximal duodenal loops and five for distal duodenal loops (each 7 cm long). After surgery, the gastrointestinal continuity was restored by end-to-end gastroduodenal or duodenojejunal anastomosis.[3] After a 6-week recovery period, the dogs were placed in cloth slings and their duodenal loops instilled either with saline solution, pH 6.0, saline solution in combination with arachidonic acid (12.5, 25, 50

[a] This work was supported by U.S.P.H.S. Grants DK 21684-10 and AA 05858-05, from the National Institutes of Health.

[b] Address for correspondence: B. L. Slomiany, Gastroenterology Research Laboratory, New York Medical College, Munger Pavilion, Valhalla, New York 10595.

and 100 μg/ml), or 50 mM HCl. The instillation procedures were continued at 30-min intervals for up to 2 hr and the bathing fluids were recovered. Each experiment was conducted without and with pretreatment with indomethacin.[3] The recovered instillates were assayed for HCO_3^- and prostaglandin,[3] and after dialysis and lyophilization they were used for mucus glycoprotein isolation.[4] Purified mucus glycoprotein was subjected to analysis of the content and composition of carbohydrates, lipids and protein.[4]

RESULTS AND DISCUSSION

The alkaline output measured under basal conditions averaged 105 μmol/30 min in the proximal duodenum and 57 μmol/30 min in the distal duodenum, whereas the corresponding values for prostaglandin were 180 and 107 ng/30 min. Topical application of arachidonic acid or HCl produced in both regions an increase in the output of HCO_3^- and prostaglandin generation, whereas pretreatment with indomethacin led to a reduction in the basal HCO_3^- output and resulted in the prevention of the increase induced by HCl and arachidonic acid (TABLE 1). Exposure of the proximal and distal duodenal mucosa to arachidonic acid or HCl produced also in each case an increase in the mucin content of the recovered instillates. With 12.5 μg/ml arachidonic acid, a two-fold increase in the glycoprotein output occurred in both areas of the duodenum, whereas a three-fold increase in mucin was observed at 100 μg/ml arachidonic acid. HCl caused a 1.5-fold increase in mucin output in the proximal duodenum and a 1.7-fold increase in the distal duodenum. Pretreatment with indomethacin prior to HCl or arachidonic acid application caused in both regions of the duodenum reduction in the mucin content of the perfusates to the levels observed with saline solution (TABLE 1).

TABLE 2 shows the composition of mucins purified from saline solution, arachidonic acid and HCl perfusates of the proximal and distal duodenal loops. The mucins from both regions exhibited similar composition and were comprised of protein, carbohydrate, associated lipids and covalently bound fatty acids. In both cases, arachidonic acid or HCl caused an increase in the content of associated lipids and covalently bound fatty acids of the glycoproteins, and a decrease in the content of protein. Less apparent differences were observed in the contents of carbohydrates. Compared to saline-treated controls, the glycoprotein from proximal duodenum showed about two-fold increase in associated lipids with arachidonic acid and 30% increase with HCl. Similar increases in associated lipids were also observed in mucin from the distal duodenum. With both proximal and distal duodenum, the changes in mucin content of proteins and lipids evoked by arachidonic acid or HCl were prevented when the dogs were pretreated with indomethacin (TABLE 2). The lipids associated with mucus glycoprotein elaborated in response to arachidonic acid or HCl, furthermore, exhibited increase in the proportions of phospholipids and decrease in neutral lipids, while the levels of glycolipids remained essentially unchanged.

The results suggest that the increase in endogenous mucosal prostaglandin level evoked by arachidonic acid or HCl causes considerable changes in the composition of elaborated mucins. The induced increases in the mucin lipid contents may be of direct relevance to the ability of duodenum to meet successfully the hostile environment to which it is continuously exposed.

TABLE 1. Alkaline Output and Prostaglandin and Mucin Content of the Proximal and Distal Duodenum in Response to HCl and Arachidonic Acid

	Proximal Duodenum			Distal Duodenum		
Type of Perfusion	HCO_3^- (μmol/30 min)	PGE_2 (ng/30 min)	Mucin (mg/100 mg perfusate)	HCO_3^- (μmol/30 min)	PGE_2 (ng/30 min)	Mucin (mg/100 mg perfusate)
Saline solution	105 ± 16	180 ± 23	8.0 ± 0.9	57 ± 12	107 ± 12	6.8 ± 0.7
Saline solution + indomethacin	56 ± 6	45 ± 7	7.2 ± 0.8	36 ± 5	42 ± 10	6.5 ± 0.7
HCl	194 ± 18	341 ± 28	11.6 ± 1.0	134 ± 11	210 ± 35	11.8 ± 1.0
Indomethacin + HCl	62 ± 8	79 ± 12	7.5 ± 0.7	35 ± 5	43 ± 12	7.2 ± 0.6
Arachidonic acid (μg/ml)						
12.5	198 ± 10	275 ± 42	16.4 ± 1.8	110 ± 6	143 ± 12	14.0 ± 1.5
25.0	195 ± 12	390 ± 49	18.5 ± 1.4	112 ± 11	225 ± 26	16.7 ± 1.5
50.0	227 ± 19	542 ± 77	23.2 ± 2.1	154 ± 12	322 ± 43	20.0 ± 2.2
100.0	283 ± 21	712 ± 84	25.1 ± 2.2	174 ± 18	514 ± 76	20.8 ± 2.1
Indomethacin + arachidonic acid (12.5 μg/ml)	61.6 ± 6	108 ± 17	8.2 ± 0.9	40 ± 5	83 ± 14	7.0 ± 0.6

NOTE: Values represent the means ± SD of triplicate analyses performed on the perfusates from five dogs.

TABLE 2. Chemical Composition of Mucins from Arachidonic Acid and HCl Perfusates of the Proximal and Distal Duodenum

Location and Type of Perfusion	Component (mg/100 mg of mucin)			
	Protein	Carbohydrate	Associated Lipids	Covalently Bound Fatty Acids
Proximal duodenum				
Saline solution	23.1 ± 2.4	59.7 ± 6.1	12.1 ± 1.2	0.3 ± 0.1
Saline solution + indomethacin	23.9 ± 2.5	59.8 ± 6.3	11.2 ± 1.1	0.2 ± 0.1
HCl	20.5 ± 2.2	61.4 ± 6.5	15.9 ± 1.6	0.3 ± 0.1
Indomethacin + HCl	23.0 ± 2.5	60.8 ± 6.1	12.0 ± 1.1	0.2 ± 0.1
Arachidonic acid (μg/ml)				
12.5	19.4 ± 2.2	62.2 ± 6.3	15.5 ± 1.3	0.3 ± 0.1
100.0	10.1 ± 1.2	65.0 ± 6.4	23.7 ± 2.4	0.6 ± 0.2
Indomethacin + arachidonic acid (12.5 μg/ml)	23.6 ± 2.4	61.1 ± 5.8	12.5 ± 1.2	0.2 ± 0.1
Distal duodenum				
Saline solution	24.6 ± 2.5	58.9 ± 6.0	11.5 ± 1.2	0.2 ± 0.1
Saline solution + indomethacin	25.3 ± 2.6	58.5 ± 5.7	10.8 ± 0.9	0.2 ± 0.1
HCl	21.1 ± 2.2	59.0 ± 5.9	14.2 ± 1.3	0.3 ± 0.1
Indomethacin + HCl	25.0 ± 2.5	58.2 ± 5.7	11.6 ± 1.0	0.2 ± 0.1
Arachidonic acid (μg/ml)				
12.5	20.8 ± 2.1	62.5 ± 6.1	14.1 ± 1.1	0.3 ± 0.1
100.0	12.3 ± 1.4	64.1 ± 6.3	22.1 ± 2.3	0.5 ± 0.2
Indomethacin + arachidonic acid (12.5 μg/ml)	24.0 ± 2.5	59.3 ± 6.1	12.0 ± 6.1	0.2 ± 0.1

NOTE: Values represent the means ± SD of triplicate analyses performed on the perfusates from five dogs.

REFERENCES

1. SLOMIANY, B. L., J. SAROSIEK & A. SLOMIANY. 1984. N.Y. Med. Quart. **4:** 124.
2. SLOMIANY, B. L., A. PIASEK, J. SAROSIEK & A. SLOMIANY. 1985. Scand. J. Gastroenterol. **20:** 1191.
3. KONTUREK, S. J., J. BILSKI, J. TASLER & J. LASKIEWICZ. 1984. Am. J. Physiol. **247:** G149.
4. SAROSIEK, J., A. SLOMIANY, A. TAKAGI & B. L. SLOMIANY. 1984. Biochem. Biophys. Res. Commun. **118:** 523.

Effect of Deglycosylation on Gastric Mucin Viscosity and Acid Impedance[a]

J. SAROSIEK, K. MIZUTA, A. SLOMIANY, AND
B. L. SLOMIANY[b]

Gastroenterology Research Laboratory
Department of Medicine
New York Medical College
Valhalla, New York 10595

INTRODUCTION

The constituent of gastric mucus primarily responsible for its viscoelastic properties is mucus glycoprotein or mucin. This densely glycosylated glycoprotein consists of a protein core and numerous O-glycosidically linked carbohydrate chains, which range in size from two to eighteen sugar units and carry various antigenic determinants.[1,2] While the studies on macromolecular organization of gastric mucin contributed significantly to understanding its polymeric structure and the changes that occur in mucus gel with disease, little is known about the contribution of the carbohydrate chains of mucin to gastric mucosal protection. Here, we present studies on the effect of deglycosylation on the viscosity of gastric mucin and on its ability to retard the diffusion of hydrogen ion.

MATERIALS AND METHODS

The undegraded gastric mucin was prepared from dog gastric mucus by the procedure involving column chromatography on Bio-Gel A-50 followed by equilibrium density gradient centrifugation in CsCl.[3] Enzymatic deglycosylation of the purified mucin was performed by incubating the glycoprotein at 37°C for 24–36 hr with specific exoglycosidases in 0.05 M sodium citrate buffer. Reactions with β-galactosidase, α-N-acetylgalactosaminidase and β-N-acetylhexosaminidase were carried out at pH 4.0, while the digestion with α-L-fucosidase was performed at pH 6.0. Removal of sialic

[a] This work was supported by U.S.P.H.S. Grant DK 21684-09 from the National Institute of Diabetes and Digestive and Kidney Diseases, National Institutes of Health.

[b] Address for correspondence: B. L. Slomiany, Gastroenterology Research Laboratory, New York Medical College, Munger Pavilion, Valhalla, New York 10595.

acid was accomplished at pH 5.5 with neuraminidase.[4] After incubation with a single enzyme or group of enzymes, the reaction mixtures were briefly heated at 100°C (3 min) and chromatographed on Bio-Gel A-1.5-m column equilibrated in 6 M urea. The glycoprotein, eluted in the excluded volume was collected, dialyzed against distilled water, and lyophilized. Such obtained intact and modified preparations of the glycoprotein were analyzed for protein and carbohydrate content.[4] For the measurements of hydrogen ion diffusion, various preparations of mucus glycoprotein were dissolved at 30 mg/ml in 0.15 M NaCl and placed in a sample port of the diffusion chamber separating the 0.15 M NaCl and 0.15 M HCl compartments.[4] The amount of H^+ diffusing through the glycoprotein was calculated in mol/sec and permeability coefficient in cm/sec. Viscosity measurements were performed with a Brookfield cone/plate digital viscometer, equipped with a 1.565° cone and constant (37°C) temperature bath.[5] Shear rates were varied from 1.15 to 230 sec^{-1} and the sample volumes were 0.5 ml.

TABLE 1. Permeability to H^+ of the Intact and Partially Deglycosylated Dog Gastric Mucus Glycoprotein

Type of Sample	Permeability ([mol/sec] $\times 10^{-10}$)	Permeability Coefficient ([cm/sec] $\times 10^{-6}$)
Intact glycoprotein	3.95 ± 0.38	8.14 ± 0.83
Glycoprotein after removal of sialic acid	2.88 ± 0.26	5.93 ± 0.52
Glycoprotein after removal of sialic acid and fucose	2.67 ± 0.22	5.50 ± 0.49
Glycoprotein after removal to sialic acid and terminal *N*-acetylgalactosamine	2.63 ± 0.23	5.42 ± 0.51
Glycoprotein after extensive (86%) deglycosylation	2.31 ± 0.19	4.76 ± 0.43
Control (0.155 M NaCl)	41.02 ± 1.56	68.77 ± 3.10

NOTE: Values represent the means ± SE of triplicate analyses.

RESULTS AND DISCUSSION

The purified dog gastric mucus glycoprotein contained 18.1% protein and 54.8% carbohydrate which consisted of fucose, galactose, *N*-acetylglucosamine, *N*-acetylgalactosamine and sialic acid arranged in a form of branched chains each containing about 10 sugar units.[4] The data on the effect of deglycosylation on the permeability of gastric mucin to hydrogen ion are summarized in TABLE 1. The results show that removal of sialic acid caused a 27% increase in the ability of asialoglycoprotein to resist the diffusion of hydrogen ion. Further 7% reduction in permeability to hydrogen ion was obtained after removal of fucose, while removal of the peripheral *N*-acetylgalactosamine decreased mucin permeability by 8.6%. Extensive deglycosylation of the glycoprotein (loss of 86% carbohydrate), however, increased the ability of mucin to impede the diffusion of hydrogen ion by 41.5%.

The effect of carbohydrate removal on the viscosity of gastric mucin is depicted in FIGURE 1. The data indicate that removal of sialic acid decreased the viscosity of resulting asialomucin by 18.5% and low shear rate and by 11% at high shear rate. Further 5% drop in viscosity was observed after treatment of the glycoprotein with α-L-fucosidase or α-N-acetylgalactosaminidase. After extensive deglycosylation of mucin (loss of 86% carbohydrate), a 40% drop in the glycoprotein viscosity at low shear rate and a 36% drop at high shear rate occurred.

The results indicate that carbohydrate component of mucus glycoprotein exerts a considerable impact on the viscoelastic and permselective properties of gastric mucins. The differences in carbohydrate chains of mucins observed along the alimentary tract of various animals may be an important factor which influence the physiological functions of the mucus gel.

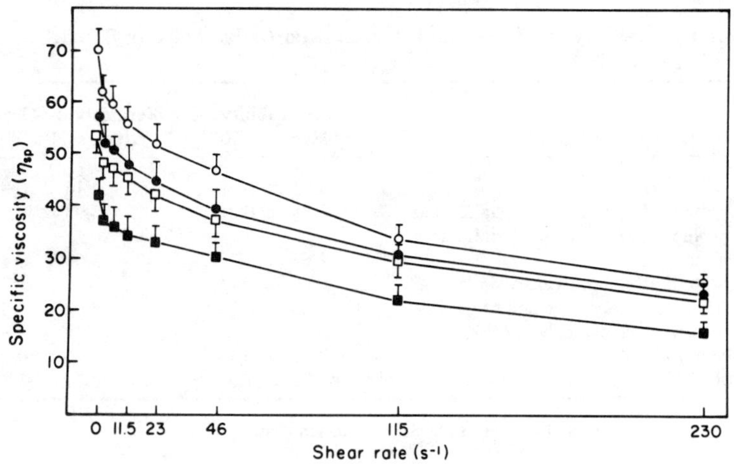

FIGURE 1. Effect of deglycosylation on the viscosity of dog gastric mucin. Intact mucin (O; mucin after removal of sialic acid (●); mucin after removal of sialic acid and fucose (□); and mucin after extensive (86%) deglycosylation (■).

REFERENCES

1. SLOMIANY, B. L., J. SAROSIEK & A. SLOMIANY. 1984. N.Y. Med. Quart. **4:** 124.
2. SLOMIANY, B. L., E. ZDEBSKA & A. SLOMIANY. 1984. J. Biol. Chem. **259:** 14743.
3. SAROSIEK, J., A. SLOMIANY, A. TAKAGI & B. L. SLOMIANY. 1984. Biochem. Biophys. Res. Commun. **118:** 523.
4. SLOMIANY, B. L., Z. BANAS-GRUSZKA, E. ZDEBSKA & A. SLOMIANY. 1982. J. Biol. Chem. **257:** 9561.
5. MURTY, V. L. N., J. SAROSIEK, A. SLOMIANY & B. L. SLOMIANY. 1984. Biochem. Biophys. Res. Commun. **121:** 521.

Biological Validation of Self-Reported AIDS Risk Reduction among New York City Intravenous Drug Users

SAMUEL R. FRIEDMAN,[a] DON C. DES JARLAIS,[b]
DONNA MILDVAN,[c] STANLEY R. YANCOVITZ,[c]
AND JONATHAN GARBER[c]

[a]Narcotic and Drug Research, Inc.
New York, New York 10013

[b]New York State Division of Substance Abuse Services
New York, New York 10027

[c]Beth Israel Medical Center
New York, New York 10003

INTRODUCTION

Much research on AIDS and other diseases depends on self-reported data about behavior. Often, such research is reported with little or no validation of the behavioral data. This paper reports the use of laboratory methods as a way to validate self-report data. This technique requires three conditions: (1) a statistically significant observed relationship between behavioral differences and biological differences; (2) mechanisms(s) that link the behavioral differences to the biological differences; and (3) evidence that the biological differences did not cause the behavioral differences.

The specified relationship and mechanism concern attempts by intravenous (IV) drug users to protect themselves against AIDS. One of our research questions was whether, among IV drug users infected by the human immunodeficiency virus (HIV), those who frequently shared needles or who reused old needles without cleaning them were more likely to develop immunosuppression. Zagury *et al.*[1] provide one mechanism to explain such a relationship. They showed *in vitro* that immunologic activation of T4 cells infected by HIV stimulates viral expression and consequent T4 cell death. Use of shared or reused needles is particularly likely to lead to immunologic stimulation through injection of allogeneic blood and/or infectious agents.[2] On the basis of these arguments, we expected seropositive IV drug users who reduced their use of shared needles or who increased the extent to which they used new or cleaned needles to have higher T4 cell counts than did seropositive IV drug users who did not report such risk reduction. Among seronegative IV drug users, on the other hand, these risk-reducing behaviors would not be related to T4 cell count.

METHODS

With informed consent, 59 methadone maintenance patients in a larger study were interviewed about how they had changed their behavior to avoid AIDS. Decreased needle sharing and increased use of cleaned or new needles were the major changes studied. (Since only two subjects reported reducing drug injection to avoid AIDS, this variable was not separately analyzed.) Medical data about AIDS-related symptoms (lymphadenopathy, unexplained fever, or night sweats lasting more than 4 weeks; unexplained weight loss of at least 10 pounds; unexplained diarrhea for more than 1 week; and oral candidiasis) were obtained by interview. Presence of HIV antibody was determined through ELISA testing with confirmation by Western blot, and T4 cell counts were determined with a Coulter EPICS C flow cytometer.

RESULTS

In the 24 seropositive subjects, protective behavior changes were associated with higher T4 cell counts. Three reported only reduced needle-sharing, four had only increased their use of new or cleaned needles, and five reported both changes. Each form of behavior change was significantly associated with higher T4 counts (see TABLE 1). (The five subjects reporting both changes had a mean of 967 T4 cells per microliter whereas the 12 with neither change had a mean of 505 cells per microliter.) The number of changes correlated 0.52 ($p < .011$) with T4 cell count. Among the 35 seronegative subjects, T4 cell count was unrelated to either behavior change (TABLE 2), and its correlation with the number of changes was only 0.09. The reported risk

TABLE 1. Mean T4 Cell Counts for HIV Antibody Seropositive Methadone Maintenance Patients with Respect to Risk Reduction[a]

Risk Reduction	T4 Count (SE)	Number of Cases
Reduced needle-sharing		
Yes	888 (91)	8
No	540 (83)	16
$p(t)$	< .02	
Increased use of new or cleaned needles		
Yes	823 (114)	9
No	555 (83)	15
$p(t)$	< .07	

[a] T4 counts in cells per microliter. Laboratory tests were conducted at the New York City Department of Health Laboratory.

TABLE 2. Mean T4 Cell Counts for HIV Antibody Seronegative Methadone Maintenance Patients with Respect to Risk Reduction[a]

Risk Reduction	T4 Count (SE)	Number of Cases
Reduced needle-sharing		
Yes	1274 (170)	9
No	1132 (80)	26
Increased use of new or cleaned needles		
Yes	1163 (169)	9
No	1170 (82)	26

[a] T4 counts in cells per microliter.

reduction did not result from medical consequences of infection leading to behavior changes. Neither the reported presence of an AIDS-related symptom nor the number of such symptoms predicts seropositive clients' (or seronegative clients') reported needle-sharing or needle-cleaning behavior.

CONCLUSIONS

These results tend to validate the behavioral data about risk reduction since there is a reasonable mechanism linking risk reduction to T4 cell count, since reported risk reduction was associated with higher T4 cell counts among seropositive but not among seronegative clients, and since the reported behavioral changes were not a result of HIV-related symptoms. The findings are consistent with the idea that immunologic stimulation leads to *in vivo* T4 cell loss among HIV-infected individuals. The data are also consistent with other reports that intravenous drug users are trying to reduce their risk for AIDS,[3] and suggest that drug users' behavioral changes may affect disease development. Finally, these data, together with prospective evidence that continued IV drug use is associated with immunosuppression,[2] indicate that seropositive IV drug users should be advised to use sterile equipment and not to share needles or other paraphernalia in order to protect themselves as well as their potential partners.

REFERENCES

1. ZAGURY, D. *et al.* 1986. Science **231**: 850-853.
2. DES JARLAIS, D. C. *et al.* 1986. AIDS **1**: 105-112.
3. DES JARLAIS, D. C., S. R. FRIEDMAN & W. HOPKINS. 1985. Ann. Int. Med. **103**: 755-759.

Internal Dose Evaluations in Standard and Nonstandard Man Using Computerized Approach

SHLOMO HOORY AND DIBYENDU
BANDYOPADHYAY

*Division of Nuclear Medicine
Long Island Jewish Medical Center, and
State University of New York at Stony Brook
New Hyde Park, New York 11042*

INTRODUCTION

A computerized approach to internal dose estimates due to intake of radionuclides in humans was recently reported by our group.[1] The dosimetry evaluations of selected human target organs are achieved through the following steps:

(1) Construction of the biophysical vector $\tilde{B}$
(2) Construction of the uptake matrix U
(3) Construction of the cumulated activity vector $\tilde{A} = U\tilde{B}$
(4) Construction of the appropriate matrix S from the absorbed dose per cumulated activity factors.[2]

Each element B_i of B is derived from the multiplication of the administered dose A_o by the mean effective life $\tau_i(\text{eff})$ of the radionuclide in organ i. Each element U_i of the diagonal matrix U represents the uptake of the organ i and U_i is greater or equal to zero and less or equal to one, while the trace of U is equal to one. Hence, the dose estimates of the target organs are obtained from $\bar{D} = S\tilde{A}$, where $\bar{D}$ is the estimated dose vector of the targets.

In the present study the algorithm was expanded to include dosimetry evaluations of target organs deviating in mass from those of the normal man. A practical case of internal dose estimates in a patient is discussed and the method of calculations in a typical problem is demonstrated.

METHODS

In reality, human organs may deviate in size from those of the normal man. Therefore, in case of an enlarged (or smaller) organ, a correction of the corresponding S factors is needed. If the mass of a normal target organ i is m_i and the mass of the

"

same organ enlarged is m'_i, the correction factor is given by $k_i = m_i / m'_i$ and $S_i' = (k_i S_{i1}, k_i S_{i2}, \ldots, k_i S_{in})$, where the elements of S'_i represent the corresponding contributions of the n source organs to the non-standard target organ i.

In case of more than one organ deviating from normal size, a similar correction is needed for each of the targets. Hence, the estimated dose of a given set of m target organs is given by

$$\overline{D}' = \begin{pmatrix} D'_1 \\ D'_2 \\ \cdot \\ \cdot \\ \cdot \\ D'_m \end{pmatrix} = \begin{pmatrix} k_1 S_{11}, k_1 S_{12}, \ldots, k_1 S_{1n} \\ k_2 S_{21}, k_2 S_{22}, \ldots, k_2 S_{2n} \\ \cdot \\ \cdot \\ \cdot \\ k_m S_{m1}, k_m S_{m2}, \ldots, k_m S_{mn} \end{pmatrix} \begin{pmatrix} \tilde{A}_1 \\ \tilde{A}_2 \\ \cdot \\ \cdot \\ \cdot \\ \tilde{A}_n \end{pmatrix} =$$

$$= \begin{pmatrix} k_1 & & O \\ & k_2 & \\ & & \cdot \\ O & & k_m \end{pmatrix} \left[\begin{pmatrix} S_{11}, S_{12}, \ldots, S_{1n} \\ S_{11}, S_{22}, \ldots, S_{2n} \\ \cdot \\ \cdot \\ \cdot \\ S_{m1}, S_{m2}, \ldots, S_{mn} \end{pmatrix} \begin{pmatrix} A_1 \\ A_2 \\ \cdot \\ \cdot \\ \cdot \\ A_n \end{pmatrix} \right] = \begin{pmatrix} k_1 & & O \\ & k_2 & \\ & & \cdot \\ O & & k_m \end{pmatrix} \begin{pmatrix} \overline{D}_1 \\ \overline{D}_2 \\ \cdot \\ \cdot \\ \cdot \\ D_m \end{pmatrix}$$

Therefore, dose estimates for target organs in the nonstandard man, $\overline{D}'$, are obtained by multiplying the correction matrix K (diagonal) by the dose vector $\overline{D}$ obtained for the standard man. The values of the correction factors of K are $k_i = 1$ or $k_i \neq 1$ depending on whether the target organ is standard or nonstandard respectively.

RESULTS AND DISCUSSION

In TABLES 1 and 2, dosimetry evaluations of a patient weighing 80kg with an enlarged liver (20%) given 5 mCi ^{99m}Tc-labeled sulfur colloid are demonstrated. Assuming a 46-hour biological half-life of technetium in the total body and an effective half-life equal to the physical half-life (6.03 hr) in all other source organs, the following biophysical vector is obtained:

$$B = 1.44 \times 5000 \, (6.03, 6.03, 6.03, 6.03, 6.03, 5.33)$$

The distribution of the radiopharmaceutical in the various source organs was chosen as follows: liver (85%), red marrow (5%), spleen (7%), thyroid (1%), and total body (2%). Arranging these uptake coefficients in a form of diagonal matrix and multiplying it by the column vector $\tilde{B}$, we can obtain the cumulated activity vector $\tilde{A}$ (TABLE 1).

TABLE 1 demonstrates dose evaluations of the target organs with respect to the normal man where the coefficients in the matrix S are not corrected for liver and total body. Such correction is achieved in TABLE 2, where the diagonal matrix K is

TABLE 1. Mean Dose Estimates of Target Organs in Normal Man after the Administration of 5 mCi ^{99m}Tc-Labeled Sulfur Colloid

| Targets | S Matrix | | | | | $\tilde{A}$ (μCi-hr) | $\bar{D}$ (rad) |
	LI	RM	SP	TY	TB		
Liver	4.6E-05	9.2E-07	9.8E-07	9.3E-08	2.2E-06	36.903E + 3	1.704
Ovaries	4.5E-07	3.2E-06	4.0E-07	4.9E-09	2.4E-06	2.171E + 3	0.029
Red marrow	1.6E-06	3.1E-05	1.7E-06	1.1E-06	2.9E-06	3.039E + 3	0.135
Spleen	9.2E-07	9.2E-07	3.3E-04	1.1E-07	2.2E-06	0.434E + 3	1.041
Lungs	2.5E-06	1.2E-06	2.3E-06	9.4E-07	2.0E-06	0.868E + 3	0.104
Thyroid	1.5E-07	6.8E-07	8.7E-08	2.3E-03	1.5E-06		1.006
Testes	6.2E-08	4.5E-07	4.8E-08	5.0E-10	1.7E-06		0.005
Total body	2.2E-06	2.2E-06	2.2E-06	1.8E-06	2.0E-06		0.095

TABLE 2. Corrected Mean Dose Estimates of Target Organs after the Administration of 5 mCi ^{99m}Tc-Labeled Sulfur Colloid

Organs	K Matrix		$\bar{D}$ (rad)		$\bar{D}'$ (rad)
Liver	0.833		1.704		1.419
Ovaries	1.000	O	0.029		0.029
Red marrow	1.000		0.135		0.135
Spleen	1.000		1.041	=	1.041
Lungs	1.000		0.104		0.104
Thyroid	1.000		1.006		1.006
Testes	O	1.000	0.005		0.005
Total body	0.875		0.095		0.079

multiplied by the $\bar{D}$ vector obtained in TABLE 1. Notice the correction factors for liver and total body appear in the first and last locations, respectively, in the diagonal matrix K.

REFERENCES

1. HOORY, S. 1986. Internal dosimetry evaluations: A computerized approach. Health Physics **50:** 195-201.
2. SNYDER, W. S., M. R. FORD, G. G. WARNER & S. WATSON. 1975. 'S' Absorbed Dose per Unit Cumulated Activity for Selected Radionuclide and Organ. MIRD Pamphlet No. 11. The Society of Nuclear Medicine, Inc. New York, NY.

Polymerization Characteristics of Divalent Cation-Free Actin[a]

LEWIS C. GERSHMAN, JAMES E. ESTES,
AND LYNN A. SELDEN

Research and Medical Services
Veterans Administration Medical Center and
Departments of Medicine and Physiology
Albany Medical College
Albany, New York 12208

At the First Colloquium in Biological Sciences three years ago, we reported that the polymerization characteristics of actin are regulated by the tightly-bound divalent cation: Actin containing tightly bound Mg^{2+} (Mg-actin) nucleates and polymerizes more readily and has a lower critical concentration than does actin containing tightly bound Ca^{2+} (Ca-actin).[1]

We have recently determined that the high-affinity binding of Ca^{2+} and Mg^{2+} to actin is much stronger than previously thought: at pH 7 the dissociation constant for Ca-actin is 2 nM and for Mg-actin about 10 nM.[2] The studies reported here on the preparation and characterization of divalent-cation-free actin (DCF-actin) were performed to further elucidate the importance of the tightly bound divalent cation to actin polymerization. However, since actin is known to denature rapidly at low divalent cation concentrations, this work was prefaced with a brief study of DCF-actin denaturation.

Ca-actin was prepared and labelled with N-1-pyrenyl-iodoacetamide (N-pyrenyl-actin) by standard procedures,[3] and Mg-actin was prepared as described previously.[4] The increase in N-pyrenyl-actin fluorescence upon polymerization was used to measure initial rates of polymerization as described in FIGURE 1. N-pyrenyl-actin fluorescence increases upon denaturation as well, though to a lesser degree than for polymerization. This N-pyrenyl fluorescence change, the change in intrinsic fluorescence of native actin on denaturation,[5] and the loss of actin polymerizability on denaturation were all used to monitor actin denaturation. DCF-actin was prepared by adding 10 mM EDTA or 5 mM BAPTA[b] (a strong Ca^{2+}-selective chelator) to monomeric Ca-actin and allowing 3-4 min for dissociation of the tightly bound Ca^{2+} ($t_{1/2}$ at pH 7 was about 45 sec).

FIGURE 1 shows the denaturation of actin at pH 7, 25°C, as measured by three methods. The denaturation rate measured by N-pyrenyl fluorescence ($t_{1/2}$ about 9 min) appears slightly faster than the denaturation rates measured by intrinsic actin fluorescence ($t_{1/2}$ about 11 min), or from initial rates of polymerization ($t_{1/2}$ about 12 min). The half-life of DCF-actin at pH 7 is thus on the order of 10 min. Since

---

[a] Supported by the Veterans Administration and by NIH Grant GM-32007.
[b] BAPTA: 1,2-bis(o-aminophenoxy)-ethane-N,N,N',N'-tetraacetic acid.

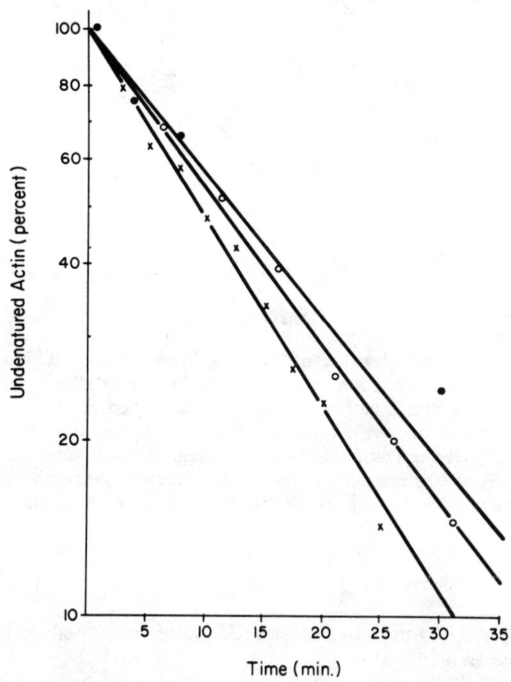

FIGURE 1. Time course of denaturation of DCF-actin at 25°C in 5 mM HEPES, 0.2 mM ATP, 0.003 mM CaCl, 1.5 mM NaN$_3$, pH 7.0 ($\times$), 10 mM EDTA added to 2 μM N-pyrenyl-actin at time t = 0 and the fluorescence intensity increase (ΔF) at 365 nm measured (excitation wavelength, 350 nm). Plotted as 100 $\times$ [1 $-$ ΔF/ΔF(∞)] versus t, where ΔF(∞) indicates end-point fluorescence change; ($\bigcirc$), 10 mM EDTA added to 2 μM native actin at time t = 0 and intrinsic fluorescence intensity *decrease*, ΔF, at 334 nm measured (excitative wavelength, 283 nm); results plotted as above after data were computer-corrected for linear drift due to bleaching; ($\bullet$), 5 mM BAPTA added at time t = 0. At subsequent times 1.5-ml aliquots were subjected to initial polymerization rate measurements by addition of 20 μl phalloidin-stabilized seeds and 20 mM KCl and monitoring of N-pyrenyl fluorescence increase at 365 nm.[3] Results were then normalized, setting the extrapolated rate at time t = 0 to 100%.

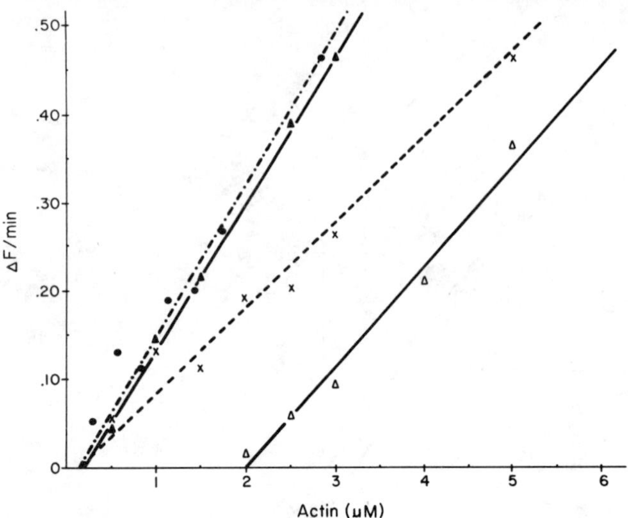

FIGURE 2. Polymerization characteristics of Ca-actin, Mg-actin and DCF-actin. Initial rate of polymerization ($\Delta F/\min$) as a function of actin concentration is shown. Solution conditions and measurement of initial polymerization rates were as described in FIGURE 1. ($\triangle$), Ca-actin; ($\blacktriangle$), Mg-actin; ($\times$), DCF-actin; ($\bullet$), DCF-actin data were corrected for denaturation; the measured initial rates of polymerization were plotted against DCF-actin concentration values calculated by subtracting the estimated amount of denaturation occurring in 6 min from the initial actin concentrations. The half-life of DCF-actin was taken as 10 min.

preparation takes only 3-4 min, studies of DCF-actin are possible, but the effects of denaturation must be considered.

FIGURE 2 shows the initial rates of polymerization as a function of actin concentration for Ca-actin, Mg-actin, and DCF-actin. Taking into account the half-life of DCF-actin of approximately 10 min estimated from FIGURE 1, we corrected the DCF-actin data in FIGURE 2 for denaturation, and the corrected curve is shown as well. It is seen that DCF-actin has a low critical concentration (x-axis intercept) of 0.2 μM, like Mg-actin, and a relative forward polymerization rate (slope) similar to that for Mg-actin and about 50% higher than for Ca-actin.

In summary, DCF-actin may be prepared using high concentrations of chelators and it has a half-life of about 10 min. Initial polymerization rate studies show that it has polymerization characteristics like those of Mg-actin. These results suggest that DCF-actin probably has a conformation unlike that of Ca-actin but similar to the conformation of "physiologic" Mg-actin.

REFERENCES

1.   GERSHMAN, L. C., J. E. ESTES & L. A. SELDEN. 1984. Ann. N.Y. Acad. Sci. **435:** 151-153.

2. GERSHMAN, L. C., L. S. SELDEN & J. E. ESTES. 1986. Biochem. Biophys. Res. Commun. **135:** 607-614.
3. SELDEN, L. A., L. C. GERSHMAN & J. E. ESTES 1986. J. Muscle Res. and Cell Motility **7:** 215-224.
4. SELDEN, L. A., J. E. ESTES & L. C. GERSHMAN. 1983. Biochem. Biophys. Res. Commun. **116:** 478-485.
5. LEHRER, S. S. & G. KERWAR. 1972. Biochemistry **11:** 1211-1217.

Growth of *Mycoplasma capricolum* with Exogenous Phosphatidylethanolamine Destabilizes the Bilayer Membrane

SANDA CLEJAN AND MOISES MALOWANY

Department of Pathology
City Hospital Center at Elmhurst and
Mount Sinai School of Medicine
Elmhurst, New York 11373

Nonbilayer lipid structures have been postulated to play important roles in membrane-mediated events like transmembrane movement of ions and enzymes.[1,2] Although natural and synthetic mixed lipids were demonstrated to undergo transitions between lamellar and hexagonal or cubic phases, there have been few examples of these structures in biological membranes.[3] Mycoplasmas were used as models to study lipid translocation because their membrane composition is easily manipulated.[4,5] Exogenous phospholipids like phosphatidylcholine and sphingomyelin were incorporated without alteration in the growth rate.[5,6] Because phosphatidylethanolamine (PE) can assume nonbilayer structures under appropriate conditions, we incorporated into the membrane of growing *M. capricolum* cells dipalmitoyl PE (DPPE), a saturated PE; dioleyl PE (DOPE), an unsaturated PE; and a natural PE, soybean PE (SPE), composed of both saturated and unsaturated PE.

METHODS

The growth of *M. capricolum,* isolation of cell membrane, preparation of small unilamellar vesicles (SUV) and measurements of [^{14}C]cholesterol exchange were described previously.[4,7] To estimate the extent of fusion between *M. capricolum* cells and SUV, we used SUV containing [^{3}H]glycerol trioleate. From the recovery of ^{3}H counts/min in the cell pellet, we estimate that during a 10-hr incubation period, about 6% fusion occurred with cells grown without PE, 10% with cells grown with DPPE and 14% with cells grown with DOPE. The recovery and leakiness of cells after incubation with and separation from acceptor SUV, monitored by supplementation of the growth media with [^{3}H]thymidine, show 80-90% retention of nucleic acids for DPPE grown cells and 75-86% ^{3}H label retained for DOPE and SPE-grown cells. The analytical procedures used for determination of protein, phospholipid and cholesterol have been described.[4]

RESULTS

FIGURE 1 shows that cells grew better with DPPE at 37°C. Cultures with SPE and DOPE at 37°C reached the stationary phase at a lower turbidity. However, at 34°C, DPPE supported growth less than SPE or DOPE. Lower amounts of SPE and DOPE were incorporated into the cell membrane (TABLE I), compared with DPPE (which differ from DOPE only in the degree of saturation of the acyl chains). There was a 50% increase in phosphatidylglycerol (PG) synthesis. However, the cardiolipin (DPG) content of the membrane remained the same. Cholesterol was also incorporated to a larger extent in the DOPE-grown cells than in the DPPE-grown cells. The structures of the membranes and of the lipids extracted from *M. capricolum* grown in media with the PEs were studied by freeze-fracture electron microscopy. The lipids

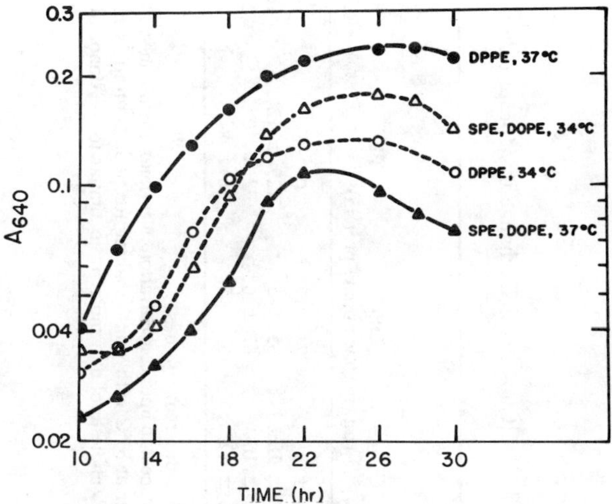

FIGURE 1. Growth of *M. capricolum* with saturated and unsaturated PE at 37 and 34°C.

extracted from cells grown in DPPE-enriched media showed bilayer-walled vesicles and the membranes consisted mainly of rigid planar bilayer. In contrast, both the cells grown on DOPE and SPE showed wide rows of flat H_{II} tubes and precursors of H_{II} phase. When membranes were isolated from cells grown in the presence of DOPE, the rate of [^{14}C]cholesterol exchange was enhanced by 15%. This is shown by the decrease in $t_{1/2}$ values (TABLE 1). We also examined the influence of unsaturated PE on the kinetics of [^{14}C]cholesterol exchange from negatively charged SUV of varying PE composition to an excess of neutral SUV acceptors. In this SUV-SUV exchange experiment, incorporation of DOPE caused the half-time of [^{14}C]cholesterol exchange to decrease by more than 26%.

TABLE 1. Lipid Composition and Kinetics of [^{14}C]Cholesterol Exchange in *M. capricolum* Cells with Saturated and Unsaturated PE Incorporated in the Membrane

Exogenous PL Added to the Growth Medium	Total PL	PG	DPG	PE	Cholesterol	Membrane SUV[a] $t_{1/2}$ (hr)	SUV-SUV[b] $t_{1/2}$ (hr)
			(μg/mg membrane protein)				
No addition	298 ± 16	154 ± 8	74 ± 7	0	190 ± 15	4.1 ± 0.3	3.7 ± 0.3
DPPE	324 ± 22	130 ± 10	65 ± 5	129 ± 12	188 ± 16	3.9 ± 0.2	3.8 ± 0.2
SPE	286 ± 20	147 ± 10	67 ± 6	72 ± 10	210 ± 14	3.6 ± 0.3	3.0 ± 0.2
DOPE	279 ± 16	196 ± 12	67 ± 5	56 ± 8	230 ± 18	3.5 ± 0.2	2.7 ± 0.1

NOTE: Values represent the mean ± SD of five different cultures.

[a] Acceptor SUV were in a 50-fold excess with respect to lipid concentration over donor membranes. Donor membranes (approximately 0.5 mg membrane protein) and SUV were incubated in STM buffer at 37°C with gentle shaking for periods up to 14 hr.

[b] Acceptor vesicles were composed of PC/PE/cholesterol at a molar ratio similar to the donor SUV.

CONCLUSIONS

(1) Introduction of PE in the *M. capricolum* lipids is not sufficient to abolish growth or to affect the rate of the logarithmic phase of growth after an initial lag phase.

(2) *M. capricolum* can be manipulated to contain many forms of nonbilayer structures. DOPE favors the formation of these structures.

(3) The rates of spontaneous [^{14}C]cholesterol exchange are increased under conditions favoring nonbilayer structures.

(4) Destabilization of membrane bilayer is a function of (a) amount of PE incorporated, (b) degree of saturation of the fatty acyl chain, (c) cholesterol content, and (d) changes in the synthesis of PG.

(5) Changes in composition of lipids in favor of bilayer-forming endogenous PG in cells grown with unsaturated PE are probably compensating the formation of nonbilayer structures which are not favorable for growth.

REFERENCES

1. CULLIS, P. R., B. deKRUIJFF, M. J. HOPE, R. NAYAR & S. L. SMID. 1980. Can. J. Biochem. **58:** 1091–1100.
2. CHENG, K.-H. & S. W. HUI. 1986. Arch. Biochem. Biophys. **244:** 382–386.
3. GOUNARIS, K., A. SEN, A. P. R. BRAUN, P. J. QUINN & W. P. WILLIAMS. 1983. Biochim. Biophys. Acta **598:** 554–560.
4. CLEJAN, S., R. BITTMAN & S. ROTTEM. 1981. Biochemistry **20:** 2200–2204.
5. CLEJAN, S. & R. BITTMAN. 1984. J. Biol. Chem. **259:** 10823–10826.
6. CLEJAN, S. & R. BITTMAN. 1984. J. Biol. Chem. **259:** 441–448.
7. CLEJAN, S. & R. BITTMAN. 1984. J. Biol. Chem. **259:** 449–455.

The Effect of Modification of the Blood-Brain Barrier to IgG on the Course of Murine Herpes Simplex Myelitis

THOMAS A. KENT[a,c,d]
AND ROBERT R. McKENDALL[a,b]

*Departments of Neurology,[a] Microbiology,[b] Psychiatry,[c] and
Pharmacology[d]
The University of Texas Medical Branch
Galveston, Texas 77550*

The blood-brain barrier (BBB) is an important factor in central nervous system (CNS) viral infection.[1] Capillary tight junctions constitute a formidable barrier to entrance of virus, unlike the more permeable peripheral capillaries. However, in addition to the effect of restricting access of the CNS to infection, the BBB may also affect immune response effectiveness once infection occurs. Clinical and animal studies demonstrate delayed and heterogeneous opening of the BBB after infection. In human herpes simplex virus (HSV) encephalitis, significant numbers of patients have normal cerebrospinal fluid (CSF) protein and brain scans at the onset of illness,[2] suggesting a relative intactness of CNS barriers. Animal studies show that Evan's blue dye fails to penetrate the BBB until severe histopathologic changes appear.[3] This may have implications for the immune response because an intact BBB will exclude neutralizing immunoglobulins. Indeed, 70% of patients with HSV encephalitis have circulating anti-HSV IgG prior to infection yet are not protected from significant illness.[2] In HSV myelitis in mice after footpad infection, the BBB does not become permeable to IgG until four days after virus has reached the spinal cord.[4] Moreover, passively transferred rabbit (Rb) anti-HSV IgG effectively clears footpad but not cord virus if given soon after cord infection.[4] These observations led us to test whether deliberate opening of the BBB combined with anti-HSV IgG may decrease illness. We injected $10^{5.5}$ pfu of HSV into the footpad of 80 mice. HSV travels intra-axonally to the dorsal root into the spinal cord and brain stem. Intra-axonal spread is also a hypothesized route in human HSV encephalitis. Half were given 0.5 cc of 1:64 Rb anti-HSV IgG 3.5 days after inoculation—a time when virus has already reached the cord, but before significant illness and IgG entrance has occurred. Half of each group was then treated with 12% CO_2 for 1.5 hours three times a day. High-dose CO_2 is known to increase the BBB to large molecules.[5] In a subset of animals, one day after CO_2 treatment was begun, footpad virus titers were obtained to see whether CO_2 had a direct antiviral effect, and CSF and serum were obtained to determine the IgG ratios as an indicator of protein transudation. A pilot study ($n=42$) was performed with 5% CO_2 for 6 hours twice a day. Both treatments reduced illness by approximately 50% with statistical significance reached in the larger 12% CO_2 group (TABLE 1). IgG appeared

TABLE 1. Effects of CO_2 and Passively Transferred Rabbit Anti-HSV IgG

Group	n	Illness Rate	Death Rate	CSF/Serum IgG $\times 10^{3a}$	Footpad Virus Titer[b]
HSV only	30	(60%)[c]	(20%)[c]		$10^{6.7}$
HSV + IgG	30	73%	47%	7.11 ± 1.79	$10^{5.4 \pm .5}$
HSV + 12% CO_2	20	35%	20%		$10^{5.7 \pm .4}$
HSV + 12% CO_2 + IgG	20	35%[e]	30%	16 ± 5.7^d	$10^{6.3 \pm .8}$
HSV + 5% CO_2	13	70%	54%		
HSV + 5% CO_2 + IgG	9	44%[f]	22%	19 ± 6.8^d	$10^{6.3 \pm .1}$

[a] Uninfected 7.7 ± 1.9 ($n = 10$); $n = 4$, HSV + IgG; $n = 9$, HSV + 12% CO_2 + IgG; $n = 7$ HSV + 5% CO_2 + IgG.

[b] $n = 3$ in each group; no significant differences.

[c] Artifactually low because some animals did not show HSV from footpads, indicating probable faulty inoculation.

[d] Combined CO_2 versus controls, $p < 0.05$ Student's t test.

[e] $p < 0.05$ versus HSV + IgG, χ^2 square.

[f] $0.1 > p > 0.05$.

to potentiate the 5% CO_2 ($0.1 > p > 0.05$), but the effect on 12% CO_2 was not as clear, possibly due to the low illness rate in the virus-only controls. This low illness rate was a result of faulty virus inoculation (some animals had no virus recoverable from footpad). 12% CO_2 alone may have had some beneficial effect, perhaps by increasing CNS levels of other immune substances. This needs to be studied further. The CSF/serum Rb IgG ratios were doubled. Neither treatment reduced footpad virus, suggesting no direct antiviral effect of CO_2. An example of the time course of illness is shown in FIGURE 1.

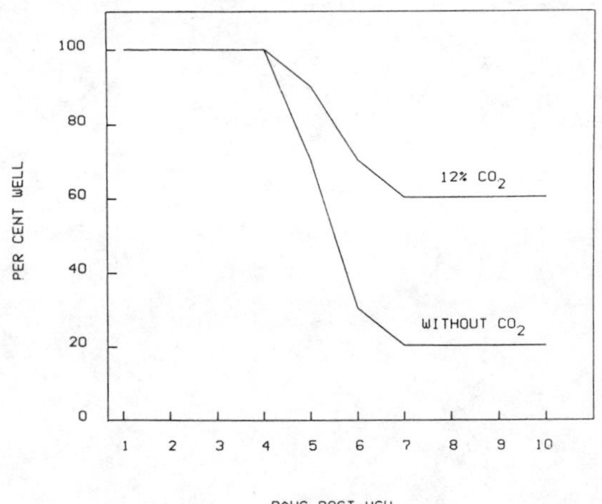

FIGURE 1. Time course of illness after footpad HSV inoculation in a representative experiment. Both groups (each $n = 10$) were treated with 0.5 cc of 1:64 rabbit anti-HSV neutralizing IgG 3.5 days after inoculation. Treatment with 12% CO_2 was then begun (*top line*). HSV 12% CO_2 + IgG significantly reduced the illness rate compared to HSV + IgG without CO_2 ($p < 0.05$, χ^2).

These results support the role of the BBB in the pathophysiology of central nervous system HSV infection in terms of clinically significant hindrance to early immune response effectiveness. Other BBB effects may also be important, such as disruption of endothelial cell transport properties. Perhaps in at least some cases of HSV infection, strategies to increase CNS levels of IgG may be beneficial. Quantification of the various BBB effects with radiotracer techniques is the immediate goal of our research.

REFERENCES

1. KENT, T. A. & R. R. McKENDALL. *In* Handbook of Clinical Neurology, Viral Diseases, Vol. 8. R. R. McKendall, Ed. Elsevier. Amsterdam. In press.
2. NAHMIAS, A. J., R. R. WHITLEY, A. N. VISINTINE, Y. TAKEI & C. A. ALFORD. 1982. J. Infect. Dis. **145:** 829.
3. KRISTENSSON, K. & P. SOURANDER. 1969. Acta Neuropathol. **14:** 38.
4. McKENDALL, R. R. 1983. J. Gen. Virol. **64:** 1965.
5. RAPOPORT, S. I. 1976. Blood-Brain Barrier in Physiology and Medicine. Raven Press. New York, NY.

Complement-Independent Stimulation of Polymorphonuclear Neutrophils by Endotoxins from the Core Polysaccharide Mutants of *Salmonella minnesota*[a]

MICHAEL D. LaSALLE[b] AND VINCENT A. DeBARI[c,d]

[b]Cook College
Rutgers University
New Brunswick, New Jersey 08903

and

[c]The Renal Laboratory
Department of Medicine
St. Joseph's Hospital and Medical Center
Paterson, New Jersey 07503

The polymorphonuclear neutrophil (PMN) plays an important role in the host-defense system of the body, especially with regard to invading bacteria. The study of the relationship between the PMN and these organisms (and their products) has long been recognized as important in the clinical approach to infectious diseases. Endotoxins, the lipopolysaccharides (LPS) from gram-negative bacteria, constitute an important aspect of the immune response to those substances from which they derive. Yet, to a large extent, the interaction of these substances with the PMN is unclear. For example, until recently, LPS was thought to "uncouple" phagocytic glucose utilization (the hexose monophosphate pathway) and free-radical generation as assessed by nitroblue tetrazolium reduction and luminol-dependent chemiluminescence (CL).[1] Recent work from our laboratory has shown this not to be the case.[2] We have, in fact, clearly demonstrated complement-independent free-radical generation in LPS from *S. minnesota* (wild-type), and from the Re mutant, R595, as well as from Lipid A. Herein, we extend these observations to the core polysaccharide mutants, Ra, Rb, Rc and Rd in an attempt to elucidate structure-function relationships between the LPS from core mutants and their ability to stimulate PMN.

[a]Supported in part by a grant-in-aid from the New Jersey Lions District 16A Charitable Foundation.

[d]Address for correspondence: Dr. DeBari, Renal Laboratory, St. Joseph's Hospital and Medical Center, 703 Main Street, Paterson, New Jersey 07503.

MATERIALS AND METHODS

Materials

The PMN used in this study was harvested from apparently healthy volunteers who gave informed consent. The protocol was approved by the Institutional Review Board of St. Joseph's Hospital and Medical Center. The method for cell harvesting has been described in detail previously.[2,3] In brief, it depends upon dextran sedimentation, followed by NH_4Cl lysis of erythrocytes, centrifugal sedimentation of PMN over fetal calf serum and, finally, buffered-saline washes of the cell suspension. The cells thus harvested always exhibited viability in excess of 85%. List Biologicals Inc. (Campbell, CA) was the source of LPS from strains R60, R345, R5 and R7, the chemotypes of which are schematically presented in FIGURE 1.

Methods

The experimental protocol used was that described earlier[2] for serum-free incubations. The LPS concentration was maintained at 1 μg/ml for all experiments. Incubation mixtures contained 10^7 PMN suspended in a total volume of 1 ml. All

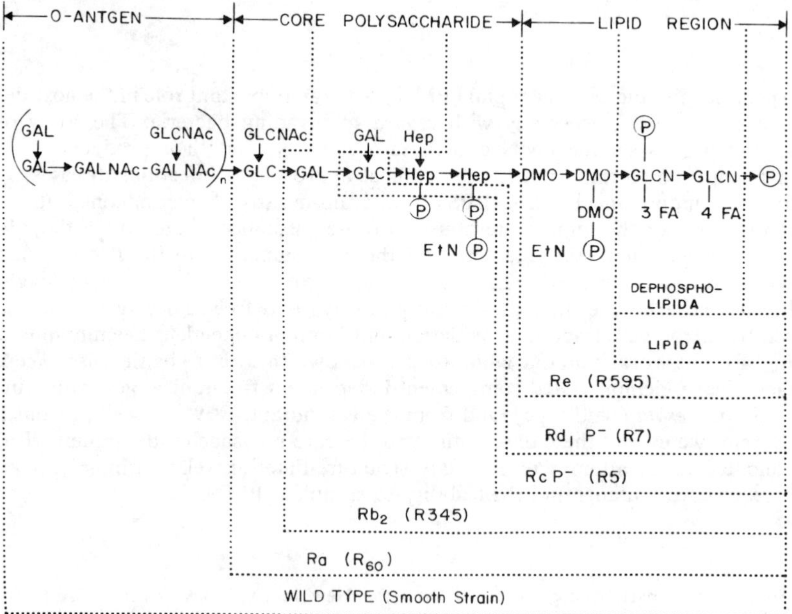

FIGURE 1. Chemotype chart for LPS from *S. minnesota.* The LPS types used in this study were from the core region mutants.

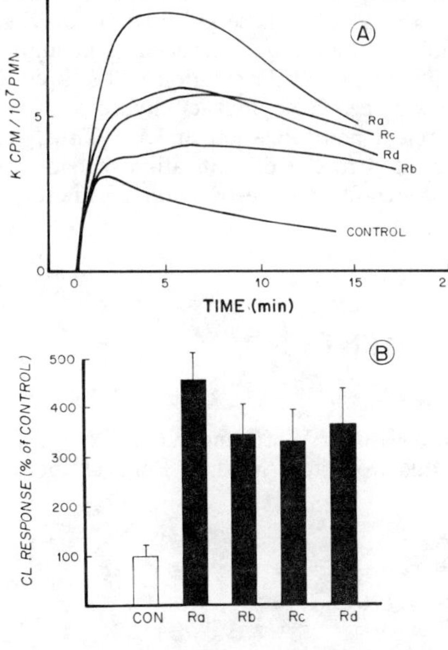

FIGURE 2. The PMN chemiluminescence response to core-mutant LPS from *S. minnesota.* (**A**) Time recording from a single experiment comparing core-mutant LPS with control. (**B**) Results of a series of experiments in which the response was integrated over 30 minutes, compared to control (100%). For this series of experiments, $n = 7$, data shown are ± 1 SEM. Control deviation is 1 SEM for mean adjusted to 100%.

LPS solutions and controls contained 0.5% triethylamine to minimize micelle formation by LPS.

RESULTS AND DISCUSSION

The time course of the CL response is shown for a single experiment in FIGURE 2A. The control curve compares favorably with those obtained previously,[2] showing a small basal response which decays rapidly. While maximal velocity, given by peak height (CPM), may not significantly exceed that of the control (note the curve obtained with Rb, which has a peak height only about 30% greater than control), the duration of the response is considerably longer and the rate of decay may be slower. The data obtained when the curves are integrated over a thirty-minute period confirm this. This is illustrated, for a series of seven experiments, in FIGURE 2B, showing the means and standard errors obtained upon incubation of PMN with the four LPS types. These data demonstrate a CL response in the range of 3-5 times that of the control. In addition to the overall stimulatory effect of the LPS, the endotoxin derived from the Ra chemotype (strain R60) appears to promote a response of greater magnitude than that of the other LPS types. In terms of structural effects, it would seem that the terminal core dimer (Glc—GlcNAc) considerably enhances the CL response. It is not clear whether this effect is one brought about by a membrane transduction mechanism or by intracellular metabolism, or by a combination of the two. Indeed, there is some disagreement about the mechanism of LPS binding to PMN.[4,5] Our data do, however, confirm that LPS can interact with PMN to stimulate oxidative metabolism in these cells and further demonstrate that structural effects may induce subtle differences in the interaction between LPS and the PMN. These interactions may, in fact, have great importance in the pathophysiology of endotoxin-induced tissue injury. The neutrophil has the ability to detoxify LPS by partial deacylation of the lipid A moiety.[6] Although this activity is carried out by a specific acyloxyacyl hydrolase, free-radical generation may enhance the physiological neutralization of LPS. Thus, the PMN may perform a number of functions related to the detoxification of bacterial endotoxins, as well as rely upon these products for the recognition of microbial invasion.

ACKNOWLEDGMENTS

We would like to thank Drs. Peter Kahn, William Ward and Keith Cooper for their thoughtful suggestions with regard to this work and Miss C. Pugliese for her technical assistance.

REFERENCES

1. Proctor, R. A. 1985. Effects of endotoxins on neutrophils. *In* Handbook of Endotoxins. Cellular Biology of Endotoxin. L. J. Berry, Ed. Vol. **3:** 248-254. Elsevier. Amsterdam.
2. DeBari, V. A., P. M. Uychich, A. J. Orsini, A. C. Ingenito & M. A. Needle. 1986. Neutrophil stimulation by lipid A and lipopolysaccharides from *Salmonella minnesota.* Trans. Am. Soc. Artif. Intern. Organs **32:** 597-600.
3. Nicotra, J., A. J. Orsini & V. A. DeBari. 1985. Chemiluminescence response of the human polymorphonuclear neutrophil to lipopolysaccharides. Cell Biophys. **7:** 283-291.
4. Gimber, P. E. & G. W. Rafter. 1969. The interaction of *Escherichia coli* endotoxin with leukocytes. Arch. Biochem. Biophys. **135:** 14-20.
5. Springer, G. F. & J. C. Adye. 1975. Endotoxin-binding substances from human leukocytes and platelets. Infect. Immun. **12:** 978-986.
6. Mumford, R. S. & C. L. Hall. 1986. Detoxification of bacterial lipopolysaccharides (endotoxins) by a human neutrophil enzyme. Science **234:** 203-205.

Restriction Enzyme Analysis of *Campylobacter* Genomic DNA

J. H. BRYNER, I. V. WESLEY, AND L. A. POLLET

United States Department of Agriculture
Agricultural Research Services
National Animal Disease Center
Ames, Iowa 50010

The genus *Campylobacter* consists of microaerophilic curved bacteria previously identified as *Vibrio* and includes pathogens of medical and veterinary importance. Few phenotypic characteristics confirm their identification. *C. jejuni* causes human enteritis and abortion in sheep whereas *C. hyointestinalis* has been associated with swine proliferative ileitis.[1] *C. coli,* the predominant *Campylobacter* colonizing the intestine of pigs, has been isolated from cases of human bloody diarrhea. *Campylobacter fetus fetus* is orally transmitted, associated with abortion in sheep and in cattle, and is responsible for bacteremia, endocarditis and enteritis in humans. *C. fetus venerealis* is transmitted venereally, and induces abortion in cattle, as does the aerotolerant *C. cryaerophila. C. sputorum bubulus* is a commensal of the genital tract of cattle and sheep.

Restriction enzyme analysis (REA) has been used to study the epidemiology of *C. coli,*[2] *C. jejuni,*[3,4] *C. pyloridis,*[5] and *C. fetus fetus.*[6] In this report we use restriction enzyme analysis of whole-cell lysate (chromosomal) DNA (1) to identify seven *Campylobacter* species, and (2) to monitor changes in the *C. fetus venerealis* genome with passage *in vivo* and *in vitro.*

Campylobacter colonies from three-day cultures grown on BHI agar supplemented with 10% defibrinated bovine blood were harvested and suspended in 0.85% NaCl, centrifuged (3020 $\times$ *g;* 30 min), and cell pellets were frozen until the time of DNA extraction. Chromosomal DNA was prepared essentially as described by Thiermann *et al.*[7] Purified *Campylobacter* DNA (1-2 µg) was digested to completion (3-4 hr) with *Bgl*II, *Bst*EII, *Eco*RI, *Hha*I, or *Xho*I in buffers and at temperatures recommended by the manufacturer.

Restriction fragments were separated by agarose gel electrophoresis (16 hr; 60 V) on a 20 $\times$ 25 cm horizontal electrophoresis box. Gels consisted of 0.7% Seakem ME agarose (Marine Colloids, FMC Corp.) in Tris-borate-EDTA buffer (pH 8.2). At the termination of electrophoresis, gels were stained with ethidium bromide for 45 minutes, visualized with shortwave UV light and photographed using a Kodak 23A red filter.

Preliminary studies with enzymes *Bgl*II, *Bst*EII, *Hha*I, *Xho*I, and *Eco*RI indicated that *Hha*I yielded the most readable DNA banding patterns. Upon *Hha*I digestion, each of the *Campylobacter* species revealed a unique DNA pattern. The restriction enzyme pattern of the aerotolerant *C. cryaerophila* after *Hha*I digestion (FIG. 1, lane 1) closely resembled but was distinguishable from two strains of *C. sputorum bubulus* (lanes 2,3). The two isolates of *C. coli* (lanes 4,5) were distinct, as were the two isolates of *C. jejuni* (lanes 6,7). *C. fetus fetus* is subdivided into serotypes A and B.

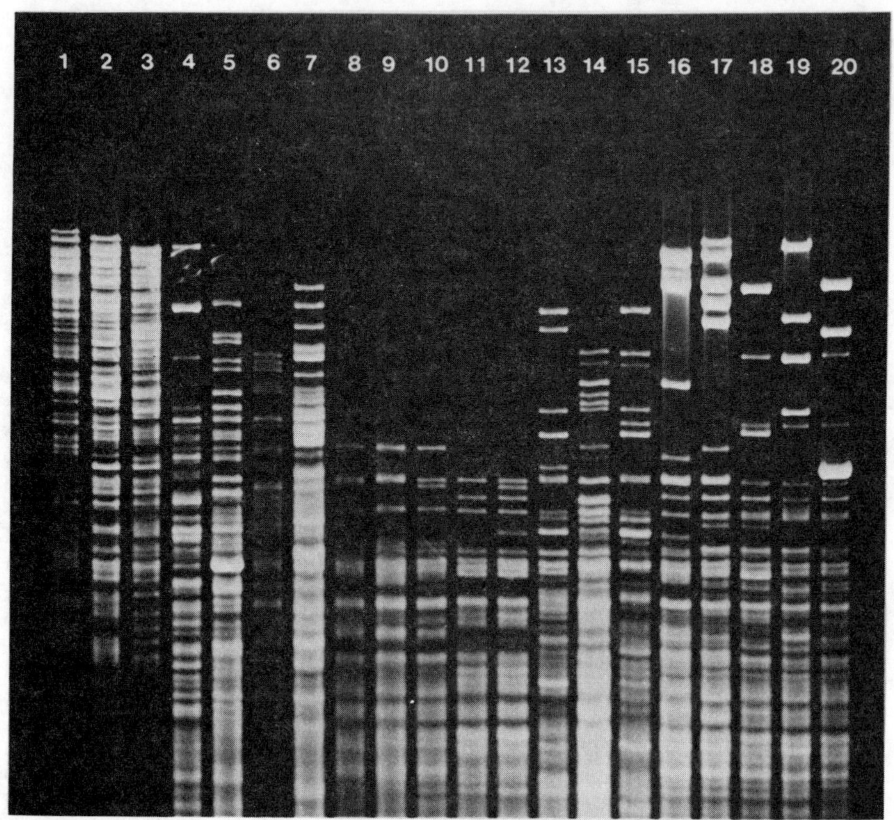

FIGURE 1. *Hha*I digests of whole-cell lysate DNA from *Campylobacter cryaerophila* (lane 1); *C. bubulus sputorum* (lanes 2,3); *C. coli* (lanes 4,5); *C. jejuni* (lanes 6,7); *C. fetus fetus* serotype A (lanes 8-10); *C. fetus fetus* serotype B (lanes 11,12); *C. hyointestinalis* (lanes 13-15); and *C. fetus venerealis* (lanes 16-20).

Digestion of chromosomal DNA with *Hha*I revealed a similar pattern for the serotype A isolates (lanes 8-10), which clearly differed from the serotype B isolates (lanes 11,12). Although clearly unique, two *C. hyointestinalis* porcine isolates were more similar to each other (lanes 13,15) than to a third isolate (lane 14). The five *C. fetus venerealis* isolates were distinguished by the presence of dense high molecular weight bands (lanes 16-20), which were not seen with other isolates nor visualized with other enzymes.

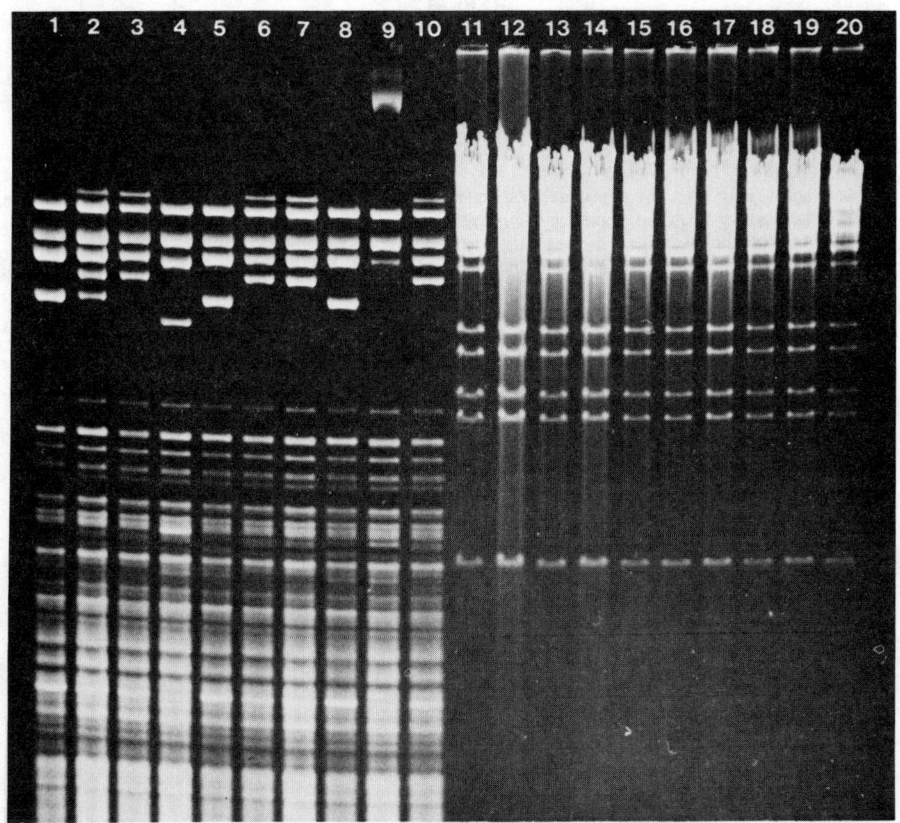

FIGURE 2. *Hha*I digestion of DNA from *C. fetus venerealis* isolates after passage *in vitro* (lanes 1-6) and in cattle (lanes 7-10). *Bst*EII digestion failed to distinguish these isolates (lanes 11-20).

Restriction enzyme analysis (REA) was used to monitor the stability of *Campylobacter* genome after passage *in vivo* and *in vitro*. Genomic changes that occur with passage in cattle were simulated by culturing *Campylobacter* with bovine leukocytes (PMNs). *Hha*I digests of 10 clones of *C. fetus venerealis* recovered from PMN cultures (FIG. 2, lanes 1-6) or from cattle (Fig. 2, 7-10) were compared. Two isolates (lanes 7,10) recovered from individual cattle were identical to two clones which were not passaged in cattle (lanes 3,6). Two pairs of identical clones were reisolated after *in vitro* passage (lanes 1,5; lanes 3,6). Four clones cultured with neutrophils resembled clones passaged in cattle (lanes 1,5,8; lanes 3,10; lanes 6,7). *Bst*EII digestion failed to discriminate any of the *C. fetus venerealis* clones (FIG. 2; lanes 11-20).

These observations indicate the sensitivity of REA for identifying *Campylobacter* isolates in epidemiologic investigations and in appraising the stability of *C. fetus venerealis* genome after *in vitro* or *in vivo* passage.

REFERENCES

1. GEBHART, C., G. E. WARD, K. CHANG & H. J. KURTZ. 1983. Am. J. Vet. Res. **44**(3): 361-367.
2. KAKOYIANNIS, C. K., P. J. WINTER & R. B. MARSHALL. 1984. Appl. Environ. Microbiol. **48**(3): 545-549.
3. PENNER, J. L., J. N. HENNESSY, S. D. MILLS, & W. C. BRADBURY. 1983. Application of serotyping and chromosomal restriction endonuclease digest analysis in investigating a laboratory-acquired case of *Campylobacter jejuni* enteritis. J. Clin. Microbiol. **18**: 1427-1428.
4. BRADBURY, W. C., A. D. PEARSON, M. A. MARKO, R. V. CONGI & J. L. PENNER. 1984. J. Clin. Microbiol. **19**(3): 342-346.
5. LANGENBERG, W., E. A. J. RAUWS, A. WIDJOJOKUSUMO, G. N. J. TYTGAT & H. C. ZANEN. 1986. J. Clin. Microbiol. **24**(3): 414-417.
6. COLLINS, D. M. & D. E. ROSS. 1984. J. Med. Microbiol. **18**: 117-124.
7. THIERMANN, A., A. L. HANDSAKER, S. L. MOSELEY & B. KINGSCOTE. 1985. J. Clin. Microbiol. **21**(4): 585-587.

Regulation of Brain Sodium Content during Hypernatremic Stress in the Brattleboro Rat

MICHAEL DePASQUALE AND HELEN F. CSERR

Section of Physiology and Biophysics
Brown University
Providence, Rhode Island 02912

Brain interstitial fluid (ISF) volume depends on total tissue content of extracellular ions, chiefly Na and Cl. In response to acute hyperosmotic stress, the brain regulates its volume by accumulating solute, principally Na, Cl, and K, thus countering the osmotic withdrawal of water.[1] In the rat, both plasma and CSF contribute to the volume-regulatory gain in electrolyte.[2] Net influx from plasma is by diffusion across the blood-brain barrier (BBB) and accounts for only a small fraction of solute gain by brain. The major fraction of Na and Cl gained by rat brain in response to acute hyperosmotic stress can be accounted for by bulk flow of CSF into brain. Following the initial osmotic withdrawal of water from brain, ISF pressure decreases,[3] and presumably this fall contributes to the driving force for bulk flow of CSF into brain.

We have investigated the possible role of vasopressin (VP) in this volume regulatory response. Experiments were conducted in the homozygous Brattleboro (DI) rat, a strain unable to synthesize VP, and for comparison, in its VP-competent congenic, the Long-Evans (LE) rat. In isosmotic DI and LE rats, brain electrolyte and water content were comparable ($p > 0.1$ or greater). After 30 min of hypernatremic stress (40-60 mosmol NaCl/kg i.p.), the DI rat brain gained 58% less Na than the LE rat ($p < 0.001$) as indicated by the slopes of the best-fit lines relating increase in brain Na content to increase in plasma osmolality (FIG. 1). To determine whether this reduction was due to an effect on electrolyte influx from plasma and/or CSF, both components of influx were evaluated.

The contribution of plasma electrolyte was assessed by measuring the unidirectional blood-to-brain transfer constant (K_1) for ^{22}Na according to the method of Ohno *et al.*[4] Values for K_1 were similar in both strains, both in isosmotic and hypernatremic animals ($p > 0.2$), showing that the difference between strains in brain Na uptake is not due to differences in influx from blood.

Volume flow of CSF into brain was estimated from the clearance of radiolabeled extracellular tracers from CSF to brain during ventriculocisternal perfusion. The tracers used were ^{125}I-labeled human serum albumin (RISA, 69,000 daltons) and ^{57}Co-labeled diethylenetriaminepentaacetic acid (DTPA, 393 daltons). Since the diffusion coefficients for these two tracers differ by a factor of 7, bulk flow can be measured as that component of tracer clearance which is independent of diffusion coefficient (i.e., the same for RISA and DTPA). In isosmotic animals, diffusion is the mechanism of tracer influx into brain and the magnitude of this influx was the same in both strains. With elevation of plasma osmolality, RISA and DTPA clearance increased similarly,

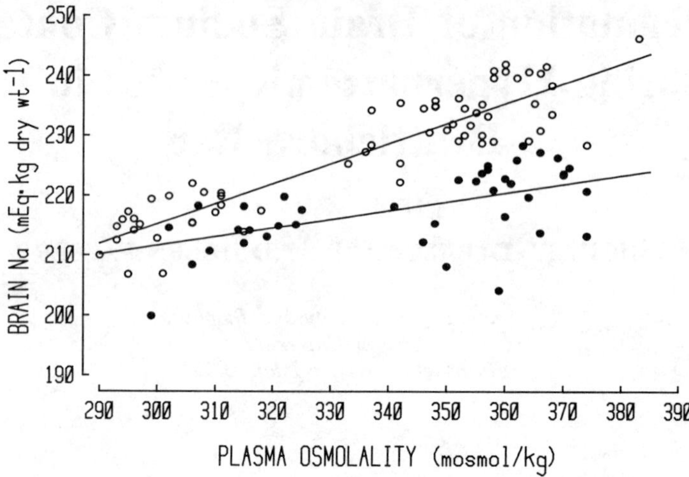

FIGURE 1. Brain sodium content as a function of plasma osmolality 30 min after a single i.p. injection of hypertonic NaCl. ($\bigcirc$) Long-Evans; ($\bullet$) Brattleboro rats. Slopes (mean $\pm$ S.E., mEq·kg dry wt^{-1}·mosmol^{-1}·kg) of the best-fit lines: 0.33 ± 0.02 ($n = 62$, $p < 0.001$) for Long-Evans and 0.14 ± 0.04 ($n = 40$, $p < 0.001$) for Brattleboro rats.

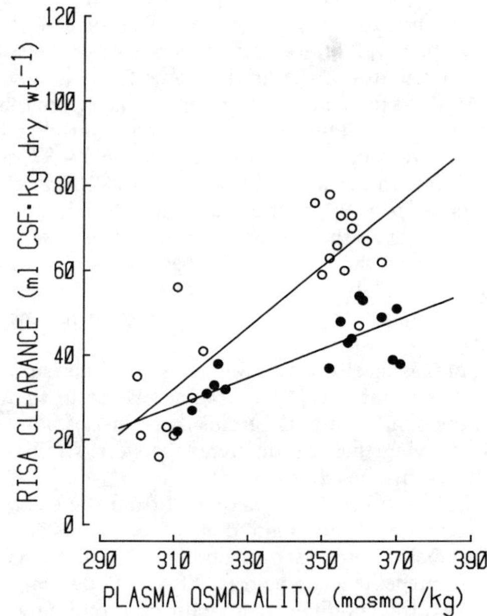

FIGURE 2. The increase in clearance of RISA from CSF to brain during 30 min of hypernatremic stress is reduced in the Brattleboro ($\bullet$) compared to the Long-Evans ($\bigcirc$) strain. Slopes (mean $\pm$ SE, ml CSF·kg dry wt^{-1}·mosmol^{-1}·kg) of the best-fit lines: 0.72 ± 0.11 ($n = 20$, $p < 0.001$) for Long-Evans and 0.34 ± 0.07 ($n = 16$, $p < 0.001$) for Brattleboro rats.

consistent with the addition of a bulk flow component of tracer influx. However, the magnitude of this osmotically stimulated bulk flow component of tracer influx was reduced by more than 50% ($p < 0.01$) in the DI strain (FIG. 2). This reduced bulk flow of CSF into brain in response to hypernatremic stress is consistent with the smaller net gain in brain Na content observed in the DI rat compared to the LE rat.

Since DI and LE rats differ primarily in their ability to synthesize VP, these results provide indirect evidence for hormonal participation in the volume-regulatory accumulation of solute by brain in response to osmotic dehydration of the interstitium.

REFERENCES

1. BRADBURY, M. W. B. 1979. The Concept of a Blood-Brain Barrier. Wiley. New York, NY.
2. CSERR, H. F., M. DePASQUALE, C. S. PATLAK & R. G. L. PULLEN. 1986. Convection of interstitial fluid and its role in brain volume regulation. Acad. of Sci. **481:** 123-134.
3. WIIG, H. & R. K. REED. 1983. Rat brain interstitial fluid pressure measured with micropipettes. Am. J. Physiol. **244:** H239-H246.
4. OHNO, K., K. D. PETTIGREW & S. I. RAPOPORT. 1978. Lower limits of cerebrovascular permeability in the conscious rat. Am. J. Physiol. **235:** H299-H307.

Changes in Systemic Blood Pressure Alter Local Cerebral Blood Flow Following Blood-Brain Barrier Injury

WALTER D. JOHNSON, THOMAS H. MILHORAT,
AND DIANA DOW-EDWARDS

*State University of New York
Health Science Center at Brooklyn
Brooklyn, New York 11203*

INTRODUCTION

Vasogenic cerebral edema (VCE) is a concomitant of a variety of CNS lesions that collectively share alterations of the blood-brain barrier (BBB). The mass effect created by this parenchymal collection of fluid often greatly exaggerates the symptoms of the underlying primary tension. In patients with rapidly evolving cerebral edema, alterations in systemic blood pressure (SBP) are common, yet there is no consensus concerning the propriety of pharmacologically controlling these changes or even the actual physiologic consequences occasioned by them. Theoretically, increasing the SBP would be desirable for enhancing perfusion in ischemic cerebral areas. However, since extracellular edema formation in other body tissues is increased by a rise in capillary blood pressure, elevation of the SBP might, in fact, be malevolent.

This study was designed to determine the effects of controlled variations in SBP on VCE formation and local cerebral blood flow (LCBF) following freeze lesions (FL) in six groups of rats: hypertensive, normotensive and hypotensive control and FL animals with five or six animals per group.

METHODS

Sprague-Dawley rats weighing 150-200 g were subjected to either sham operation of FL.[1] Evan's blue (EB) dye, given prior to FL, defined plasma albumin extravasation. Following a 30-min recovery period, 0.1% Neosynephrine or 2.5% Arfonad was infused intravenously to achieve mean arterial pressures 20-30% above and below baseline, respectively, for 60 min. Normotensive animals received normal saline solution at an equivalent rate.

LCBF was determined during the last minute of drug or saline infusion by the ^{14}C-antipyrine technique of Sakurada.[2] Brain images were obtained on film according to standard autoradiographic procedure and analyzed densitometrically to determine local tissue ^{14}C concentrations, which together with the plasma ^{14}C concentrations were used to calculate LCBF.

All statistical determinations were made using Student's t test.

RESULTS

In addition to the FL site, six brain areas were examined bilaterally: cortex, ventral thalamus, hypothalamus, hippocampus, corpus callosum, and internal capsule.

In FL animals, the area of extravasated EB dye was exactly congruent with the area of decreased LCBF shown autoradiographically. In hypertensive animals, this area was significantly enlarged ($p < 0.05$) from an average width of 1.85 mm to 3.25 mm. Hypotension decreased this area by a comparable degree. Excluding the FL site, no significant differences in LCBF were seen between ipsilateral and contralateral structures. Small but significant differences were noted between hypotensive and normotensive controls in bilateral cortex ($p < 0.01$), thalamus ($p < 0.01$), hippocampus ($p < 0.05$) and ipsilateral internal capsule ($p < 0.01$). However, all values were within LCBF ranges previously published.

DISCUSSION

This study clearly demonstrates that following BBB disruption, a rise in SBP increases the rate of formation and the extent of spread of VCE, producing a simultaneous and congruent decrease in LCBF. In contrast, a decrease in SBP reduces the congruent zone of edema/ischemia without compromising perfusion. These findings may have important clinical implications in the management of patients with VCE.

REFERENCES

1. KLATZO, I., A. PIRAUX & E. J. LASKOWSKI. 1958. Exp. Neurol. **17:** 548-564.
2. SAKURADA, O., C. KENNEDY, J. JEHLE, J. BROWN, G. CARBIN & L. SOKOLOFF. 1978. Am. J. Physiol. **234(1):** H59-H66.

Delivery of Solutes in Cerebrospinal Fluid to Central Neurons via "Paravascular" Fluid Pathways in the Central Nervous System[a]

M. L. RENNELS,[b] O. R. BLAUMANIS,[b] P. A. GRADY,[b]
AND K. FUJIMOTO[c]

[b]*Departments of Neurology and Anatomy*
University of Maryland School of Medicine
Baltimore, Maryland 21201

[c]*First Division, Department of Anatomy*
Shimane Medical University
Izumo City, Shimane, Japan

Blood vessels throughout the brain and spinal cord are surrounded by longitudinal fluid pathways that are in continuity with cerebrospinal fluid (CSF) in the subarachnoid space (SAS).[1,2] These pathways can be demonstrated light microscopically by infusion of the tracer protein, horseradish peroxidase (HRP), into the SAS and subsequent localization of the enzyme in brain sections using the highly sensitive chromogen, tetramethylbenzidine (TMB).[3] HRP enters the perivascular spaces (PVS) around penetrating arterioles and, within minutes, is distributed throughout the CNS along the basal laminae (BL) of capillaries. This rapid, unidirectional fluid/tracer movement along the vascular network appears to be facilitated by arteriolar pulsations.[2] From these "paravascular pathways," HRP spreads into the tissue spaces and also selectively delineates localized groups of neurons.

A needle was inserted into the cisternae magna of adult cats anesthetized with sodium pentobarbital, and CSF efflux was measured. An equal volume (0.5–1.0 ml) of 4.0% HRP (Sigma Type II) in Hanks' balanced salt solution was infused by gravity flow. After HRP circulation periods of 4 min (6 cats) or 10 min (9 cats), fixation was carried out by perfusion of aldehyde solutions.[2] Vibratome sections (100 μm) of the forebrain and brainstem were prepared and HRP was localized using TMB[3] or diaminobenzidine (DAB)[4,5] as substrates.

CNS intraparenchymal vessels were outlined, *in toto*, by paravascular reaction product after 10 min HRP circulation in the SAS, with TMB incubation. In the brainstem, vascular outlining occurred after as little as 4 min HRP circulation, and localized groups of tracer-containing neurons were observed in the hypothalamus and in lateral tegmental zones of the midbrain, pons and medulla (FIG. 1). This rapid cellular uptake was not observed elsewhere. The soma or dendrites of HRP-filled cells

[a]This work was supported by Grant No. 2P50 NS16332-04A2 from NINCDS, National Institutes of Health.

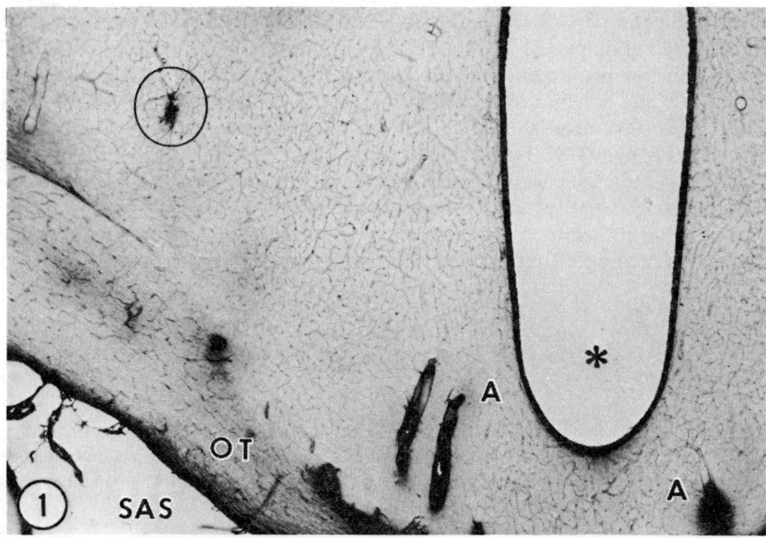

FIGURE 1. Cat hypothalamus fixed 4 min after infusion of 4.0% HRP solution into the subarachnoid space. Neurons within the outlined area (see FIGURE 2) contain dense HRP reaction product, unlike cells elsewhere in the section. The perivascular spaces of penetrating arterioles (A) contain dense tracer material. HRP migration from these spaces along the basal laminae of capillaries has also resulted in the outlining of these vessels. The ependyma of the third ventricle (*) is labelled owing to tracer reflux from the subarachnoid space (SAS) into the ventricular system. OT, optic tract; no counterstain; × 12.

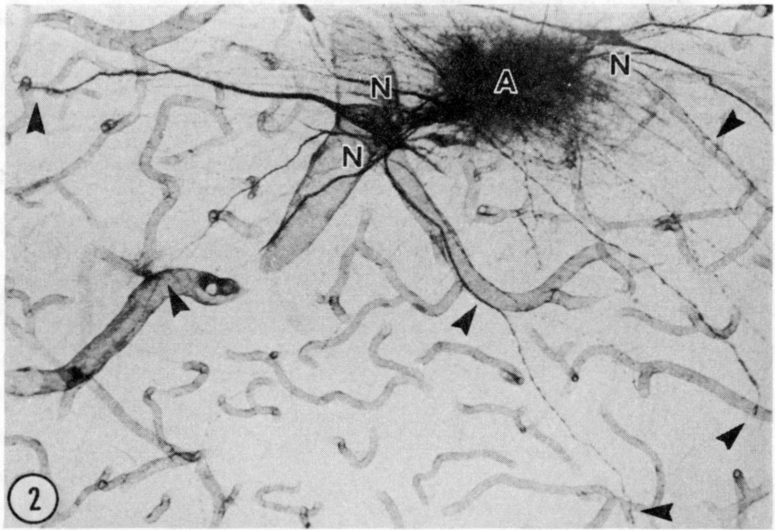

FIGURE 2. A penetrating arteriole (A) is surrounded by diffuse tracer in the extracellular spaces and a dense meshwork of HRP-filled neurons (N) and fibers. Larger vessels of lesser density are probably venules. Dendrites and radially distributed varicose axons appear to contact the walls of microvessels at numerous sites (*arrowheads*). The vascular network is completely outlined by reaction product in the paravascular pathways; × 123.

adjoined penetrating arterioles, and their varicose axons appeared to contact the walls of neighboring capillaries (FIG. 2). Tracer accumulation by these hypothalamic and brainstem neurons presumably occurs via the close association of their neurites with arteriolar PVS and/or by direct contact of these processes with the BL of capillaries. The latter sites have been shown in electron microscopic studies.[6,7] The paravascular pathways thus convey CSF-borne solutes to neurons "behind" the blood-brain barrier. The rapid and predictable access of HRP to these cells implies a specialized relationship to the paravascular pathways of nearby microvessels that is not shared by adjacent neurons. Axonal projections to neighboring capillaries suggest that these cells play a local neurovascular role that may be neurosecretory, afferent or regulatory in nature.

REFERENCES

1. GREGORY, T. F., M. L. RENNELS, O. R. BLAUMANIS & K. FUJIMOTO. 1985. A method for microscopic studies of cerebral angioarchitecture and vascular-parenchymal relationships, based on the demonstration of "paravascular" fluid pathways in the mammalian central nervous system. J. Neurosci. Methods **14:** 4-14.
2. RENNELS, M. L., T. F. GREGORY, O. R. BLAUMANIS, K. FUJIMOTO & P. A. GRADY. 1985. Evidence for a "paravascular" fluid circulation in the mammalian central nervous system, provided by the rapid distribution of tracer protein throughout the brain from the subarachnoid space. Brain Res. **326:** 47-63.
3. MESULAM, M. M. 1978. Tetramethylbenzidine for horseradish peroxidase neurohistochemistry: a non-carcinogenic blue reaction product with superior sensitivity for visualizing neural afferents and efferents. J. Histochem. Cytochem. **26:** 106-117.
4. GRAHAM, R. C. & M. J. KARNOVSKY. 1966. The early stages of absorption of injected horseradish peroxidase in the proximal tubule of the mouse kidney. Ultrastructural correlates by a new technique. J. Histochem. Cytochem. **14:** 291-302.
5. RENNELS, M. L., T. F. GREGORY & K. FUJIMOTO. 1983. Innervation of capillaries by local neurons in the cat hypothalamus: A light microscopic study with horseradish peroxidase. J. Cereb. Blood Flow Metab. **3:** 535-542.
6. RENNELS, M. L., K. FUJIMOTO & E. NELSON. 1979. Capillary innervation in the hypothalamus: A fluorescence histochemical and electron microscopic study. Acta Neurol. Scand. **60** (Suppl. 72): 92-93.
7. RENNELS, M. L. & E. NELSON. 1975. Capillary innervation in the mammalian central nervous system: An electron microscopic demonstration. Amer. J. Anat. **144:** 233-241.

HTLV-III Can Cross the Blood–Brain Barrier[a]

P. SHAPSHAK,[b,c,d,e] L. RESNICK,[f] M. A. OSBORNE,[b]
W. W. TOURTELLOTTE,[b,d] P. SCHMID,[b,d] M. LEE,[g]
G. RUBINSTEIN,[b] D. T. IMAGAWA,[c] R. MITSUYASU,[h]
M. GOTTLIEB,[h] AND R. C. GALLO[i]

AIDS Dementia Complex is a distinct clinicopathologic entity characterized by altered mental status and/or "subcortical dementia" and neuropathologic involvement. Increasing evidence supports the hypothesis that HTLV-III is the cause of this disorder.

The Tourtellotte equation corrects for diffusion of IgG from the blood across the BBB and also for possible damage to the BBB. Therefore, an elevated intra-BBB IgG synthesis rate (>3 mg/day) provides evidence for *de novo* IgG synthesis within the CNS, and this evidence is supported by the presence of unique CSF oligoclonal bands. Intra-BBB IgG synthesis was found in patients with ADC, patients with known active CNS infections, and patients with no apparent signs of neurologic disease.

HTLV-III may not be detected easily by standard virologic techniques as evidenced by the fact that only 7 of 39 CSF specimens from seropositive patients were HTLV-III culture-positive. In addition, the virus may be more easily isolated in early infection, as demonstrated by the higher incidence of positive cultures in patients with no neurologic signs than in patients with neurologic disease (5/23 compared to 2/16) and by the fact that antibodies against p24 were present by Western Blot in all seven patients from whom positive cultures were obtained. Studies are under way to determine factors that may be predictive of impending neurologic disease in HTLV-III-infected patients.

[a] Supported in part by Key Pharmaceuticals Medical Research and Educational Foundation of Miami, Florida; U.S. Army Grant Medical Research and Development Command, Order No. 86 PP 6854; Veterans Administration Merit Funds; Arthur L. Swim Foundation; and California Universitywide Task Force on AIDS (R86LA015).

[b] Neurology and Research Service, Veterans Administration Medical Center West Los Angeles, Wadsworth Division, Los Angeles, California.

[c] Department of Pediatrics, Harbor-University of California at Los Angeles Medical Center, Torrance, California.

[d] Department of Neurology, Reed Neurological Research Center, University of California at Los Angeles School of Medicine, Los Angeles, California.

[e] Address for correspondence: Dr. Paul Shapshak, Neurology Service (W127A), Veterans Administration Medical Center Wadsworth, Los Angeles, California 90073.

[f] Mount Sinai Medical Center, Miami, Florida.

[g] Department of Immunology, University of Texas School of Medicine, Denton, Texas.

[h] Department of Medicine, University of California at Los Angeles School of Medicine, Los Angeles, California.

[i] National Cancer Institute, National Institutes of Health, Bethesda, Maryland.

SUMMARY

In order to determine the etiologic role of HTLV-III in the AIDS Dementia Complex (ADC), a common but as yet unexplained occurrence in AIDS patients,[1-3] intra-blood-brain barrier (BBB) total and HTLV-III-specific IgG synthesis were determined in patients with AIDS or ARC, or seropositive for HTLV-III. HTLV-III isolation from CSF was also attempted.[4] Eighty-six percent of patients with AIDS Dementia Complex had evidence of intra-BBB IgG synthesis by an elevated rate according to the Tourtellotte equation,[6] and 43% had CSF oligoclonal bands by isoelectric focusing (IEF). Sixty percent of AIDS/ARC patients with known opportunistic infections of the central nervous system (CNS) had elevated intra-BBB IgG

TABLE 1. Summary of Intra-BBB Synthesis and IEF Data for HIV Seropositive Individuals

Patient's Neurologic Diagnosis	No. of Patients with Elevated Intra-BBB IgG Synthesis Rate	No. of Patients with CSF Oligoclonal Bands	No. of Patients with Abnormal Serum Bands	No. of Patients with Abnormal Intra-BBB Albumin Leakage Rate
ADC but no CNS opportunistic infections	12/14 (86%)	6/24 (43%)	7/24 (50%)	6/14 (43%)
Known CNS opportunistic infections	3/5 (60%)	2/5 (40%)	3/5 (60%)	4/5 (80%)
No neurological symptoms	5/7 (71%)	1/5 (20%)	4/5 (80%)	1/7 (14%)

synthesis and 40% had oligoclonal bands. Seventy-one percent of HTLV-III-seropositive patients with no neurologic signs of disease had elevated intra-BBB IgG synthesis and 20% had oligoclonal bands. Sixty percent ($^3/_5$) of patients had ADC and intra-BBB IgG synthesis to HTLV-III ranging from 3-10% of the total IgG synthesized, while the remaining two, one with cerebral toxoplasmosis and one with no neurologic disease, were negative. Twenty-two percent ($^5/_{23}$) of neurologically asymptomatic patients were culture-positive for HTLV-III in the CSF compared to 14 percent ($^2/_{16}$) of neurologically symptomatic patients. All seven of the CSF viral isolations were associated with both the p24 major viral core protein and the gp41 transmembrane envelope protein.

[NOTE ADDED IN PROOF: Additional results have been published confirming evidence for HTLV-III-caused intra-BBB IgG synthesis in seropositive asymptomatic individuals.[7]]

TABLE 2. Tourtellotte Equations

Intra-BBB Total IgG Synthesis

Intra-BBB IgG SYN (mg/day) =

$$\left[\left(\text{IgG}_{\text{CSF}} - \frac{\text{IgG}_{\text{serum}}}{369}\right) - \left(\text{ALB}_{\text{CSF}} - \frac{\text{ALB}_{\text{serum}}}{230}\right) \times \left(\frac{\text{IgG}_{\text{serum}}}{\text{Alb}_{\text{serum}}}\right)(0.43)\right] \times 5$$

Intra-BBB HTLV-III IgG Synthesis

Intra-BBB HTLV-III IgG Synthesis (mg/day) =

$$\left[\left(\text{HTLV-III IgG}_{\text{CSF}} - \frac{\text{HTLV-III IgG}_{\text{serum}}}{369}\right) - \left(\text{ALB}_{\text{CSF}} - \frac{\text{ALB}_{\text{serum}}}{230}\right) \times \left(\frac{\text{HTLV-III IgG}_{\text{serum}}}{\text{Alb}_{\text{serum}}}\right)(0.43)\right] \times 5$$

Intra-BBB Albumin Leakage

Intra-BBB Albumin Leakage (mg/day) =

$$\left[\text{Alb}_{\text{CSF}} - \frac{\text{Alb}_{\text{serum}}}{230}\right] \times 5$$

NOTE: Concentrations are in mg/dl for IgG and albumin; 369 and 230 are the average normal serum/CSF ratios for IgG and albumin, respectively; 0.43 is the molecular weight ratio of albumin/IgG; and 5 is the daily CSF production in deciliters.
Normal intra-BBB IgG synthesis rate is < 3 mg/day.
Normal intra-BBB albumin leakage is < 75 mg/day.

REFERENCES

1. NAVIA, B. A., B. D. JORDAN & R. W. PRICE. 1986. The AIDS dementia complex: I. Clinical Features. Ann. Neurol. **19:** 517-524.
2. NAVIA, B. A., E.-S. CHO, C. K. PETITO & R. W. PRICE. 1986. The AIDS dementia complex: II. Neuropathology. Ann. Neurol. **19:** 525-535.
3. PRICE, R. W., B. A. NAVIA & E.-S. CHO. 1986. AIDS encephalopathy. Neurol. Clin. **4:** 285-301.
4. RESNICK, L., F. DE MARZO-VERONESE, J. SCHUPBACH, W. W. TOURTELLOTTE, *et al.* 1985. Intra-blood-brain barrier synthesis of HTLV-III-specific IgG in patients with neurologic symptoms associated with AIDS or AIDS-related complex. N. Engl. J. Med. **313:** 1498-1504.
5. STAUGAITIS, S., P. SHAPSHAK, W. TOURTELLOTTE, M. LEE & H. REIBER. Isoelectric focusing of unconcentrated CSF: Applications to ultrasensitive analysis of oligoclonal immunoglobulin G. Electrophoresis **6:** 287-291.
6. TOURTELLOTTE, W. W., M. J. WALSH, R. W. BAUMHEFNER, S. STAUGAITIS & P. SHAPSHAK. 1984. The current status of multiple sclerosis intra-blood-brain barrier IgG synthesis. Ann. N.Y. Acad. of Sci. **436:** 52-67.
7. RESNICK, L., J. B. BERGER, P. SHAPSHAK & W. W. TOURTELLOTTE. 1988. Early penetration of the blood-brain barrier by HIV. Neurology **38:** 9-14.

Blood-Brain Barrier Ultrastructure: Beyond Tight Junctions

P. A. STEWART,[b] C. R. FARRELL,[a] AND
B. L. COOMBER [a]

Departments of [a] Anatomy and [b] Medicine
University of Toronto
Toronto, Ontario, Canada

It is well known that the structural basis of the blood-brain barrier (BBB) is the continuous interendothelial tight junctions; however, the significance of clefts in the junctional zone and of other structural features to BBB function is not as well understood. Because descriptive studies are insensitive and vulnerable to both subjectivity and perceptual errors, we have quantified BBB ultrastructure in a variety of species (FIGURE 1).

BBB capillary walls are 30% thinner than walls in other continuous capillaries in the mouse.[1] Since important nutrients cross the endothelial membrane via membrane-bound transport molecules and then diffuse across the endothelial cytoplasm, a thinner capillary wall would provide for a shorter diffusion path and therefore may be an advantage in vessels with restricted permeability. In the human, gray matter capillary walls are thinner than those in white matter. This may reflect the higher metabolic demands of gray matter.

In the developing mouse, BBB capillary walls are thick in the fetus and become thin in parallel with the maturation of the BBB.[2] In the aging human, capillary walls in both gray and white matter become thinner. This is due to both thinning of the endothelial layer and a decline in both size and number of pericytes.[3]

If pericytes form a "second line of defense" in the BBB by phagocytosing molecules that escape the endothelial barrier, then their gradual loss during aging suggests that the BBB in the elderly may not be as able to compensate for transient leaks as in younger brains. We have serially reconstructed pericytes from normal human brain and have shown that at least 95% of them are granular ($p < 0.02$).[4]

In non-BBB capillaries serial reconstruction of interendothelial junctions has shown that the junctional clefts form channels, some of which span the junction from lumen to ablumen.[5] Our studies have shown that in the normal BBB, junctional clefts constitute 26% of the junctional length in both mouse and human (FIGURE 2). In the developing mouse this "cleft index" is high in the fetus and declines during the maturation of the BBB in parallel with the declining permeability.[2] Furthermore, the clefts enlarge and can be filled with vascular tracer under some pathological conditions. A possible explanation for these observations is that in the normal BBB the channels formed by the clefts are not continuous across the entire junction, but that when the BBB is permeable, some of the clefts anastomose and form complete channels.

Endothelial vesicles are much more numerous in permeable endothelium than in blood-brain barrier endothelium and so have been implicated in vascular permeability. Barrier vessels have densities of vesicles ranging from 3 to 11 per μm^2 cytoplasm.[1,3]

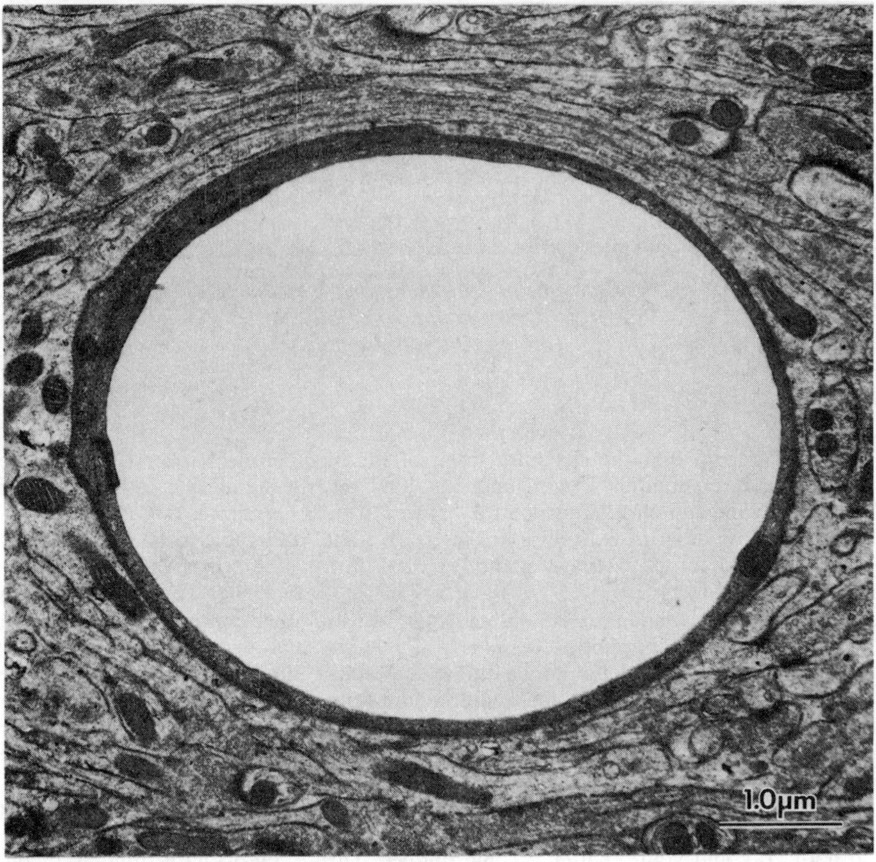

FIGURE 1. A typical capillary profile from a 4-month-old mouse cerebellum. Note the thin wall, tight junctions (two), and few vesicles.

FIGURE 2. Profile of an interendothelial junction from a capillary of a 4-month-old mouse cerebellum. The junctional clefts (*arrowheads*) constitute 26% of the total junctional length.

In contrast, the density of vesicles in normally permeable vessels ranges from 37 (area postrema) to 80 per μm^2 (skeletal muscle).[1] The densities of vesicles in developing mouse brain[2] and aging human brain[2] are not significantly different from those of young adult brain of the same species. It is unlikely, therefore, that they are involved in the changes in cerebrovascular permeability that occur during development, or that are postulated to occur during aging.

Because the three-dimensional arrangement of vesicles within the endothelium is important in understanding their role in permeability, we reconstructed vesicles in brain and in permeable endothelium. We found that vesicles in permeable endothelium were arranged mostly into invaginating clusters, but that few such clusters existed in BBB vessels.[6] These data support the hypothesis that vesicles may act in vascular permeability by forming transient channels spanning the endothelial cells.

REFERENCES

1. COOMBER, B. L. & P. A. STEWART. 1985. Microvasc. Res. **30:** 99-115.
2. STEWART, P. A. & E. M. HAYAKAWA. 1987. Dev. Brain Res. **32:** 271-282.
3. STEWART, P. A. *et al.* 1987. Microvasc. Res. **33:** 270-282.
4. FARRELL, C. F. & P. A. STEWART. 1987. Anat. Rec. **218:** 466-469.
5. BUNDGAARD, M. 1984. J. Ultrastr. Res. **88:** 1-17.
6. COOMBER, B. L. & P. A. STEWART. 1986. Anat. Rec. **215:** 256-261.

Use of a Chemical Redox System for Delivery of Drugs to the Brain: Ethinyl Estradiol

M. E. BREWSTER, K. S. ESTES, AND N. BODOR[a]

Pharmatec, Inc.
Alachua, Florida 32615

Center for Drug Design and Delivery
University of Florida
Gainesville, Florida 32610

The normally protective function of the blood-brain barrier (BBB) can hinder delivery of therapeutic drug concentrations to the central nervous system (CNS). In addition, several CNS active drugs which can readily penetrate the BBB owing to their lipophilic properties have short biological half-lives and require frequent dosing. A chemical redox-based system of drug-enhanced delivery through the BBB has recently been developed in our laboratories.

The method is designed to biomimic the $NAD^+ \rightleftarrows NADH$ redox system and has been shown to both enhance delivery of drugs to the CNS and to maintain sustained drug levels in this tissue.[1,2] The system utilizes a lipophilic dihydropyridine carrier which is covalently attached to a relatively hydrophilic drug. The drug carrier complex enhances drug penetration through the BBB due to augmented lipophilicity; however, the complex is designed to be metabolically labile. *In vivo* oxidation to the corresponding hydrophilic quaternary salt of the drug complex impedes efflux through the BBB. Excretion from peripheral tissues is conversely facilitated by this oxidation. The resulting "locked-in" drug carrier complex is hydrolyzed to release free drug and the naturally occurring nontoxic carrier metabolite, trigonelline. The rate of hydrolysis, and therefore the duration of drug activity, can be altered by chemical design of the carrier linkage.[1,2]

The present studies applied this chemical delivery system (CDS) to ethinyl estradiol (EE). The EE-CDS was synthesized by reacting EE with nicotinic anhydride, followed by preferential phenolic hydrolysis, quaternization and reduction to yield final product. This method is similar to that previously described from this laboratory.[3,4]

In vivo distribution studies verified that the system was capable of enhanced and sustained brain drug concentrations compared to measured circulating levels. FIGURE 1 illustrates the results of this study. Sprague-Dawley male rats were treated with 7.5 mg/kg EE-CDS via tail vein injection. Blood and brain samples were collected at 0.25, 1, 3, 6, 24 and 48 hr post treatment and analyzed for concentrations of the "locked-in" quaternary salt form of the drug (EQ) and free EE using HPLC methods similar to those previously described.[3,4] As shown in FIGURE 1, EQ levels were

[a] Address for correspondence: Dr. N. Bodor, Center for Drug Design and Delivery, Box J-497 JHMHC, University of Florida, Gainesville, Florida 32610.

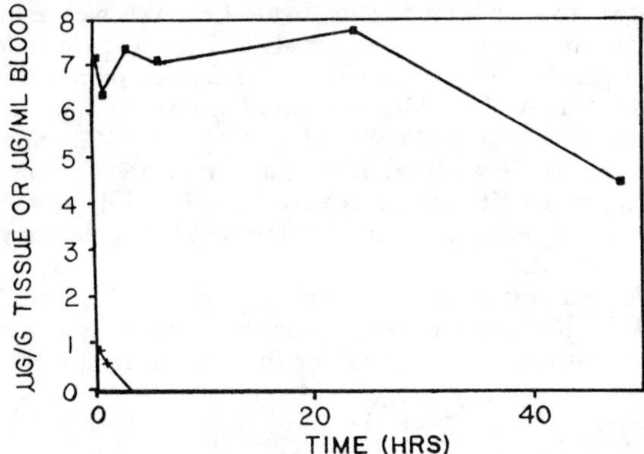

FIGURE 1. Concentrations of EQ in brain (■) and blood (+) after 7.5 mg/kg EE-CDS via tail vein injection. Samples were extracted in acetonitrile with further deproteinization by saturated sodium chloride and directly analyzed by HPLC using UV detection. Pooled samples were assayed for free EE.

sustained for 24 hr and only decreased to 60% of their peak concentrations by 48 hr. In contrast, blood levels were undetectable by 3 hr. Small but detectable free EE brain tissue concentrations of 22.0 and 7.3 ng/g were measured at 6 and 48 hr post treatment, while no EE was detected in blood samples.

The duration of luteinizing hormone (LH) inhibition in ovariectomized (OVX) Sprague-Dawley rats was used to evaluate biological action of CDS-EE.[3] Two weeks after ovariectomy, rats were treated with 1.0, 0.3, 0.1 or 0.03 mg/kg of EE-CDS or dimethyl sulfoxide (DMSO) 0.5 ml/kg vehicle control via tail vein injection. Blood samples were obtained on day 12 and 18 post treatment via heart puncture, and trunk

TABLE 1. Effect of EE-CDS and EE Dose on Serum LH 12, 18 and 25 Days post treatment in Ovariectomized Rats

Treatment Group	n	Day 12	Day 18	Day 25
			Mean and SEM for Serum LH (ng LHRP-2/ml)	
DMSO (0.5 ml/kg)	8	11.4 ± 1.2^a	15.6 ± 2.9^b	$11.6 \pm 1.0^{a,b}$
EE-CDS				
1.0 mg/kg	7	3.8 ± 0.4^c	3.7 ± 0.6^c	2.6 ± 0.5^c
0.3 mg/kg	7	8.4 ± 1.2^c	$8.4 \pm 1.2^{a,c}$	10.4 ± 1.0^a
0.1 mg/kg	5	6.4 ± 0.8^c	$13.1 \pm 2.0^{a,b}$	21.2 ± 2.2^d
0.03 mg/kg	7	9.8 ± 0.9^a	15.9 ± 1.0^b	14.6 ± 0.7^b
EE				
0.7 mg/kg	6	11.1 ± 0.9^a	$12.0 \pm 1.3^{a,b}$	$12.7 \pm 2.2^{a,b}$
0.2 mg/kg	7	10.5 ± 1.7^a	16.9 ± 1.7^b	$13.7 \pm 0.9^{a,b}$

[a-d] Values with different superscripts are significantly different, $p < 0.05$.

blood was collected at sacrifice on day 25. Serum LH levels were estimated by routine radioimmunoassay procedures using materials and methods included in NIAMDD kits. Values are expressed in terms of LH-RP-2 reference preparation and the results are shown in TABLE 1. A clear dose and duration response were found. On day 12 post treatment LH was dampened 67%, 53%, 44% and 14% compared to controls, and by day 18 decreased 76%, 46%, 16% and increased 2% in these groups by 1.0, 0.3, 0.1, and 0.03 mg/kg EE-CDS, respectively. LH was decreased on day 25 post treatment only in the highest dose group. Equimolar EE was ineffective at these distant time points post treatment.

The enhanced and sustained brain concentrations of EQ and EE compared to blood levels and the prolonged dose and duration activity response shown in these studies support the therapeutic potential for the CDS to treat CNS impairments.

REFERENCES

1. BODOR, N., H. FARAG & M. BREWSTER. 1981. Site-specific, sustained release of drugs to the brain. Science **214:** 1370.
2. BODOR, N. & M. E. BREWSTER. 1983. Problems of delivery of drugs to the brain. Pharm. Ther. **19:** 337.
3. SIMPKINS, J., J. MCCORNACK, K. ESTES, M. BREWSTER, E. SHEK & N. BODOR. 1986. Sustained brain-specific delivery of estradiol causes long-term suppression of LH secretion. J. Med. Chem. **29:** 1810.
4. BODOR, N., J. MCCORNACK & M. E. BREWSTER. 1987. Improved delivery through biological membranes XXII. Synthesis and distribution of brain-selective estrogen delivery systems. Int. J. Pharm. **35:** 47.

Increased ACTH Plasma Levels with Prenatal and Postnatal Nicotine Administration in Rats[a]

JEAN A. KING AND FLEUR L. STRAND[b]

Biology Department
New York University
New York, New York 10003

Nicotine mediates the release of ACTH in adult rats,[1,3] but the resultant effects on development remain unknown. Acute administration of nicotine results in a rise in ACTH which is followed by an increase in plasma corticosterones.[1,3] This nicotine-mediated ACTH release is accompanied by a release of beta-endorphin.[3] In addition, corticotropin-releasing factor (CRF), has been elicited from the hypothalamus in a dose-dependent manner in response to varying concentrations of nicotine.[9] Fabro and Sieber[4] demonstrated that nicotine administered to pregnant females can be subsequently localized in the implanted blastocyst during gestation. Other investigations indicate that the alkaloid is present in high concentrations in the milk of dams postpartum.[7] Taken together, it is not unreasonable to postulate that nicotine administration to pregnant females may cause changes in fetal ACTH levels either indirectly through the mothers' circulatory system or directly by penetration of the placenta. The present study was designed to determine the effects of nicotine on ACTH levels of neonates during development.

METHOD

Pregnant Sprague-Dawley rats were used in all experiments. The pregnant females were maintained on a 12-hr light/dark cycle at a room temperature of 22°C. Each dam was housed in an individual cage and supplied with rat chow and water *ad libitum*. All treatment of pregnant females began on the third day of gestation and continued through parturition. Nicotine tartrate (0.25 mg/kg, 2× daily i.p.) was given to pregnant females. The control animals received a corresponding volume of 0.9% saline solution. The pups were divided into four treatment groups. One group of pups received nicotine prenatally only; another group of pups received nicotine prenatally and postnatally (0.05 mg/kg/day s.c.); and the third experimental group received nicotine during postnatal life only. Control rats received saline solution during

[a] This work was supported by the Council for Tobacco Research.

[b] Address for correspondence: Professor Fleur L. Strand, Department of Biology, 1009 Main Building, New York University, Washington Square, New York, New York 10003.

both prenatal and postnatal life (saline/saline). All animals received last injections 24 hours prior to the experiment.

ACTH plasma levels were determined in all nicotine-treated neonates and saline-treated controls on days 7, 15 and 21. Blood was obtained by means of cardiac puncture according to the method of Waynfork.[8] Blood was drawn in pre-chilled syringes containing EDTA. The blood was transferred to microcentrifuge tubes coated with EDTA. The plasma obtained from these samples was stored at $-70°C$ prior to use. Each plasma sample was thawed once on the day of assay. ACTH plasma level determinations were carried out by means of radioimmunoassay (RIA), using RIA kits obtained from Radioassay Systems Laboratories Inc., Carson, California. This assay is less than 0.01% reactive with the beta-endorphins and ACTH fragments smaller than ACTH 1-24.

RESULTS

During the first week of life, only those pups exposed to nicotine both prenatally and postnatally had significantly higher titers of ACTH. However, by postnatal day 15 all neonates treated with nicotine at different time periods during ontogeny had significantly higher levels of the peptide in their plasma. The plasma ACTH levels in the rats receiving prenatal treatment only or postnatal treatment only were not significantly different from each other and they represented a 3-4-fold increase as compared to saline-treated controls. Exposure to nicotine during gestation and neonatal life was the most effective combination, resulting in dramatic increases in plasma ACTH levels at day 15. While ACTH levels decrease by day 21, they are still significantly higher than are levels in saline-treated controls (TABLE 1.)

DISCUSSION AND CONCLUSIONS

These results provide some evidence that nicotine may modulate the hypothalamic-pituitary-adrenal axis in rats during development. They show that on postnatal day 7, when the pituitary gland is still unresponsive to endogenous CRF,[1] only high levels of nicotine (administered prenatally and postnatally) elevate ACTH titers, probably acting directly on the pituitary gland and bypassing the hypothalamic link. By day 15 the pituitary is responsive to CRF stimulation,[5,6] which thus can evoke ACTH release in a dose-related manner. Since nicotine causes dose-dependent CRF release in adult animals,[3] a similar mechanism may be involved here. The dramatically high ACTH levels in the 15-day-old pre- and postnatally treated rats indicate that the negative feedback mechanisms initiated by corticosteroids are not yet functional. The subsequent modulation of the ACTH response of all nicotine-treated animals at day 21 implies that the hypothalamic-pituitary-adrenal axis has matured three weeks after birth.

TABLE 1. Nicotine-Evoked ACTH Release in Neonatal Rats

Postnatal Day	Treatment (prenatal/postnatal)[a]	ACTH (pg/ml)
7	Saline/saline	58
7	Saline/nicotine	57
7	Nicotine/saline	73
7	Nicotine/nicotine	224[b]
15	Saline/saline	75
15	Saline/nicotine	220[b]
15	Nicotine/saline	269[b]
15	Nicotine/nicotine	633[b]
21	Saline/saline	95
21	Saline/nicotine	140
21	Nicotine/saline	110
21	Nicotine/nicotine	235[b]

[a] prenatal = nicotine, 0.25 mg/kg i.p., given twice daily to pregnant females; postnatal = nicotine, 0.05 mg/kg/daily s.c., given to pups from birth.
[b] $p < 0.05$ vs. saline/saline-treated controls of same age.

REFERENCES

1. CAM, G. R. & J. R. BASSETT. 1984. Effect of prolonged exposure to nicotine and stress on the pituitary adrenocortical response: The possibility of cross adaptation. Pharmacol. Biochem. Behav. **20:** 221-226.
2. CARNES, M., M. S. BROWNFIELD, N. H. KALIN, S. LENT & C. M. BARKSDALE. 1986. Episodic secretion of ACTH in rats. Peptides **7:** 219-223.
3. CONTE-DEVOLX, B., C. OLIVER, P. GIRAUD, P. GILL, P. CASTANAS, E. LISSITSKY, F. BOUDOURESQUE & Y. MILLET. 1981. Effect of nicotine on *in vivo* secretions of melanocorticotropic hormones in rat. Life Sci. **28:** 1067-1073.
4. FABRO, S. & S. M. SIEBER. 1969. Caffeine and nicotine penetrate the preimplantation blastocyst. Nature **23:** 410-411.
5. GUILLET, R. & S. M. MICHEALSON. 1978. Corticotropin responsiveness in the neonatal rat. Neuroendocrinology **27:** 119-125.
6. GUILLET, R., M. SAFFRAN & M. MICHEALSON. 1980. Pituitary-adrenal response in neonatal rats. Endocrinology **106:** 991-994.
7. HILL, P. & E. L. WYNDER. 1979. Nicotine and cotinine in breast fluid. Cancer Lett. **6:** 251-254.
8. WAYNFORK, H. G. 1983. Obtaining body fluids. *In* Experimental and Surgical Techniques in the Rat. H. Waynfork, Ed.: 62-86. Academic Press. New York, N.Y.
9. WEIDENFELD, J., R. SEIGEL, N. CONFORTI, R. MIZACHI & T. BENNER. 1983. Effect of intracerebroventricular injection of nicotine acetylcholine receptor antibodies on ACTH, corticosterone and prolactin secretion in the male rat. Brain Res. **265:** 152-156.

Benzodiazepine Stimulation of Gamma-Aminobutyric Acid Receptor Desensitization in Chick Spinal Cord Cell Cultures

DANIEL MIERLAK AND DAVID H. FARB[a]

Department of Anatomy and Cell Biology
The State University of New York
Health Science Center at Brooklyn
Brooklyn, New York 11203

In many receptor systems, the response initiated by agonist binding fades, or desensitizes, upon prolonged agonist presence.[1] The role for desensitization in neurotransmitter receptors coupled to ion channels remains poorly understood. We have been investigating the process of desensitization for the γ-aminobutyric acid (GABA) receptor. GABA, a major inhibitory neurotransmitter in the vertebrate CNS, causes an increase in membrane chloride permeability. In addition, the GABA receptor-chloride ionophore complex is thought to be the site of action for benzodiazepines, such as Valium and Librium. Benzodiazepines are known to potentiate GABA-mediated conductance increases[2] through specific binding sites.[3] In light of this neuromodulation of GABA action, we were interested in determining the effect of benzodiazepines on the process of GABA receptor desensitization.

Electrophysiological experiments were performed on embryonic chick spinal cord neurons grown in primary monolayer culture as previously described.[4] Cell conductance was determined from Ohm's Law after measuring the voltage response to constant, hyperpolarizing current pulses. GABA and benzodiazepine solutions were applied directly to neuronal soma by pressure ejection.

Application of GABA rapidly produces a peak conductance increase (peak g_{GABA}) which then desensitizes to a steady-state level (g_{ss}). To quantitate GABA-mediated desensitization, both extent (%D) and apparent rate constant for desensitization (k_{app}) were determined (see TABLE 1).

In the presence of saturating chlordiazepoxide (CDPX, 300 μM), a water-soluble benzodiazepine, the peak response to 10 μM GABA is potentiated and both %D and k_{app} increase. In control experiments, %D and k_{app} were shown to be proportional to peak GABA response. Since CDPX potentiates the response to 10 μM GABA, both %D and k_{app} would be expected to increase in the presence of CDPX. We determined whether the increase in %D and k_{app} seen with CDPX could be explained simply by the increased peak response produced by CDPX. Several concentrations of GABA were applied to individual neurons. In most cases, plots of %D or k_{app} against peak

[a] Address for correspondence: Dr. David H. Farb, Department of Anatomy and Cell Biology, Box 5 SUNY-HSCB, 450 Clarkson Avenue, Brooklyn, New York 11203.

response were linear. Thus, the values of %D and k_{app} for 10 μM GABA in the presence of CDPX were compared to the corresponding values predicted for an equivalent peak GABA response elicited by GABA alone. In five of six neurons, the increase in %D and k_{app} in the presence of CDPX was significantly greater than predicted, based on the increase in peak response (TABLE 1).

To determine whether the benzodiazepine binding site mediates this effect, we tested a very weak benzodiazepine modulator of GABA action,[4] Ro15-1788, on desensitization of the GABA response. When applied at saturating concentrations, Ro15-

TABLE 1. CDPX Stimulates GABA-Mediated Desensitization over Predicted Levels

Condition	k_{app} (s^{-1})	%D	Percentage Increase in k_{app}	Percentage Increase in %D
10 μM GABA	0.052 $\pm$ 0.006	69 $\pm$ 4	—	—
10 μM GABA + 300 μM CDPX	0.187 $\pm$ 0.024	94 $\pm$ 0.7	260[a]	36[b]
GABA response equivalent to 10 μM GABA + 300 μM CDPX	0.096 $\pm$ 0.011	82 $\pm$ 2	85[a]	19[b]

NOTE: Summary of results from five neurons. %D is defined as [(peak g_{GABA} − g_{ss}) / peak g_{GABA}] $\times$ 100%. k_{app} is determined by first subtracting g_{ss} from all g_{GABA} values; (g_{GABA} −g_{ss}) is then plotted semi-logarithmically against time. The best straight line defines k_{app} as follows: k_{app} = 0.693 / $t_{1/2}$, where $t_{1/2}$ is the time for any (g_{GABA} − g_{ss}) value to decay by 50%. %D and k_{app} for 10μM GABA alone and in the presence of 300μM CDPX were directly measured. The predicted values of %D and k_{app} for a GABA response of equal magnitude to 10 μM GABA + 300 μM CDPX were determined by interpolation from control curves of %D or k_{app} against peak g_{GABA} for several GABA concentrations. Percentage increases are expressed with respect to 10 μM GABA alone. Data are means $\pm$ SEM.

[a] Significantly different (p < 0.005) by two-tailed Student's t test.
[b] Significantly different (p < 0.05) by two-tailed Student's t test.

1788 had little effect on %D and k_{app} for 10 μM GABA. Importantly, however, when both CDPX and Ro15-1788 were applied simultaneously, the CDPX stimulation of desensitization was partially antagonized.

The results demonstrate a specific stimulation of GABA-mediated desensitization by CDPX. The inability of Ro15-1788 to stimulate desensitization while antagonizing the CDPX effect strongly suggests that the benzodiazepine binding site mediates the stimulation of desensitization. These results may prove important in understanding the mechanism of action of benzodiazepines as well as understanding neuromodulation of ion channel-coupled neurotransmitter receptors.

REFERENCES

1. LEVITSKI, A. 1986. Trends Pharmacol. Sci. **7:** 3-6.
2. CHOI, D. W., D. H. FARB & G. D. FISCHBACH. 1977. Nature **269:** 342-344.
3. FARB, D. H., L. A. BORDEN, C. Y. CHAN, C. M. CZAJKOWSKI, T. T. GIBBS & G. D. SCHILLER. 1984. Ann. N.Y. Acad. Sci. **435:** 1-31.
4. CHAN, C. Y. & D. H. FARB. 1985. J. Neurosci. **5:** 2365-2373.

Sex-Hormone-Binding Globulin (SHBG) Is a Normal Constituent of Human Cerebrospinal Fluid (CSF)

PETER POHL,[a] REINHARD FÄSSLER,[b] AND
SIEGFRIED SCHWARZ[b]

[a]*Department of Neurology*
[b]*Institute for General and Experimental Pathology*
University of Innsbruck
A-6020 Innsbruck, Austria

Important functions in the regulation of the bioactivity of sex steroid hormones have been ascribed to plasma sex-hormone-binding globulin (SHBG),[1-3] which is synthesized in the liver. Its presence in the interstitial fluid compartment may regulate the bioavailability of sex steroids at their target cells. SHBG was recently shown to occur in human amniotic fluid[4] and in human breast fluid.[3] In this study we describe the characteristics of SHBG in human cerebrospinal fluid (CSF) and, in addition, present data on SHBG levels in patients with normal and disturbed blood-CSF barrier function.

Cerebrospinal fluid samples were analyzed in 49 patients: 26 females aged 44 ± 15 years (mean ± SD; range: 8-82 years) and 23 males aged 48 ± 15 years (range: 15-75 years). According to conventional CSF investigations (albumin, IgG, IgG index, CSF protein profile, red and white blood cell counts), patients could be divided into 31 cases with undisturbed and 18 cases with disturbed blood-CSF barrier function.

Sex-hormone-binding globulin was measured by a highly sensitive radioligand saturation assay that has been recently described.[5] In this assay tritiated 5α-dihydrotestosterone was used as the specific radioligand. The data obtained were subjected to Scatchard plot analysis.[6] The molecular similarity of cerebrospinal fluid SHBG with plasma SHBG was substantiated by a number of experiments in which the CSF protein displayed the same properties as plasma SHBG with respect to thermolability, affinity, specifity, and sedimentation rate (data not shown).

In the 31 patients with undisturbed blood-CSF barrier function there were characteristic differences in mean values of cerebrospinal fluid SHBG levels between different groups of subjects (TABLE 1). The lowest CSF values were seen in men (0.083 ± 0.03 nmol/L, mean ± SD) and the highest in a pregnant woman (0.415 nmol/L). SHBG levels in the CSF of normal women were found to be 0.139 ± 0.04 nmol/L. The CSF/plasma ratios of SHBG are constant in the entire physiological range. The ratio, in turn, is similar for IgG but not for albumin.

In the 18 cases with a blood-CSF barrier impairment the cerebrospinal SHBG levels correlated with total protein in a similar way as did albumin and IgG (FIG. 1), indicating an influx rate of SHBG not different from the other proteins under pathologic conditions. Accordingly, one patient with normal CSF albumin, but elevated

TABLE 1. SHBG Levels in CSF and Plasma, and CSF/Plasma Ratios of SHBG, IgG, and Albumin from 31 Patients with Undisturbed Blood-CSF Barrier

| | SHBG | | CSF/Plasma Ratio ($\times$ 100) | | |
| | CSF | Plasma | | | |
	(nmol/L)		SHBG	IgG	Albumin
Males (n = 12)	0.083 ± 0.03	16[a] (n = 1/12)	0.27[a]	0.23	0.50
Females (n = 18	0.139[b] ± 0.04	52 ± 2 (n = 7/18)	0.27	0.18	0.41
Pregnant subject (N = 1)	0.415	n.d.	0.21[a]	0.26	0.67

[a] Since no plasma was available from the pregnant subject or from the 12 men, plasma SHBG determination was performed in only one man. These ratios were calculated from the cerebrospinal fluid SHBG levels found in this study and the reference values for plasma SHBG established in a previous study[5]: 30 ± 8 nmol/L (mean ± SD) for normal men; 200 ± 45 nmol/L for second-trimester pregnant subjects.

[b] Statistically significant difference ($p < 0.005$) from males as determined by unpaired t test.

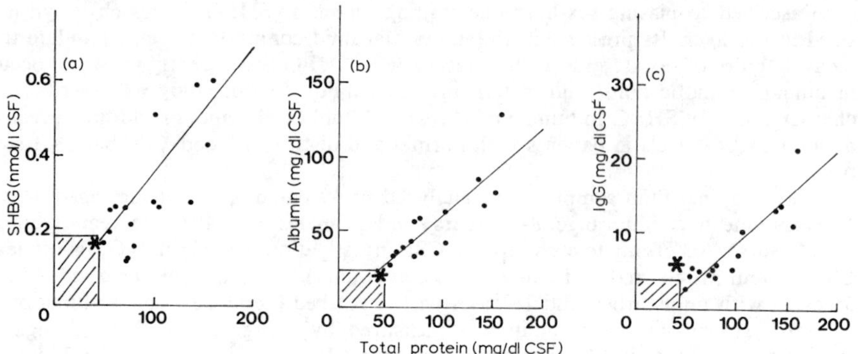

FIGURE 1. Correlation of (**a**) SHBG, (**b**) albumin and (**c**) IgG with total protein content in CSF of 18 patients with disturbed blood-CSF barrier function. *Hatched areas* show reference limits. The *asterisk* indicates one patient with normal barrier function displaying only elevated IgG, yet normal SHBG, albumin and total protein content. r = coefficient of correlation as determined by linear regression analysis using a standard computer program. (**a**) r = 0.616, 0.005 < p < 0.01; (**b**) r = 0.855, p < 0.001; (**c**) r = 0.905, p < 0.001.

cerebrospinal fluid IgG concentrations due to intrathecal IgG synthesis, had normal cerebrospinal fluid SHBG (asterisk in FIG. 1).

SUMMARY

This study shows that there is evidence that:

(1) a protein exists in the CSF that binds 5α-dihydrotestosterone with the same affinity and specificity as plasma-SHBG;

(2) cerebrospinal fluid SHBG shares similar molecular properties with plasma SHBG with respect to thermolability and sedimentation rate; and

(3) under normal conditions (i.e., functioning blood-CSF barrier) SHBG leaks in the same manner as do other proteins from plasma into the CSF compartment.

At present, the physiological function of steroid-binding globulins within the CSF is unknown. Data of previous studies suggest that the entry of steroids into the CSF seems to be primarily governed by their affinities to plasma-binding globulins[7-9]: molecules that are less firmly bound to plasma-binding globulins can transgress the blood-CSF barrier more readily than do others. It therefore becomes conceivable that the rates of entry of sex steroids into the target cells within the central nervous system are modulated or controlled, in the same way as their transfer from plasma into CSF, by their affinity to cerebrospinal fluid SHBG, or reciprocally, by the amount of SHBG in the cerebrospinal fluid.

REFERENCES

1. ANDERSON, D. C. 1974. Sex-hormone-binding globulin. Clin. Endocrinol. **3:** 69-96.
2. HEYNS, W. 1977. The steroid binding β-globulin of human plasma. Adv. Steroid Biochem. Pharmacol. **6:** 59-79.
3. SIITERI, P. K., J. T. MURAI, G. L. HAMMOND, J. A. NISKER, W. J. RAYMORE & R. W. KUHN. 1982. The serum transport of steroid hormones. Rec. Progr. Horm. Res. **38:** 457-510.
4. HAMMOND, G. L., P. LEMONEN, N. J. BOLTON & R. VIHKO. 1983. Measurement of sex hormone binding globulin in human amniotic fluid: Its relationship to protein and testosterone concentration, and fetal sex. Clin. Endocrinol. **18:** 377-384.
5. SCHWARZ, S., H. HINTNER & G. TAPPEINER. 1981. Hormone binding globulin levels in a patient with hereditary angiooedema during treatment with danazol. Clin. Endocrinol. **14:** 563-570.
6. SCHWARZ, S. 1979. A simple computer program for SCATCHARD plot analysis of steroid receptors including non-specific binding correlation on a low cost desk top calculator. J. Steroid Biochem. **11:** 1641-1646.
7. BACKSTRÖM, T., H. CARSTENSEN & R. SÖDERGARD. 1976. Concentration of estradiol, testosterone and progesterone in cerebrospinal fluid compared to plasma unbound and total concentrations. J. Steroid Biochem. **7:** 469-472.
8. MARYNICK, S. P., G. B. SMITH, M. H. EBERT & D. L. LORIAUX. 1977. Studies on the transfer of steroid hormones across the blood-cerebrospinal fluid barrier in the Rhesus monkey. II. Endocrinology **101:** 562-567.
9. WOOD, J. H. 1982. Neuroendocrinology of cerebrospinal fluid: Peptides, steroids and other hormones. Neurosurgery **11:** 293-305.

Neurons in Entorhinal Cortex of Etherized Rats Exhibit Rhythmic Bursting Correlated with the Hippocampal EEG[a]

GREGORY J. QUIRK AND MARK STEWART

Department of Physiology
State University of New York
Health Sciences Center at Brooklyn
Brooklyn, New York 11203

The EEG recorded from the hippocampus is marked by a large-amplitude, sinusoidal pattern with a frequency of 4 to 10 Hertz, termed "theta" rhythm. Recordable from several species, theta rhythm may reach a peak-to-peak amplitude of 0.5 millivolts or more. The frequency of the theta rhythm is associated with the behavioral state of the animal.[1] The higher frequencies tend to occur during behaviors such as walking, while lower frequencies are more often recorded from animals anesthetized with urethane or ether. The rhythmic hippocampal EEG seen in anesthetized rats is abolished by systemic administration of the muscarinic receptor antagonist, atropine, while that seen during walking persists.[2] The presence of the theta rhythm, a rhythmic extracellular field potential, suggests that the activity of many neurons is synchronous. It is apparent from the rhythmic firing of single cells in the hippocampus[3] and the medial nuclei of the septum[4,5] that at least the commissural, associational, and septal afferent inputs to the hippocampus could produce rhythmic postsynaptic currents which underlie the field potentials of the theta rhythm.

In addition to the hippocampus proper, the theta rhythm has been recorded from several related brain regions, including the entorhinal cortex,[6] a major source of input to the hippocampus. Our aim was to determine if, during walking and anesthesia, the theta rhythm recorded from the entorhinal cortex reflected the phasic firing of neurons there. Such synchronous activity could contribute to the production of the extracellular theta rhythm both locally and in the hippocampus. Vanderwolf has made lesions of the entorhinal cortex in rats and found that the hippocampal theta rhythm was still present; however, the theta rhythm associated with walking became atropine-sensitive.[7] This implies that the contribution of the entorhinal cortex to the production of atropine-sensitive, hippocampal theta rhythm is not a critical one. However, a demonstration of phasic firing by output neurons of the entorhinal cortex during atropine-sensitive theta rhythm would suggest a contribution of entorhinal cortex to both types of hippocampal theta rhythm.

Rats were implanted with a movable fine-wire electrode in medial entorhinal cortex for recording multiunit activity and a fixed wire electrode in hippocampus (CA1)

[a]This work was supported in part by NIH Grants NS17095 and NS14497.

which served as the EEG reference. Multiunit activity is particularly useful for generalizing about the behavior of a population of neurons. The hippocampal EEG was filtered (0.3 to 100 Hz), frequency modulated, and stored on video tape along with the filtered (500 Hz high pass) multiunit activity from the entorhinal cortex. Data were recorded during two conditions: (1) while the animal was awake and unrestrained, and (2) following the administration of ether sufficient to produce a deep level of anesthesia. Recordings were analyzed off line with a PDP 11/45 computer programmed to detect cycles of theta rhythm and correlate them with the multiunit activity (for details of the analysis see Ref. 3).

Seven experiments were performed in five rats. FIGURE 1 shows a sample of data recorded in one experiment. In FIGURE 1A the data were recorded during walking. The multiunit activity in this and all other experiments was grouped around the negative peak of the simultaneously recorded hippocampal (CA1) theta rhythm. This confirms a previous result from this laboratory examining single units in the entorhinal

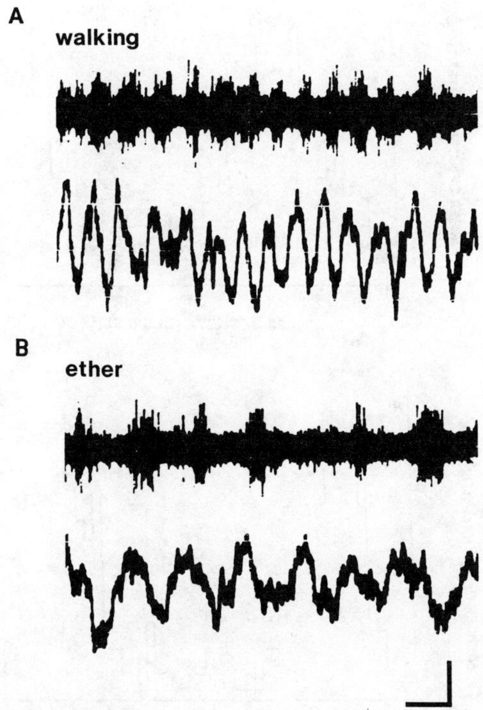

FIGURE 1. (A) Simultaneously recorded entorhinal multiunit activity and hippocampal (CA1) EEG in an awake, walking rat. *Top trace* is the filtered multiunit activity recorded from a chronically implanted fine-wire electrode located in the medial entorhinal cortex. Note that the bursts of action potentials occur near the negative peaks of the CA1 theta rhythm (*lower trace*). In both traces, negativity is indicated by an upward deflection. (B) Similar recordings during ether anesthesia. The rat whose activity was recorded above was given ether until withdrawal reflexes in response to pinching of the foot were abolished. Note that the action potentials are grouped near the same phase as the units were during walking theta rhythm. Calibration: 50 μV top, 150 μV bottom, 0.2 sec.

cortex of freely moving rats.[8] During ether anesthesia, the multiunit activity in six of the seven experiments remained phase-locked to the negative peak of the lower frequency hippocampal theta rhythm. A sample of data is shown in FIGURE 1B. In the seventh experiment, the multiunit activity was essentially randomly distributed over the theta rhythm cycle. Atropine (100 mg/kg i.p.) abolished this ether-induced theta rhythm in two of three experiments. FIGURE 2 presents histograms of the phase of occurrence of action potentials relative to an averaged theta wave. In this form, the tendency for action potentials to cluster around the negative peak of the hippocampal theta rhythm is obvious. The histology indicated that the recordings had been made in layers II and III of the medial entorhinal cortex, which are known to project to the hippocampus.

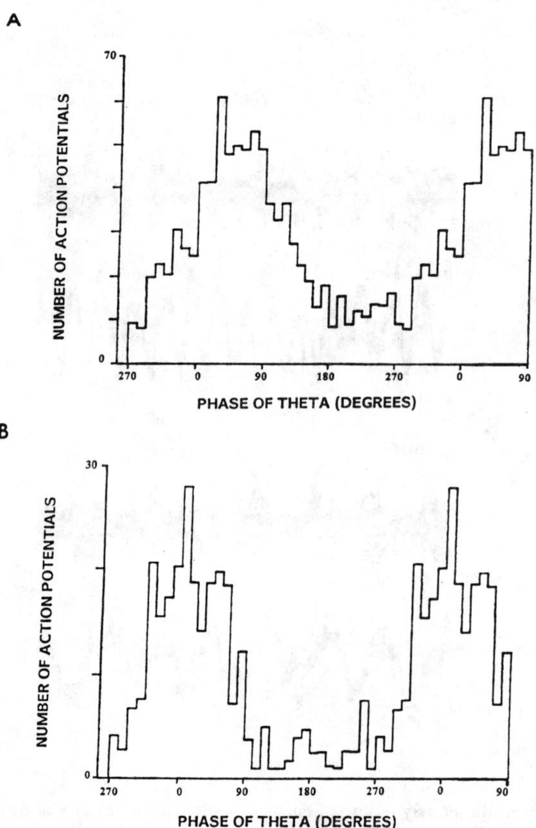

FIGURE 2. Phase of occurrence of action potentials on an averaged theta cycle. The *abscissa* of these representative histograms consists of one average theta rhythm cycle divided into 32 bins, each 11.25 degrees wide. Zero degrees indicate the peak negativity of the hippocampal (CA1) theta rhythm. On the ordinate is the number of action potentials per bin. Note that during both walking (A) and ether anesthesia (B), the action potentials tend to cluster around the negative peak (0°). For the grouped ether data, $p < 0.01$, as determined by the Rayleigh test.

These data suggest that the entorhinal input to the hippocampus is rhythmic during atropine-sensitive theta rhythm. This implies that the entorhinal cortex may make a contribution to the generation of the hippocampal theta rhythm. The fact that atropine-sensitive, hippocampal theta rhythm can still be recorded after the elimination of the entorhinal cortex suggests that the hippocampus and the entorhinal cortex are phasically modulated by a common "pacemaker," presumably the medial septal nuclei. Direct projections to both the hippocampus and the entorhinal cortex from these nuclei have been demonstrated anatomically, and contain cholinergic fibers.[9]

ACKNOWLEDGMENTS

We are indebted to Drs. Steven E. Fox and James B. Ranck, Jr. for their assistance throughout this study.

REFERENCES

1. VANDERWOLF, C. H. 1969. Hippocampal electrical activity and voluntary movement in the rat. Electroencephalog. Clin. Neurophysiol. **26:** 407-418.
2. KRAMIS, R., C. H. VANDERWOLF & B. H. BLAND. 1975. Two types of rhythmical slow activity in both the rabbit and the rat: relations to behavior and effects of atropine, diethyl ether, urethane, and pentobarbital. Exp. Neurol. **49:** 58-85.
3. FOX, S. E. 1986. Hippocampal theta rhythm and the firing of neurons in walking and urethane anesthetized rats. Exp. Brain Res. **62:** 495-508.
4. PETSCHE, H., C. STUMPF & G. GOGOLAK. 1962. The significance of the rabbit's septum as a relay station between the midbrain and the hippocampus 1. The control of hippocampus arousal activity by the septum cells. Electroencephalog. Clin. Neurophysiol. **14:** 202-211.
5. STEWART, M. & S. E. FOX. 1986. Two populations of rhythmically bursting neurons in the septal nuclei are revealed by atropine. Soc. Neurosci. Abst.**12:** 1527.
6. MITCHELL, S. J. & J. B. RANCK, JR. 1980. Generation of theta rhythm in medial entorhinal cortex of freely moving rats. Brain Res. **189:** 49-66.
7. VANDERWOLF, C. H. & L.-W. S. LEUNG. 1983. Hippocampal rhythmical slow activity: a brief history and the effects of entorhinal lesions and phencyclidine. *In* Neurobiology of the Hippocampus. W. Seifert, Ed.: 275-302. Academic Press. London.
8. QUIRK, G. J. & J. B. RANCK, JR. 1986. Firing of single cells in entorhinal cortex is location specific and phase locked to hippocampal theta rhythm. Soc. Neurosci. Abst. **12:** 1524.
9. AMARAL, D. G. & J. KURZ. 1985. An analysis of the origins of the cholinergic and non cholinergic septal projections to the hippocampal formation of the rat. J. Comp. Neurol. **240:** 37-59.

Prolongation of Cardiovascular Actions of Methionine Enkephalin by *d*-Phenylalanine in Intact Rabbits

H. M. RHEE

Department of Pharmacology
Oral Roberts University
School of Medicine
Tulsa, Oklahoma 74137

The enkephalins produce a variety of cardiovascular responses depending upon the species of animals used, the route of administration and the state of anesthesia.[1] In pentobarbital-anesthetized rabbits, met-enkephalin decreased systemic blood pressure, heart rate and renal sympathetic nerve activity. The cardiodepressive effect of met-enkephalin is in part due to the central sympatho-inhibitory action of the peptide.[2,3] *d*-Phenylalanine is known to decrease blood pressure. This hypotensive mechanism is poorly understood. *d*-Phenylalanine is hydroxylated to *d*-tyrosine and eventually converted to *d*-norepinephrine, which may act as a false transmitter. At the same time, *d*-phenylalanine inhibits enkephalinase, which is one of the enzymes that digest enkephalin. Thus, the main purpose of this work was to test whether *d*-phenylalanine increases the duration and effect of met-enkephalin on the cardiovascular system by inhibiting the enkephalinase system.

In pentobarbital-anesthetized (30 mg/kg) male rabbits (2 kg), blood pressures (systolic, mean, diastolic and pulse), heart rate and renal nerve activity were monitored as reported previously.[3] The effects of met-enkephalin (3 to 300 μg/kg i.v.) were evaluated in this animal model. Separately, the same dose of met-enkephalin was evaluated in the presence of *d*-phenylalanine (100 mg/kg i.v.), which was given 10 min before the peptide. Blood pressure, heart rate and renal nerve activity were quantitated as reported.[3] Time to peak action (Tpa) of met-enkephalin, the time required to decrease 50% of peak effect of met-enkephalin (Tpa$_{50}$) and time to 50% recovery from the maximal effect (Tr$_{50}$) on blood pressure were calculated for statistical analysis.

In anesthetized rabbit met-enkephalin (30 μg/kg i.v.) produced a significant reduction in heart rate, blood pressure and renal nerve activity (FIG. 1), which confirms the previous studies from this laboratory.[2,3] Administration of met-enkephalin 10 min after the pretreatment of *d*-phenylalanine prolonged the duration of met-enkephalin action (FIG. 1, panel B), which is clearly manifested by the fact that Tr$_{50}$ was increased by almost 100% (TABLE 1). *d*-Phenylalanine also potentiated the duration of RNA suppression, particularly at the low doses of met-enkephalin. At any dose of met-enkephalin tested in this study, the treatment of *d*-phenylalanine increased Tr$_{50}$ significantly ($p < 0.001$). A single low dose of *d*-phenylalanine did not decrease blood pressure significantly in this study, although the hypotensive trend was noted after *d*-phenylalanine.

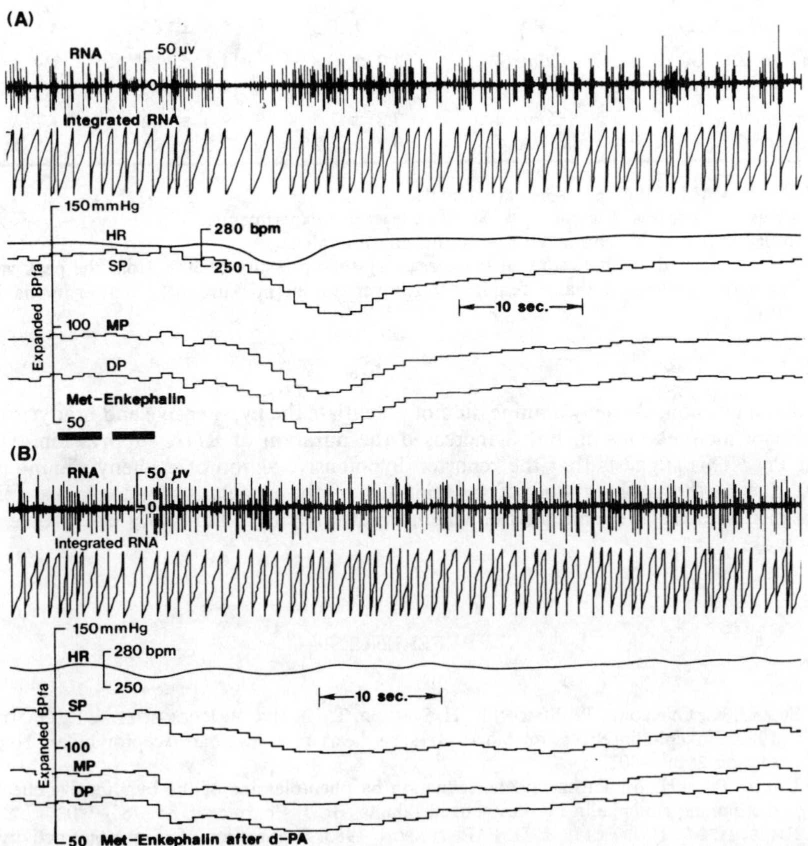

FIGURE 1. Interaction between met-enkephalin and *d*-phenylalanine (*d*-PA) in cardiovascular system of anesthetized rabbits. (**A**) In pentobarbital-anesthetized rabbit (30 mg/kg) met-enkephalin (30 µg/kg) was injected (*first bar*) and subsequently flushed (*second bar*). (**B**) The same treatment was repeated 10 min after the administration of *d*-PA (100 mg/kg i.v.). Renal nerve activity (RNA) was monitored and analyzed by an integrator. HR, SP, MP and DP stand for heart rate and systolic, mean and diastolic pressure, respectively. BPfa indicates femoral arterial blood pressure.

TABLE 1. Effect of d-Phenylalanine on Cardiovascular Action of Met-Enkephalin in Anesthetized Rabbit.[a]

Parameters	Experiments	Intravenous Dose of Met-Enkephalin		
		3 μg/kg	10 μg/kg	30 μg/kg
Duration of RNA	Control	3.3 ± 0.4[b]	5.7 ± 1.1	9.7 ± 1.8
suppression (sec)	d-PA	7.7 ± 0.3[c]	8.5 ± 2.9	14.1 ± 3.5
Time to 50% of	Control	11.3 ± 0.8	11.1 ± 0.4	11.9 ± 0.6
peak action (Tpa$_{50}$)[d]	d-PA	13.6 ± 1.2	12.8 ± 0.3[c]	13.4 ± 1.1
Time to 50% of	Control	7.8 ± 1.5	9.0 ± 0.2	13.7 ± 1.0
recovery (Tr$_{50}$)[e]	d-PA	13.7 ± 1.2[c]	27.0 ± 4.4[c]	23.6 ± 4.3[c]

[a] See the text for the details of experiments.
[b] Values are expressed as mean ± SE of at least five experiments.
[c] Indicates $p < 0.05$, compared to relevant control value.
[d] Tpa$_{50}$ is defined as time required to decrease systolic pressure to 50% from the peak value.
[e] Tr$_{50}$ is defined as time that is required to recover systolic pressure to 50% after its maximal reduction.

In conclusion, d-phenylalanine did not potentiate the hypotensive and bradycardiac action of met-enkephalin, but it increased the duration of RNA suppression, Tpa$_{50}$ and Tr$_{50}$. This suggests that the reported hypotensive action of d-phenylalanine is in part due to the prolongation of the half life of the circulating endogenous opioid peptides, such as met-enkephalin, which results from its inhibitory effect of enkephalinase.

REFERENCES

1. SCHAZ, K., G. STOCK, W. SIMON, K. H. SCHLÖR, T. UNGER, R. ROCKHOLD & D. GANTEN. 1980. Enkephalin effects on blood pressure, heart rate, and baroreceptor reflex. Hypertension **2:** 395-407.
2. EULIE, P. & H. M. RHEE. 1984. Reduction by phentolamine of the hypotensive effect of methionine enkephalin in anesthetized rabbits. Br. J. Pharmacol. **83:** 783-790.
3. RHEE, H. M., P. J. EULIE & D. F. PETERSON. 1985. Suppression of renal nerve activity by methionine enkephalin in anesthetized rabbits. J. Pharmacol. Exp. Ther. **234:** 534-537.
4. FREDERICKSON, R. C. & L. E. GEARY. 1982. Endogenous opioid peptides: Review of physiological, pharmacological and clinical aspects. Prog. Neurobiol. **19:** 19-69.

Dopamine Metabolism Increases Levels of Oxidized Glutathione in Rat Striatal Synaptosomes[a]

MARY BETH SPINA AND GERALD COHEN

Department of Neurology and Neurobiology Center
Mount Sinai School of Medicine of the
City University of New York
New York, New York 10029

A major cellular mechanism for the detoxification of hydrogen peroxide is through the glutathione peroxidase/glutathione reductase enzyme system. The oxidation of reduced glutathione (GSH) by glutathione peroxidase results in the formation of oxidized glutathione (GSSG). Increased levels of GSSG can provide an index of oxidative stress. Dopamine (DA), a prominent monoamine neurotransmitter in brain, is metabolized by monoamine oxidase (MAO), a mitochondrial enzyme, and H_2O_2 is produced as a product. H_2O_2 is an oxidant that is potentially hazardous to the cell. In these experiments, we have linked DA metabolism by MAO in isolated nerve terminals to an increase in GSSG. If normal catecholamine metabolism induces an oxidative stress within monoamine neurons, then increased neurotransmitter turnover may evoke changes that alter neuronal activity.

We studied DA-rich striatal synaptosomes. Increased DA catabolism was provoked by treatment with reserpine. Reserpine prevents the storage of DA in synaptic vesicles and thereby increases the concentration of free DA in the cytoplasm.

Sprague-Dawley rats weighing 200–250 grams (Perfection Breeders) were used. To prepare synaptosomes, striata were dissected and homogenized at 4°C in 0.32 M sucrose. The homogenate was centrifuged at $1000 \times g$ for 10 min to clear cellular debris. The supernatant fraction was recentrifuged at $12,000 \times g$ for 40 min. The resultant P2 pellet was resuspended in Krebs-Ringer phosphate buffer (pH 7.4). Aliquots were incubated at 37°C with and without the appropriate drugs. Subsequently, the synaptosomes were repelleted and homogenized in 0.4 M perchloric acid. After centrifugation at $11,000 \times g$ for 15 min, the resultant supernatant was analyzed for GSSG.

In order to remove GSH, the supernatant was reacted with N-ethylmaleimide (NEM) at pH 7.5. After a 20-min incubation at room temperature, the excess NEM was removed by Sep-Pak chromatography according to the method of Adams.[1] GSSG was measured by the method of Tietze[2] as modified by Cooper *et al.*[3] The assay utilized GSSG reductase and 5,5′-dithiobis(2-nitrobenzoic acid) (DTNB). The rate of color formation was monitored at 412 nm for 5 min and compared to standards that were subjected to the same procedure; the blank rate, consisting of the rate of reduction of DTNB by GSSG reductase, was subtracted from the observed values. Data were evaluated with a paired t test.

[a] This work was supported by USPHS Grant NS-23017.

The steady-state level of GSSG in striatal synaptosomes was increased by 22.9 ± 2.6% (SEM) during incubation with 10 μM reserpine and 10 μM DA (TABLE 1). When 10 μM clorgyline, an MAO inhibitor, was added, the rise in GSSG was suppressed to only 3.5 ± 1.4% ($p < 0.001$ compared to reserpine + DA).

In separate experiments, synaptosomes were incubated with 10 μM reserpine and 1 mM L-dopa. L-dopa is transformed to DA by dopa decarboxylase. Therefore, incubation with L-dopa will increase the intraneuronal level of DA. In the presence of reserpine and L-dopa, the level of GSSG was increased to 30.0 ± 2.5% (TABLE 1). Carbidopa is an inhibitor of the decarboxylase enzyme. In the presence of carbidopa, the rise in GSSG was suppressed to 12.7 ± 2.4% ($p < 0.05$ compared to reserpine + L-dopa).

Exposure of whole organs or tissues to H_2O_2 or organic peroxides produces a measurable elevation in the GSSG level.[4] The rise in GSSG is an indicator of oxidative stress. We increased the mitochondrial production of H_2O_2 in striatal nerve terminals by increasing the turnover of DA. Metabolism of DA by MAO induced a substantial increase in the steady-state level of GSSG. Increased DA turnover has been implicated in a variety of disease states, such as Parkinson's disease and schizophrenia. Our experiments show that increased turnover of DA can evoke an oxidative stress. These observations provide insights into mechanisms that may be responsible for neuronal senescence or alterations in neuronal function in CNS disorders that affect monoamine neurons.

TABLE 1. GSSG Levels in Rat Striatal Synaptosomes after Treatment with Reserpine plus Dopamine, or Reserpine plus L-DOPA

Sample (n)	Increase in GSSG from Control Levels[a]	
	pmol/mg of Original Tissue	% Increase
Reserpine + DA (11)	4.07 ± 0.46[b]	22.9 ± 2.6
Reserpine + L-Dopa (7)	5.89 ± 0.56[b]	30.0 ± 2.5

[a] GSSG levels in control samples were 19.01 ± 1.06 pmol/mg tissue (mean ± SEM, $n = 18$).

[b] $p < 0.05$ compared to controls (paired t test).

REFERENCES

1. ADAMS, J. D., B. H. LAUTERBURG & J. R. MITCHELL. 1983. Plasma glutathione disulfide in the rat: Regulation and response to oxidative stress. J. Pharm. Exp. Ther. **227:** 749-754.
2. TIETZE, F. 1969. Enzymic method for quantitative determination of nanogram amounts of total and oxidized glutathione: Applications to mammalian blood and other tissues. Anal. Biochem. **27:** 502-522.
3. COOPER, A. J. L., W. A. PULSINELLI & T. E. DUFFY. 1980. Glutathione and ascorbate during ischemia and postischemic reperfusion in rat brain. J. Neurochem. **35:** 1242-1245.
4. SIES, H. 1983. Hydroperoxides and thiol oxidants in the study of oxidative stress in intact cells and organs. *In* Oxidative Stress. H. Sies, Ed.: 73-90. Academic Press. London.

Effects of Methionine Enkephalin on Contractile Force of Isolated Blood Vessels

H. M. RHEE

Department of Pharmacology
Oral Roberts University
School of Medicine
Tulsa, Oklahoma 74137

Methionine enkephalin (met-enkephalin) has a modulatory effect on the responsiveness of peripheral tissues to catecholamines or neuropeptides.[1,2] Met-enkephalin also suppresses centrally mediated sympathetic nerve activity and thereby reduces heart rate and systemic blood pressure in anesthetized rabbits.[3] This suggests that met-enkephalin crosses the blood-brain barrier to produce its cardiovascular effects. However, the hypotensive effect of the peptide might be related to its ability to dilate peripheral blood vessels directly. Therefore, the purpose of this study was to examine a potential direct effect of met-enkephalin on the vascular contractile force of the descending aorta and femoral artery of rabbits.

The descending aorta was removed from anesthetized rabbits and helically sectioned muscle strips were prepared as described by Furchgott.[4] One end of the preparation was fastened to a force-displacement transducer and the strip was allowed to equilibrate for 30 minutes in modified Krebs-Henseleit (K-H) solution[5] and bubbled with 95% O_2 and 5% CO_2 at 37°C. The other end of the strip was fixed to an organ bath and was maintained under 0.1 to 0.5 gm resting tension for an additional 30 minutes. Histamine or *l*-norepinephrine was added to the organ bath to make an appropriate final concentration and the resultant tension generated by the muscle was analyzed. The identical procedure was repeated in the presence of indicated dose of met-enkephalin.

FIGURE 1 shows the effects of met-enkephalin (10^{-4}M) on the force of contraction induced by histamine in rabbit aortic strip. Histamine did not produce a significant effect at the dose below 10^{-9}M, but a good dose-response curve for histamine on resting tension and contractile force was obtained at concentrations from 10^{-7} to 10^{-4} M (FIGS. 1 and 2). Met-enkephalin did not alter the dose-response curve of histamine at 10^{-5}M or 10^{-4}M. Met-enkephalin also had little or no effects on contraction induced by *l*-norepinephrine in the rabbit aortic smooth muscle.[6] At the same time, met-enkephalin had minimal effect on the heart rate, the left ventricular pressure and its dp/dt, including the right ventricular contractile tension.[7] Therefore, it may be safely said that the direct peripheral vasodilatory effect of met-enkephalin does not play a significant role in the hypotensive effect of the peptide in rabbits. This further suggests that the hypotensive action of met-enkephalin is related to its ability to enter into the brain in a time frame that matches with the onset of its peripheral cardiovascular actions.

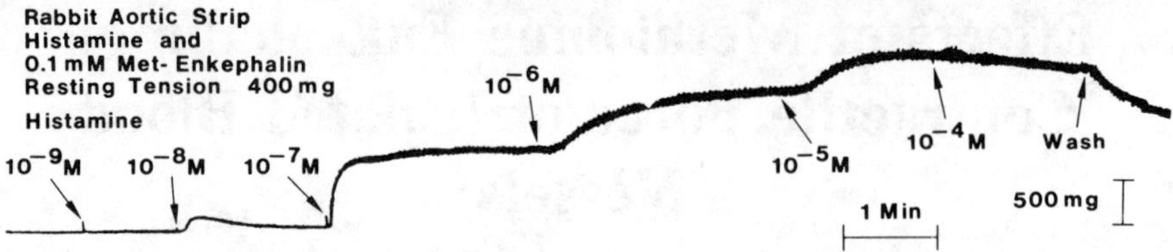

FIGURE 1. Effects of met-enkephalin (10^{-4}M) on the force of contraction induced by histamine in rabbit aortic strip. Helical strips of rabbit aortic muscle were prepared according to the method of Furchgott.[4] After an appropriate control period, met-enkephalin (10^{-4}M) was added to the muscle bath at 37°C. Indicated dose of histamine was added cumulatively to the incubation muscle bath. Time and weight scales are indicated, and resting tension for the muscle was 400 mg.

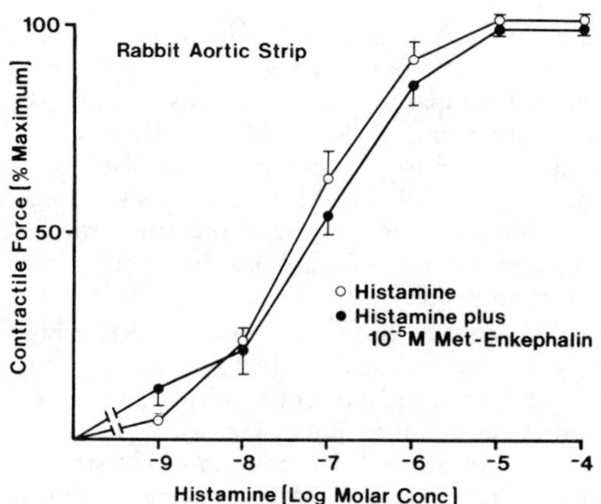

FIGURE 2. Lack of met-enkephalin effect on the dose-response curve of histamine in rabbit aortic strips. Rabbit aortic muscle strips were prepared as in FIGURE 1 and a dose-response curve of histamine was obtained as described in the text. The dose-response analysis was repeated in the presence of met-enkephalin (10^{-5}M) under identical conditions. Data are expressed as percentage of the maximal contraction obtained with 10^{-4}M histamine. *Vertical bars* indicate SE and each *point* represents mean of at least four experiments.

REFERENCES

1. EIDEN, L. E. & J. A. RUTH. 1982. Enkephalins modulate the responsiveness of rat atria *in vitro* to norepinephrine. Peptides **3:** 475-478.
2. GANTEN, D. 1985. Opioid peptides and cardiovascular control. Presented at a meeting of the Federation of the American Society of Experimental Biologists.
3. RHEE, H. M., P. J. EULIE & D. F. PETERSON. 1985. Suppression of renal nerve activity by methionine enkephalin in anesthetized rabbits. J. Pharmacol. Exp. Ther. **234:** 534-537.
4. FURCHGOTT, R. F. 1960. Spiral-cut strip of rabbit aorta for *in vitro* studies of responses of arterial smooth muscle. Meth. Med. Res. **8:** 177-186.

5. DUTTA, S., H. M. RHEE & B. H. MARKS. 1972. Effects of metabolic inhibitors on the accumulation of digitaloids by the isolated guinea-pig heart. J. Pharmacol. Exp. Ther. **180:** 351-358.
6. RHEE, H. M. & P. J. EULIE. 1985. Characteristics of stress induced by methionine enkephalin in intact rabbit. *In* Stress and Heart Diseases, R. E. Beamish *et al.,* Eds: 298-311. Martinus Nijhoff. Boston.
7. EULIE, P. & H. M. RHEE. 1985. Mechanism of cardiodepressant actions of enkephalin. Ann. N.Y. Acad. Sci. **435:** 408-411.

Index of Contributors